Lactic Acid Bacteria

Wei Chen

Editor

Lactic Acid Bacteria

Bioengineering and Industrial Applications

Editor
Wei Chen
Jiangnan University
Wuxi, China

The print edition is not for sale in The Mainland of China. Customers from The Mainland of China please order the print book from: Science Press.

ISBN 978-981-13-7282-7   ISBN 978-981-13-7283-4   (eBook)
https://doi.org/10.1007/978-981-13-7283-4

This Springer imprint is published by the registered company Springer Nature Singapore Pte Ltd.
The registered company address is: 152 Beach Road, #21-01/04 Gateway East, Singapore 189721, Singapore

Contents

Chapter 1
Lactic Acid Bacteria and γ-Aminobutyric Acid and Diacetyl

Shunhe Wang, Pei Chen, and Hui Dang

1.1 Lactic Acid Bacteria and γ-Aminobutyric Acid

1.1.1 Introduction of γ-Aminobutyric Acid

1.1.1.1 γ-Aminobutyric Acid

γ-Aminobutyric acid (GABA), or γ-ammonia butyric acid, is a kind of nonprotein amino acid which widely exists in vegetables and animals. It exists in the seeds, roots, and tissue fluid of many plants, such as *Glycine* L., *Panax*, herbal, and in animals almost exclusively present in nervous tissues. The content of GABA in brain tissue is 0.1~0.6mg/g, and immunology research show that substantia nigra contain the highest concentration of GABA (Krajnc et al. 1996). Meanwhile GABA is also present in microorganisms, such as yeast, *Lactobacillus*, and *Escherichia coli*.

GABA was discovered in the metabolite of plants and microorganisms in 1883. Awapara et al. (1950), Roberts and Frankel (1950), and Udenfriend (1950) found that the content of GABA was relatively high in the brain of mammals, but the biological significance was unknown. Roberts reported that GABA was a product of glutamate through α-decarboxylation. Krnjević and Schwartz (1966) researched that GABA played a part in the brain of mammals as an endogenous neurotransmitter and GABA was confirmed as an inhibitory neurotransmitter in the Second International Symposium on GABA in 1975.

S. Wang (✉)
Jiangnan University, Wuxi, China
e-mail: wangshunhe@jiangnan.edu.cn

P. Chen
Shaanxi Radio & TV University, Xi'an, China

H. Dang
Shaanxi Normal University, Xi'an, China

© Springer Nature Singapore Pte Ltd. and Science Press 2019
W. Chen (ed.), *Lactic Acid Bacteria*,
https://doi.org/10.1007/978-981-13-7283-4_1

Many studies in recent years have reported the physiological function of GABA, including delaying senescence of nerve cells, lowering blood pressure, repairing the skin, treating mental illnesses, and regulating cardiac arrhythmia and hormone secretion (He Xipu et al. 2007). Scientists from all around the world studied the synthesis technology and its related product. GABA could be produced by chemical synthesis, plant enrichment, and microbiological fermentation. Sawai et al. (2001) researched that the content of GABA in tea could be enhanced through a variable anaerobic-aerobic process. Meanwhile, it also could be produced by controlling the germination condition of brown rice and rice germ. Xia Jiang et al. reported that *Lactobacillus brevis* CGMCCNO.1306 was isolated from unpasteurized raw milk and the capacity of GABA was 76.36 g/L after the mutation breeding (Xia Jiang 2006; Xia Jiang et al. 2006).

1.1.1.2   General Physicochemical Property

The chemical formula is $C_4H_9NO_2$ and molar mass is 103.120 g/mol. It appears as a white microcrystalline powder, easily soluble in water, slightly soluble in ethanol, and insoluble in cold ethanol, benzene, and ether. The dissociation constant of pKCOOH and pKNH3 were 4.03 and 10.56, respectively. GABA has no optical rotation, the melting point was 203~204°C. Its products of decomposition were pyrrolidone and water (Chen Lilong and Shi 2010). In most part, GABA exists in the form of amidogen with positive charge or carboxyl with negative charge. The state of GABA determined its molecular conformation. In the gaseous state, the molecular conformation is highly folded due to the electrostatic interaction of two charged groups; in solid state, the molecular conformation is extended due to the intermolecular interaction between the two groups. When in liquid state, molecular conformations exist in both states. The various conformations of GABA are combined with different receptor proteins, which play significant physiological functions.

1.1.2   GABA Production of Lactic Acid Bacteria

The production of GABA by microbial fermentation is to use the glutamate decarboxylase in organisms as a catalyst, which GABA is produced from L-glutamate or its sodium salt by α-decarboxylation reaction. Glutamate decarboxylase is the single enzyme to generate GABA from L-glutamate or its sodium salt, which was discovered in bacteria, archaea, and eucaryotic microbes. Compared with chemical synthesis and plant enrichment, GABA from microorganism fermentation, which has low cost and high yield, could be used in food safely.

Lactic acid bacteria, as a safe microorganism (GRAS), are widely used in foods, such as yogurt, cheese, and pickles. Many studies show that lactic acid bacteria have glutamate decarboxylase which generates GABA from L-glutamate by decarboxylation reaction. Nomura et al. (1998) reported that the content of GABA was 383 mg/kg in cheese, in which a strain of *Lactococcus lactis* 01–7 with high GABA was isolated. Komatsuzaki et al. (2005) researched that a strain of *Lactobacillus paracasei* NFRI7415 with high GABA was isolated from traditional fermented food and the capacity of GABA was 302 mmol/L. Xu Jianjun et al. (2002) isolated a strain of *Lactococcus lactis* with high GABA for the first time interiorly, and the amount of GABA in the fermentation broth was 250 mg/100 ml, which was fermented in 25 L tank for 72 h. It could be used in the development of yogurt or as an ingredient in other foods. Cui Xiaojun et al. (2005) also reported the screening and fermentation conditions of high GABA-producing *Lactobacillus*, and the capacity of GABA was 5.4 g/L in the fermentation broth. Xia Jiang et al. (2006) isolated a strain of *Lactococcus lactis* hjxi-01, and the maximum accumulation concentration of GABA was 7 g/L in the medium containing 5% L-sodium glutamate (Xia Jiang 2006). And on this basis, the initial strain was mutagenized by ultraviolet rays and radiation. Finally, a high-yield mutant hjxj-80,119 was obtained, which was passed on for 12 consecutive generations and had stable genetic characters. Compared with the original strain hjxj-01, the average yield increased by 142.9%, and the average product was 17 g/L. Using neural network and particle swarm optimization algorithm, the researchers optimized the fermentation conditions of lactic acid bacteria in the shake flask, and the accumulated amounts of GABA was 33.4 g/L. After the optimization of indirect fermentation conditions, the concentration of GABA reached 107.5 g/L. Siragusa et al. (2007) isolated 12 different species of lactic acid bacteria with high GABA from 22 kinds of Italian cheese. Five species of *Lactobacillus*, including *Lactobacillus casei* PF6, *Lactobacillus bulgaricus* PR1, *Lactococcus lactis* PU1, *Lactobacillus plantarum* C48, and short lactic acid bacteria PM17, had strong ability to synthesize GABA. And the result of PCR acquired the glutamate decarboxylase gene. The homology was analyzed by amino acid analyzer with 85%~99%. Ji Linli et al. (2008) isolated nine strains of lactic acid bacteria with high GABA from traditional fermented milk. The strains WH11–1 and M10–4-3 were identified as *Lactococcus lactis* subsp. *lactis* through their traditional morphological, physiological, biochemical, and chemical characteristics and 16S rDNA gene identification. Fan et al. (2012) isolated *Lactobacillus brevis* CGMCC1306 with high glutamate decarboxylase active from fresh and unpasteurized milk and further analyzed the glutamate decarboxylase, which provided a basis for an approach that could use glutamate decarboxylase and high-density culture to produce GABA in the future. Xian Qianlong (2013) isolated 14 strains of lactic acid bacteria producing GABA, and two strains, including QL-14 (0.39 g/L) and QL-20 (0.584 g/L), had relatively high GABA content. Through morphological features, physiological and biochemical experiment, and 16S rDNA gene identification, strains of QL-14 and QL-20 were identified as *Lactobacillus plantarum* and *Lactobacillus casei*, respectively.

1.1.3   Physiological Function of γ-Aminobutyric Acid

The current research about function of γ-aminobutyric acid mainly focus on the treatment and improvement of the nervous system disease, high blood pressure, liver function, and renal function, the impact on the physiology and reproduction, and many other aspects.

1.1.3.1   Delaying the Senescence of Nerve Cells

Leventhal et al. (2003) researched that GABA could make the brain of aged macaque in peak form, while as the growth of age, the brain GABA would reduce gradually. Macaque visual function was similar with human, which would be deteriorated with the increase of age. When brain in peak form, macaque could choose signal to respond. Lack of GABA could lead to the degeneration of nerve cell in macaque, and that would decline the ability of visual selection. Compared with younger macaque, after injecting GABA and its agonist in an aging macaque, visual cortex cells could be a greater degree to restore the ability of visual stimulus orientation and directional selectivity, making the "senescence" cells show the characteristics of being "young."

Leventhal et al. (2003) studied the effects of electrophoretic application of GABA by using multibarreled microelectrodes, and the agonist and antagonist of GABA A-type receptor were used in individual V1 cells in old monkeys. Compared with old monkeys, GABA A-type receptor antagonist played an important role in neuronal responses in young monkeys, which indicated the degradation of GABA-mediated inhibition with the increasing monkey age. On the other side, the treatment of GABA resulted in improved visual function. The treated cells in area V1 of old animals displayed the special characteristic of young cells.

1.1.3.2   Lower Blood Pressure

High blood pressure is a common human disease and a major risk factor for coronary heart disease, stroke, or other cardiac-cerebral vascular diseases. According to statistics, cardiac-cerebral vascular diseases, which were caused by high blood pressure, caused about more than 12 million deaths yearly. Takahashi et al. (1959) first reported the effect of GABA on the cardiovascular system, which found that GABA had the strongest effect on cardiovascular activity in the γ-amino acid group. Later, they also confirmed that 10 mg/kg GABA could reduce blood pressure and cause bradycardia in rabbit, dog, and cat experiments. But its effect could be prevented by resecting sympathetic nerve or blocking ganglion. In addition, this effect didn't need the vagus nerve, aortic nerve, and sinus nerve. Therefore, GABA might regulate blood pressure and heart rate by interacting central nervous system. Antonaccio and Taylor (1977) injected 3~1000 g/kg GABA into the encephalocoele of anesthetized cats, which decreased the blood pressure and heart rate with a dose-effect relationship. It confirmed that GABA adjusted blood pressure and heart rate by central nervous system for the first time. Kazami et al. (2002) also reported that compound flavor containing GABA could significantly reduce the blood pressure of patients with mild hypertension, which reduced systolic blood pressure by 6 mmHg and diastolic blood pressure by 4 mmHg. The blood pressure of the normal group remained unchanged, and no adverse reactions were observed. Hayakawa et al. (2004) reported that GABA-enriched skim milk fermented by lactic acid bacteria was fed to normal rat with spontaneous hypertension. The result showed that the GABA-enriched milk had significantly lowered the blood pressure. Further study found that the fermented milk didn't contain antihypertensive peptides and the functional component for lowering the blood pressure was GABA, which was similar to the antihypertensive active ingredient in *Astragalus membranaceus* and manyinflorescenced sweetvetch root.

1.1.3.3   Effects on Physiological Reproduction

The main inhibitory neurotransmitter in the brain was γ-aminobutyric acid. And it could participate in the secretion regulation of anterior pituitary hormone at the level of the hypothalamus or pituitary, which indirectly affected the function of the ovary, oviduct, male gonads, and accessory organs and was closely related to the movement of sperm and the production of steroid hormones. Murashima and Kato (1986) found that the capacity of GABA in the tubal mucosa of rats was ten times higher than that in the brain, and the concentration of GABA decreased in a gradient from the oviduct fimbria to the connection of uterine and oviduct. Meanwhile, research also showed that GABA is present in the male gonads and accessory organs. GABA receptor was found on the surface of sperm membrane, suggesting that GABA was related to sperm function. Roldan et al. (1994) reported that GABA could significantly induce acrosome reaction in capacitated spermatozoa of mice and human and had significant dose-effect relationship. Progesterone had promoting effect to the reaction. Bian Shuling et al. (2002) reported that GABA could increase the sperm acrosin activity and significantly increased Na^+-k^+-ATPase and superoxide dismutase activity in normal and positive antisperm antibody (AsAb) men and found that GABA could increase the sperm acrosin activity and significantly increased Na^+/K^+-ATPase and superoxide dismutase activity, so it has important theoretical significance and broad application outlooks in reproductive physiology.

1.1.3.4   Treatment of Mental Illness

GABA was the most important inhibitory neurotransmitter in the central nervous system of mammal, shellfish and certain parasitic worms and mediated over 40% of inhibitory nerve conduction. The abnormality of GABA induced various nervous system diseases, including excitotoxic reaction, epilepsy, insomnia, etc. (Mombereau et al. 2004). Since Tower first proposed that the occurrence of epilepsy was related to GABA in the brain in the early 1960s, a large number of studies had shown that the content of GABA in the spinal fluid of patients with epilepsy was significantly lower than that of normal people and the extent of the decline was related to the type of seizure. Many research showed that the content of GABA in the spinal fluid of epileptic patients was significantly lower than that of normal people, and the decline level was related with the type of epilepsy (Song Wei et al. 2008). GABA was a special biochemical drug to treat intractable epilepsy. Experimental epilepsy could be induced by GABA inhibitors, such as allylglycine, GABA A-type receptor, and receptor antagonists, while GABA receptor agonist had anticonvulsive and antiepileptic effects (Usuki et al. 2007). Okada et al. (2000) used GABA rice germ food orally and found that it had the role of sedation, promoting sleep and resistance to anxiety, and the total improvement rate of woman menopause syndrome and early mental disorders was 75% in the elderly.

$$1 \text{ mmHg} = 133.322 \text{ Pa}$$

1.1.4 Application of γ-Aminobutyric Acid

The development of GABA-enriched foods began in 1986. GABA-enriched teas were first successfully developed by Japan's Ministry of Agriculture, Forestry, and Fisheries in tea testing grounds for sale. In 1994, Japan's Ministry of Agriculture, Forestry, and Fisheries successfully developed GABA-enriched rice germs and rice bran at China agricultural testing ground. Then, Japan developed the GABA-enriched germinated brown rice and the healthy food materials with high GABA concentration which were fermented by *Lactobacillus* and yeast. In recent years, the research and development of GABA-enriched healthy food, including plants (rice germ, rice bran, green tea, pumpkin, etc.) and microorganisms (*Lactobacillus*, yeasts), have become the research focuses at home and abroad (Cao Jiaxuan et al. 2008).

Lactobacillus starter is essentially used in fermenting milk, which ferments glucose into lactic acid, providing the best environment for the formation of clots, texture, and flavor. The anaerobic fermentation of *Lactobacillus* increased acidity, which gave it the acid-resistant ability to maintain vitality at low pH, The low pH environment maintained the activity of glutamate decarboxylase in *Lactobacillus*. And glutamate decarboxylase consumed hydrion and reduced acidity, so that *Lactobacillus* could produce the functional dairy products with GABA activity. Nomura et al. (1998) reported that three of the seven commercial *Lactobacillus* starters could produce GABA in skim milk culture and cheese-ripening process, in which the yield was 383 mg/kg (Huang Yahui et al. 2005).

Fermented soybean food mostly used natural fermentation rich in microbial species. Lactic acid bacteria widely existed in various kinds of fermented soybean food. Lactic acid bacteria, producing GABA in fermented soybean food, improved the flavor of soybean fermentation food, and the glutamic acid and its sodium salt could be used to synthesize GABA, which provided a new idea for future research on increasing GABA content and enhancing the function of soybean fermentation food (Geng Jingzhang 2012).

1.2   Lactic Acid Bacteria and Diacetyl

1.2.1   Introduction of Diacetyl

Diacetyl, or butanedione (chemical name, 2,3-butanedione), is a yellow or light green liquid with the molecular formula $CH_3COCOCH_3$. When diluted to 1 mg/L, it has intensely buttery flavor and easily soluble in ethanol, ether, propylene glycol, glycerol, and water (Guo Zheng 1998; Xie Haiyan and Yin Dulin 2000; Ai 1984). Diacetyl has a relatively low flavor threshold (1.5~5.0 μg), so it can give the product a strong flavor at a low concentration. Acetoin is the reductive form, and the fragrance is far inferior to diacetyl. Natural diacetyl existed in Foeniculum vulgare, narcissus, tulip, raspberry, strawberry, lavender, citronella, rockrose, and cream (Xie Haiyan and Yin Dulin 2000). In the food industry, as a flavor enhancer, diacetyl was mainly used in soft drinks, cold drinks, baked food, candy, and other products. In addition, diacetyl was also one of the taste and flavor compounds that play an important role in fermented dairy products (Yang Lijie and Wang Junhu 2004). It was also an important unpleasant odor source of beer and wine (Yang Lijie and Wang Junhu 2004).

At present, *Lactobacillus* could be used alone or in combination with starter culture to produce yogurt with the diacetyl as main flavor compounds in european and american countries. Meanwhile, diacetyl was an essential flavor ingredient in butter, yogurt, cream, cheese, and many nondairy products that require milk flavor. And it was widely used in the preparation of flavoring butter essence, strawberry essence, essence of suckling pig feed, and synthetic fats flavor enhancer and also could be used as gelatin hardener and photographic adhesive. Although there are many methods to synthesize diacetyl, it was only manufactured by few producers in China. And because of the limitations of productive technology, the production of diacetyl had low yield, poor quality, and high cost and had the difficulty in meeting the requirements of replacing imported products. Diacetyl, which was used to produce essence and other products by perfumery plant, was imported from the Netherlands, Japan, and other countries. And the import price was around 200 yuan/kg. Domestic demand all year round was 30~40 t (Han Guangdian 1978; Wang Zhen 1992).

In addition, diacetyl could inhibit the growth of many microorganisms, including pathogenic bacteria and spoilage bacteria (Dan Tong and Zhang Heping 2013). Trace amounts of diacetyl inhibited the growth of *E. coli*. And with lower system pH, the bacteriostatic activity was strengthened (Jay et al. 1992), which indicated that lower pH enhanced the bacteriostatic activity of diacetyl (Meng Xiangchen 2009; Jay et al. 1992). On the contrary, glucose, acetate, and Tween-80 had an antagonistic effect on the antibacterial activity of diacetyl (Meng Xiangchen 2009). The reaction between diacetyl and arginine-binding protein interfered the use of arginine by gram-negative bacteria and inhibited bacterial growth (Zheng Yingfu et al. 2005).

1.2.2   Diacetyl Production of Lactic Acid Bacteria

Many microorganisms could produce diacetyl, such as *Streptococcus, Leuconstoc, Lactobacillus, Pediococcus, Oenococcus,* and other microbial species (Bartowsky and Henschke 2004). The leavening agents, including *Lactobacillus bulgaricus* and *Streptococcus thermophilus,* were used to ferment milk in China. The two strains had a symbiotic relationship and could promote the growth of each other in culture. Generally, *Streptococcus thermophilus* was considered as the major strain to produce diacetyl and had larger contribution to the flavor of fermented milk (Meng Xiangchen 2009). Scholars from the United States, the Netherlands, Denmark, and other countries used genetic engineering and mutant breeding technology to select and transform *Lactobacillus* with high diacetyl (Hua Chaoli and Zhao Zheng 2004). However, it was still in the beginning in China. Song Huanlu (2002) studied the production of diacetyl by *Lactobacillus* D21 and *Lactococcus lactis* NCIMB 8763 in milk and MRS medium. Yu Peng et al. (2006) used nitrosoguanidine to mutate and breed the strain with high diacetyl. Liu Fang et al. (2006) conducted a large-scale screening of *Streptococcus thermophilus* that produced high diacetyl. Benson et al. (1996) overexpressed the coding genes for α-acetolactic acid synthetase in α-acetolactic acid decarboxylase-defective *Lactococcus* and made the diacetyl production increased significantly. Boumerdassi et al. (1997) overexpressed the coding genes for NADH oxidase (nox gene) in α-acetolactic acid decarboxylase-defective *Lactococcus.* The production of diacetyl was increased by more than ten times, and the growth of engineered bacteria remained unchanged.

1.2.3   Production Way and Metabolic Regulation of Diacetyl in Lactobacillus

1.2.3.1   Production Way of Diacetyl in *Lactobacillus*

1.2.3.1.1   Production of Diacetyl by Citric Acid Metabolic Pathways

Diacetyl is the key aroma of many fermented dairy products and also produces unpleasant odor in wine and lemon juice (Jordan and Cogan 1988). The microorganism used to produce diacetyl was mainly *Lactobacillus* which could use citric acid. Metabolism of citric acid by lactic acid bacteria was shown in Fig. 1.1. As was well

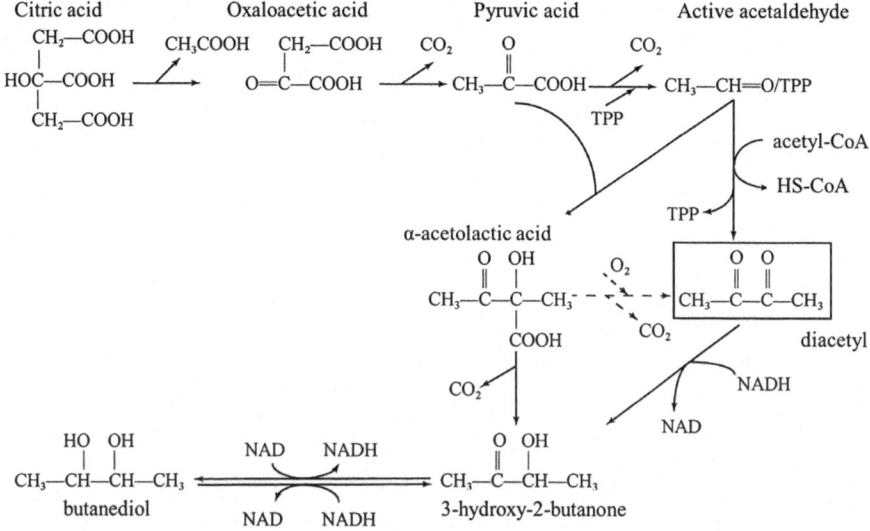

Fig. 1.1 Citric acid metabolic pathways in *Lactobacillus*

known, the two features of lactose fermentation and lactic acid production were important to lactic acid bacteria in dairy products. And screening the strain that could use citric acid to produce diacetyl was an important industry characteristic in production of dairy products (Ma Guihua 1989; Borts 1963; Rodríguez et al. 2012). Lactic acid bacteria which were able to take advantage of citric acid to produce acetoin could be used to produce diacetyl. Therefore, lactic acid bacteria that produced diacetyl included *Leuconostoc dextranicum*, *Leuconostoc citreum*, *Streptococcus lactis*, and some *Bacterium lacticum* (Yang Jiebin et al. 1996).

Citric acid was one of the most important substrates in the generation of diacetyl by *Lactobacillus*. Citric acid was transferred into the cell by enzymes and then lysed into acetic acid and oxaloacetic acid by lyase. Oxaloacetic acid was decomposed into CO_2 and pyruvic acid, while the process of pyruvate to diacetyl had not been clearly explained (Hemme and Foucaud-Scheunemann 2004; Mcsweeney and Sousa 2000). It is currently believed to be in two ways: The first was proposed by Speckman and Collins (1968). It illustrated that active acetaldehyde was synthesized from pyruvate and pyrithiamine, then condensed with acetyl-CoA, and split into diacetyl by diacetyl synthetase. However, so far there was no direct evidence of diacetyl synthetase. The second was proposed by Deman. The active acetaldehyde was condensed with other pyruvate to form α-acetolactic acid by its synthetase (Monnet et al. 1994), and then α-acetolactic acid was oxidized to form diacetyl (Cogan et al. 1981). When in hypoxia, the outcome was 3-hydroxy-2-butanone and then reduced to 2,3-butanediol with no obvious smell. Both of 3-hydroxy-2-butanone and diacetyl had fragrant smell. The process of oxidative decarboxylation had been supported by the data of biochemistry, genetics, and nuclear magnetic resonance (Marth and Steele 1998). Diacetyl could be reduced by reductase to acetoin (Cogan et al. 1981; Seitz et al. 1963). The catalytic activity of the enzyme had great difference between different species or strains and, to a certain extent, was affected by pH. Alpha-acetolactic acid was converted to acetoin by its decarboxylase. O'Sullivan et al. (2001) had successfully separated and purified the α-acetolactic acid decarboxylase in *Leuconostoc*, determined the amino acid sequence, and confirmed the existence of the oxidative decarboxylation.

1.2.3.1.2 Production of Diacetyl by Glucose Metabolic Pathways

The metabolic pathways of diacetyl by lactic acid bacteria were mainly citric acid metabolic and glycolytic pathways. Many *Lactobacillus*, especially *Lactococcus*, could use lactose and citric acid to produce pyruvate and synthesize diacetyl (Fig. 1.2). During the production of diacetyl from lactose, the excessive pyruvate was used to synthesize α-acetolactic acid by its synthetase, and then diacetyl was generated by oxidative decarboxylation reaction under the acidic condition. Diacetyl was very unstable and could be reduced by reductase to acetoin. In the metabolic pathways of citric acid to diacetyl, pyruvate and α-acetolactic acid were the most important intermediates. Citric acid was lysed into oxaloacetic acid by lyase and synthesizes to α-acetolactic acid from oxaloacetic acid directly or produced to pyruvate by decarboxylic reaction. Moreover, pyruvate decarboxylation reaction generated active acetaldehyde, which could be used with pyruvate to synthesize α-acetolactic acid by its synthetase. And then diacetyl was generated by oxidative decarboxylation reaction. Alpha-acetolactic acid was converted to acetoin by its decarboxylase. Because of the low content of citric acid in milk, in general, diacetyl was generated by glycolytic pathway in the fermented milk (Aymes et al. 1999; Hugenholtz 1993; Curic et al. 1999).

During the process of glucose metabolism in *Lactococcus*, pyruvate was dehydrogenized by lactic dehydrogenase, became lactic acid, and transformed to acetyl CoA by pyruvate dehydrogenase and formate lyase. Furthermore, acetyl CoA became ethanol and acetaldehyde. Pyruvate was synthesized to α-acetolactic acid

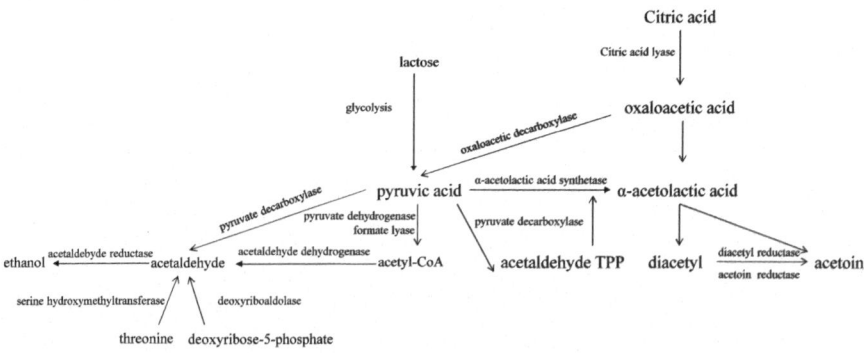

Fig. 1.2 Synthesis pathways of acetaldehyde and diacetyl in *Lactobacillus*

by its synthetase. And diacetyl was generated by nonenzymatic natural oxidative decarboxylation reaction under the acidic condition. Diacetyl could be reduced by reductase to acetoin, which had been supported by a large number of biochemical and molecular biological data. In recent years, more and more studies confirmed that the main pathway was the nonenzymatic natural oxidative decarboxylation reaction which made α-acetolactic acid to become diacetyl in *Lactococcus* (Snoep et al. 1992).

1.2.3.2   Metabolic Regulation of Diacetyl in Lactic Acid Bacteria

1.2.3.2.1   Regulation of Culture Conditions

Production and stability of diacetyl depended on the growing environment of lactic acid bacteria, including nutrient composition, temperature, pH, and the presence of oxygen. Due to the established dairy technology, many of the external factors could not be controlled in the fermented dairy products processing. Even in lactic acid bacteria that could ferment citric acid, the capacity of diacetyl was relatively low in dairy products. Researcher focused on the regulation of culture conditions to increase the content of diacetyl in cheese and sour milk. In the production of dairy products, citric acid fermentation strains were often mixed with acid-produced strains to improve oxygen concentration and production of diacetyl (Boumerdassi et al. 1997). In addition, oxygen could oxidize the NADH and decreased the reductive ratio of diacetyl to acetoin.

1.2.3.2.2   Regulation of α-Acetolactic Acid

The metabolism of pyruvate and citrate in *Lactococcus* and *Leuconostoc* confirmed that diacetyl was produced with α-acetolactic acid by nonenzymatic natural oxidative decarboxylation reaction. Alpha-acetolactic acid was transformed to acetoin by α-acetolactic acid decarboxylase in *Lactococcus*, which irreversibly decreased α-acetolactic acid. And the metabolism regulation of α-acetolactic acid was a possible way to improve the content of diacetyl. Based on the understanding of diacetyl generation pathways, several metabolic engineering strategies have been designed to increase the content of diacetyl in lactic acid bacteria. Because of the very low content of citric acid in milk, researches all focused on the transformation of lactose to diacetyl. Researchers had tried several methods as follows: devitalized the gene of lactate dehydrogenase (LDH) (Gasson et al. 1996); overexpressed the genes of ilvBN or alsS of α-acetolactic acid in anabolism and catabolism; devitalized the genes of pyruvate lyase (Pfl2) and α-acetolactic acid decarboxylase (aldB) (Monnet et al. 1997; Yang and Wang 1996); and overexpressed the gene of NADH oxidase (nox) (Benson et al. 1996; Felipe et al. 1998). These strategies also could be combined to make lactose and glucose efficiently transform into acetone, especially in the inactivation of lactate dehydrogenase (LDH).

The inactivation of LDH led to excessive accumulation of pyruvate. In theory, the mutant strain produced more diacetyl than the wild strain. Studies had confirmed that site-specific mutation of LDH in *Lactococcus* was feasible to change the pathway of glycometabolism. And the pyruvate couldn't be converted to lactate but to a mixed acid fermentation pathway with formic acid and alcohol as the main product. Under aerobic conditions, the accumulation of pyruvate led to a large amount of acetoin and butanediol.

Increasing the copy number of α-acetolactic acid gene and NADH oxidase gene improved the content of α-acetolactic acid and eventually enhanced the production of diacetyl (Benson et al. 1996; Felipe et al. 1998). Marugg et al. (1994) cloned and expressed the gene of α-acetolactic acid synthase (als) and analyzed the characteristics in diacetyl subspecies of *Lactococcus lactis* (Marugg et al. 1994). When the gene of α-acetolactic acid synthase was devitalized or the enzyme was deficient in *Lactococcus lactis*, more than 80% of the pyruvate was converted under the appropriate conditions.

One of the most promising ways to increase the production of diacetyl was to inactivate the aldB gene, which leads to the accumulation of α-acetolactic acid as the precursor of diacetyl (Goupil et al. 1996). The inactivation of α-acetolactic acid decarboxylase directly led to the conversion of α-acetolactic acid to acetoin, therefore, increasing the diacetyl (Monnet et al. 1997; Yang and Wang 1996). Due to the redox balance, sugar was converted to pyruvic and then to lactic acid. Meanwhile, citric acid was converted to α-acetolactic acid and then was decarboxylated into acetone. In dairy fermentation, through the chemical decarboxylation of α-acetolactic acid, these bacteria could produce a large amount of diacetyl directly. New screening methods and genetic engineering had made the use of these bacteria possible. The aldB mutant strain of *Lactococcus* could grow in leucine medium without adding other branched-chain amino acids to achieve separation. The theoretical basis was that the activity of Ald was regulated by allosteric modification of leucine, which could directly activate Ald, and that α-acetolactic acid was the precursor of leucine and valine. Most of lactococci irreversibly lost the ability to synthesize branched-chain amino acids (Godon et al. 1993), and the growth was restricted in the screening medium (Godon et al. 1993).

1.2.3.2.3 Regulation of Mixed Acid

Under different fermentation conditions, researchers studied the production of diacetyl of *Lactobacillus* or its precursor, α-acetolactic acid, and the results showed that the production of diacetyl was related to the utilization of citric acid. Because of the low content of citric acid in foods, diacetyl could be produced efficiently with sugars such as lactose or fructose.

In the production of dairy products, *Lactobacillus* had the potential to produce various terminal products in the sugar fermentation, such as lactic acid, acetic acid, diacetyl, acetaldehyde, or extracellular polysaccharide, which was beneficial to the flavor and structure formation of dairy products and had important economic value.

Generally, *Lactobacillus* was the homolactic fermentation, and metabolic changes have been found under certain conditions. For example, when *Lactobacillus* used glucose as the sole carbon source and energy, under restrictive carbon sources or aerobic conditions, the terminal products included other mixed acids besides lactic acid. Recent studies had indicated that the decrease of glycometabolism rate led to the variation of homolactic fermentation to mixed acid fermentation. Compared with control, when high glycolysis was applied during metabolism under aerobic conditions, it had been observed that metabolism flowed to mixed acid fermentation. Direct oxidation of NADH, which was required for pyruvate reduction, led to a decrease of flowing to lactic acid via LDH. As an indicator of cellular redox, the ratio of NAD^+ to NADH was directly affected by the activity of NADH oxidase, which determines the flow direction of metabolism.

Lactobacillus usually conducted glycometabolism through homolactic fermentation and heterolactic fermentation. The terminal products of glycometabolism could be produced by a variety of ways, not only included lactic acid in homozygotic strain, the terminal products of glycometabolism could be produced by a variety of ways, not only included lactic acid (Fig. 1.3). Usually these fermentation products were formed by the accumulation of pyruvate and the ratio of NADH to NAD+. When the concentration of intracellular pyruvate exceeded lactic acid that was produced by lactate dehydrogenase, it should supply other pathways of pyruvate removal, providing NADH oxidation. These alternate pathways also provided the cells with the means to produce extra ATP. What conditions and circumstances did the pyruvate will be accumulated? When substrate fermentation is limited, the lack of fructose-1,6-diphosphate, low activity of isomerases, lactic dehydrogenase, pyruvate would be accumulated. With high oxygen content in the environment, NADH, which acted on the reduction of pyruvate, could be directly oxidized, preventing lactate dehydrogenase reaction.

The enzymes and pathways were identified, which were involved in making pyruvate to form lactic acid or other products in *Lactococcus* and other *Lactobacillus* (Cocaign-Bousquet et al. 1996; Garrigues et al. 1997). Under anaerobic conditions, when carbohydrates were limited, low growth rate led to mixed acid fermentation, forming alcohol, acetic acid, and formate. Pyruvate formate lyase was activated to break down pyruvate into formate and acetyl CoA, which could be converted into ethanol and/or acetic acid. ATP also could be formed from acetic acid by acetokinase.

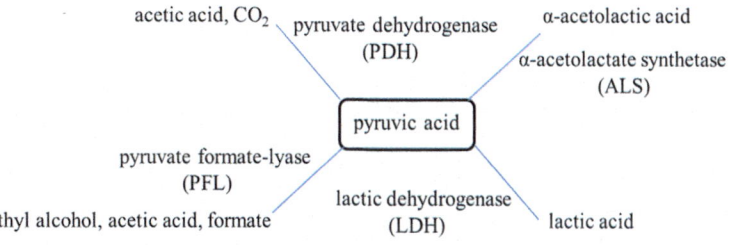

Fig. 1.3 Mixed acid fermentation of pyruvate

Under aerobic conditions, the inactivation of pyruvate formate lyase led to the conversion of pyruvate into acetic acid and CO_2 by pyruvate dehydrogenase. Finally, excess pyruvate was able to produce α-acetolactic acid by its synthase, and the reaction had other important implications, because α-acetolactic acid was the precursor of diacetyl.

These alternative pathways of pyruvate metabolism were greatly influenced by environmental conditions, and mutant strains that could not produce lactic dehydrogenase must use excess pyruvate to produce other terminal products. Under certain conditions, excess pyruvate might be converted to other products in cells, especially flavor substance, such as diacetyl. Diacetyl was usually produced from citric acid, but if the right conditions were established or the cells were generally modified, cells that did not ferment citric acid could use lactose to form diacetyl. For example, the overexpression of NADH oxidase in *Lactococcus* resulted in the reduction of lactic acid from pyruvate while forming the precursor α-acetolactic acid of diacetyl (Felipe et al. 1998). Studies had also found that glycometabolism of *Lactobacillus* could be manipulated by changes in the concentration of pyruvate, an intermediate product of metabolic engineering.

1.2.4   Strategy for Increasing Production of Diacetyl in Lactobacillus

1.2.4.1   Optimized the Culture Conditions to Increase the Yield of Diacetyl

The yield and stability of diacetyl were affected by many external factors in the fermentation of *Lactococcus*, such as the type and proportion of carbon and nitrogen sources in the culture medium, growth temperature, pH, oxygen, and so on. Studies have found that under aerobic conditions, the yield of diacetyl was much higher than that under anaerobic conditions (Curic et al. 1999). Under aerobic conditions, the supplementation of citric acid in the medium could increase the yield of diacetyl. The reason might be that under aerobic conditions, oxygen acted as an electron receptor to promote the production of diacetyl during the process of the formation through decarboxylation of α-acetolactic acid. And the supplementation of citric acid led to further accumulation of pyruvate, resulting in the increased production of diacetyl (Levata-Jovanovic and Sandine 1996). Therefore, in the fermentation industry, in order to prevent the reduction reaction of diacetyl to form acetoin, making fermentation dairy products lose their required flavor, the method of adding citric acid or sodium citrate was often used to strengthen the milk to guarantee the synthesis of diacetyl with constant and high yield. When it had achieved the desired acidity and flavor, the product would be immediately cooled to inhibit the activity of diacetyl reductase and decrease its damage to the diacetyl. In addition, it should prevent the contamination of psychrophilic bacteria with high diacetyl reductase activity. Optimization of culture conditions could improve the diacetyl production capacity of *Lactococcus* to some extent but at the same time might affect the proportion of other nutrients in the fermentation products and might produce some harmful substances (Yang and Wang 1996).

1.2.4.2   Application of Genetic Engineering and Metabolic Engineering to Increase the Yield of Diacetyl

The yield of diacetyl in yogurt fermentation was closely related to the activity of several key enzymes, including the citrate lyase, oxaloacetic decarboxylase, α-acetyl synthase, NADH oxidase, α-acetolactate decarboxylase, diacetyl reductase, and so on. These enzymes could be inactivated or overexpressed by using molecular biology technology, which could increase the accumulation of the intermediate product pyruvate, α-acetolactic acid, change the metabolic pathway of diacetyl, and increase the yield of diacetyl. In order to change the metabolic pathway of α-acetolactic acid to acetoin, some domestic and foreign scholars had focused on the screening and construction of some strains with α-acetolactic acid decarboxylase deficiency. Aymes et al. (1999) used random mutagenesis to screen the subspecies of *Lactococcus lactis* butanedione with α-acetolactic acid decarboxylase deficiency, and under anaerobic conditions, the yield of diacetyl was much higher than that of the wild strain. In order to increase the yield of the diacetyl in *streptococcus thermophiles*, Monnet and Corrieu (2007) used the method of directional mutation to select the *Streptococcus thermophilus* with α-acetolactic acid decarboxylase deficiency in the medium containing ketobutyric acid, leucine, isoleucine. Under the microaerobic conditions, the yield of diacetyl in mutant strain was nearly 3 times higher than that of the wild strain. On the basis of screening and construction of the strain with α-acetolactic acid decarboxylase deficiency, some researchers overexpressed the key genes which regulated the production of diacetyl, and further attempts were made to improve the yield of diacetyl by means of joint regulation. Hugenholtz et al. (1993) found that the overexpression of nox gene regulating NADH oxidase under aerobic conditions significantly increased the content of diacetyl in the *Lactococcus lactis* with α-acetolactic acid decarboxylase deficiency. At present, the gene regulation and metabolism of diacetyl in most studies focused on the regulation of some key genes such as glycolysis, citric acid metabolic pathway. With detailed studies, more new products should be produced that meet the demands of consumers.

1.2.5   Application of Diacetyl in Foods

Diacetyl inhibited many microorganisms, including spoilage and pathogenic bacteria. Low concentrations (several milligrams per liter) of diacetyl inhibited the growth of *E. coli*, and the antibacterial activity of diacetyl increased as pH decreased. Some scholars had confirmed the synergistic effect of diacetyl antibacterial activity at low pH (Jay et al. 1992). Diacetyl inhibited the growth of gram-negative bacteria by interfering with the use of arginine by reacting with the arginine-binding protein of gram-negative bacteria.

Diacetyl was considered to be one of the most important ingredients in the flavor of dairy products, as well as an important flavor in butter, cream, cheese, and many nondairy products that require milk flavor (Hugenholtz et al. 2000). Most *Lactobacillus* such as *Lactococcus lactis, Streptococcus lactis,* and *Streptococcus thermophilus* could produce diacetyl (Aymes et al. 1999). However, diacetyl was the main factor affecting the maturation period of beer flavor in beer fermentation. When the content exceeded the threshold ($0.15 \times 10{-}6$), it would produce an unpleasant sour beer taste (Bartowsky and Henschke 2004).

References

Ai M (1984) Oxidation of methyl ethyl ketone to diacetyl on V2O5-P2O5 catalysts. J Catal 89:413–421

Antonaccio MJ, Taylor DG (1977) Involvement of central GABA receptors in the regulation of blood pressure and heart rate of anesthetized cats. Eur J Pharmacol 46:283–287

Awapara J, Landua AJ, Fuerst R et al (1950) Free gamma-aminobutyric acid in brain. J Biol Chem 187:35–39

Aymes F, Monnet C, Corrieu G (1999) Effect of alpha-acetolactate decarboxylase inactivation on alpha-acetolactate and diacetyl production by Lactococcus lactis subsp. lactis biovar. Diacetylactis. J Biosci Bioeng 87:87–92

Bartowsky EJ, Henschke PA (2004) The 'buttery' attribute of wine-diacetyl-desirability, spoilage and beyond. Int J Food Microbiol 96:235–252

Benson KH, Godon JJ, Renault P et al (1996) Effect of ilvBN-encoded α-acetolactate synthase expression on diacetyl production in Lactococcus lactis. Appl Microbiol Biotechnol 45:107–111

Bian Shuling, Zhang Wei, Zhu Hui et al (2002) Effect of γ-aminobutyric acid on the sperm acrosin activity. Natl J Androl 8:326–328

Borts IH (1963) Dairy bacteriology. Am J Public Health Nations Health 200:529

Boumerdassi H, Monnet C, Desmazeaud M et al (1997) Isolation and properties of Lactococcus lactis subsp. lactis biovar. Diacetylactis CNRZ 483 mutants producing diacetyl and acetoin from glucose. Appl Environ Microbiol 63:2293–2299

Cao Jiaxuan, Li Yuping, Xiong Xiangyuan et al (2008) Applications of γ-aminobutyric acid in functional foods. J Hebei Agric Sci 12:52–54

Chen Lilong, Jiang Qingyan, Xiao Shi (2010) Biological function of γ-aminobutyric acid and its application as a novel feed additive. Feed Ind 31:1–3

Cocaign-Bousquet M, Garrigues C, Loubiere P et al (1996) Physiology of pyruvate metabolism in Lactococcus lactis. Antonie Van Leeuwenhoek 70:253–267

Cogan TM, O'Dowd M, Mellerick D (1981) Effects of pH and sugar on acetoin production from citrate by Leuconostoc lactis. Appl Environ Microbiol 41:1–8

Curic M, Richelieu MD, Henriksen CM et al (1999) Glucose/citrate cometabolism in Lactococcus lactis subsp. lactis biovar. Diacetylactis with impaired α-acetolactate decarboxylase. Metab Eng 1:291–298

Dan Tong, Zhang Heping (2013) Classification, biosynthesis and their applications of bacteriocins produced from lactic acid Bacteria. Zhonggue Rupin Gongye 41:29–32

Fan E, Huang J, Hu S et al (2012) Cloning, sequencing and expression of a glutamate decarboxylase gene from the GABA-producing strain Lactobacillus brevis CGMCC 1306. Ann Microbiol 62:689–698

Felipe FLD, Kleerebezem M, Vos WMD et al (1998) Cofactor engineering: a novel approach to metabolic engineering in Lactococcus lactis by controlled expression of NADH oxidase. J Bacteriol 180:3804–3808

Cui Xiaojun, Jiang Bo, Feng Biao (2005) Optimization of fermentation conditions for GABA (γ-aminobutyric acid) production by lactobacillus SK005. Food Res Dev 26:64–69

Garrigues C, Loubiere P, Lindley ND et al (1997) Control of the shift from homolactic acid to mixed-acid fermentation in Lactococcus lactis: predominant role of the NADH/NAD+ ratio. J Bacteriol 179:5282

Gasson MJ, Benson K, Swindell S et al (1996) Metabolic engineering of the Lactococcus lactis diacetyl pathway. Dairy Sci Technol 76:33–40

Geng Jingzhang (2012) Research on use of gamma-amino butyric acid (GABA) in food industry. Beverage Ind 15:11–14

Godon JJ, Delorme C, Bardowski J et al (1993) Gene inactivation in Lactococcus lactis: branched-chain amino acid biosynthesis. J Bacteriol 175:4383–4390

Goupil N, Corthier G, Ehrlich SD et al (1996) Imbalance of leucine flux in Lactococcus lactis and its use for the isolation of diacetyl-overproducing strains. Appl Environ Microbiol 62:2636–2640

Guo Zheng (1998) Research of Butanedione synthesis technology. Zhejiang Chem Ind 2:22–23

Han Guangdian (1978) Handbook of organic preparation chemistry. Chemical Industry Press, Bei Jing

Hayakawa K, Kimura M, Kasaha K et al (2004) Effect of a gamma-aminobutyric acid-enriched dairy product on the blood pressure of spontaneously hypertensive and normotensive Wistar-Kyoto rats. Br J Nutr 92:411–417

He Xipu, Zhang Min, Li Junfang et al (2007) The physiological function of γ -aminobutyric acid and the general research about γ -aminobutyric acid. J Guangxi Univ Nat Sci Ed 32:464–466

Hemme D, Foucaud-Scheunemann C (2004) Leuconostoc, characteristics, use in dairy technology and prospects in functional foods. Int Dairy J 14:467–494

Hua Chaoli, Zhao Zheng (2004) Studies on a nes ketone flavor yogurt co-fermented by Lactobacillus helveticus and Streptococcus diacetylactis. Zhonggue Rupin Gongye 32:17–20

Huang YH, Zheng HF, Liu XL, Wang X et al (2005) Studies of the variation of GABA and Glu in Gabaron tea process. Food Sci 26:117–120

Hugenholtz J (1993) Citrate metabolism in lactic acid bacteria. FEMS Microbiol Rev 12:165–178

Hugenholtz J, Kleerebezem M, Starrenburg M et al (2000) Lactococcus lactis as a cell factory for high-level diacetyl production. Appl Environ Microbiol 66:4112–4114

Jay JM, Loessner MJ, Golden DA (1992) Modern food microbiology. Chapman & Hall, New York

Ji Linli (2008) The screening and identification of LAB strains isolated from traditional dairy products with γ-amino butyric acid producing and optimizing their fermentation conditions. Inner Mongolia agricultural university, Hu He Hao Te

Jordan KN, Cogan TM (1988) Production of acetolactate by Streptococcus diacetylactis and Leuconostoc spp. J Dairy Res 55:227–238

Kazami D, Ogura N, Fukuchi T et al (2002) Antihypertensive effect of Japanese taste seasoning containing γ-amino butyric acid on mildly hypertensive and high-normal blood pressure. Nippon Shokuhin Kagaku Kogaku Kaishi 49:409–415

Komatsuzaki N, Shima J, Kawamoto S et al (2005) Production of γ-aminobutyric acid (GABA) by Lactobacillus paracasei isolated from traditional fermented foods. Food Microbiol 22:497–504

Krajnc D, Neff N, Hadjiconstantinou M (1996) Glutamate, glutamine and glutamine synthetase in the neonatal rat brain following hypoxia. Brain Res 707:134–137

Krnjević K, Schwartz S (1966) Is gamma-aminobutyric acid an inhibitory transmitter? Nature 211:1372–1374

Levata-Jovanovic M, Sandine WE (1996) Citrate utilization and diacetyl production by various strains of Leuconostoc mesenteroides ssp. Cremoris 1. J Dairy Sci 79:1928–1935

Leventhal AG, Wang Y, Pu M et al (2003) GABA and its agonists improved visual cortical function in senescent monkeys. Science 300:812–815

Liu Fang, Wang Yutang, Huo Guicheng (2006) Screening and identification of S. Thermophiles producing diacetyl. J Dairy Sci Technol 29:272–275

Ma Guihua (1989) Lactobacillus and human health. Food Herald:10–12

Marth EH, Steele JL (1998) Applied dairy microbiology. Marcel Dekker, New York

Marugg JD, Goelling D, Stahl U et al (1994) Identification and characterization of the alpha-acetolactate synthase gene from Lactococcus lactis subsp. lactis biovar. Diacetylactis. Appl Environ Microbiol 60:1390–1394

Mcsweeney PLH, Sousa MJ (2000) Biochemical pathways for the production of flavour compounds in cheeses during ripening: a review. Lait 80:293–324

Meng Xiangchen (2009) Lactic acid Bacteria and dairy starter culture. Science Press, Bei Jing

Mombereau C, Kaupmann K, Froestl W et al (2004) Genetic and pharmacological evidence of a role for GABA (B) receptors in the modulation of anxiety-and antidepressant-like behavior. Neuropsychopharmacology 29:1050–1062

Monnet C, Corrieu G (2007) Selection and properties of alpha-acetolactate decarboxylase-deficient spontaneous mutants of Streptococcus thermophilus. Food Microbiol 24:601–606

Monnet C, Schmilt P, Divies C (1994) Diacetyl production in milk by an α-acetolactic acid accumulating strain of Lactococcus lactis ssp. lactis biovar. Diacetylactis. J Dairy Sci 77:2916–2924

Monnet C, Schmitt P, Divies C (1997) Development and use of a screening procedure for production of alpha-acetolactate by Lactococcus lactis subsp. lactis biovar. Diacetylactis strains. Appl Environ Microbiol 63:793–795

Murashima YL, Kato T (1986) Distribution of gamma-aminobutyric acid and glutamate decarboxylase in the layers of rat oviduct. J Neurochem 46:166–172

Nomura M, Kimoto H, Someya Y et al (1998) Production of gamma-aminobutyric acid by cheese starters during cheese ripening. J Dairy Sci 81:1486–1491

O'Sullivan SM, Condon S, Cogan TM et al (2001) Purification and characterisation of acetolactate decarboxylase from Leuconostoc lactis NCW1. FEMS Microbiol Lett 194:245–249

Okada T, Sugishita T, Murakami T et al (2000) Effect of the defatted rice germ enriched with GABA for sleeplessness, depression, autonomic disorder by oral administration. J Jpn Soc Food Sci Technol Nippon Shokuhin Kagaku Kogaku Kaishi 47:596–560

Roberts E, Frankel S (1950) Gamma-aminobutyric acid in brain: its formation from glutamic acid. J Biol Chem 187:55–63

Rodríguez A, Martínez B, Suárez J (2012) Dairy starter cultures. CRC Press, Boca Raton

Roldan ER, Murase T, Shi QX (1994) Exocytosis in spermatozoa in response to progesterone and zona pellucida. Science 266:1578–1581

Sawai Y, Yamaguchi Y, Miyama D et al (2001) Cycling treatment of anaerobic and aerobic incubation increases the content of gamma-aminobutyric acid in tea shoots. Amino Acids 20:331–334

Seitz EW, Sandine WE, Elliker PR et al (1963) Distribution of diacetyl reductase among bacteria. J Dairy Sci 46:186–189

Siragusa S, Angelis MD, Cagno RD et al (2007) Synthesis of gamma-aminobutyric acid by lactic acid bacteria isolated from a variety of Italian cheeses. Appl Environ Microbiol 73:7283–7290

Snoep JL, Mj TDM, Starrenburg MJ et al (1992) Isolation, characterization, and physiological role of the pyruvate dehydrogenase complex and alpha-acetolactate synthase of Lactococcus lactis subsp. lactis biovar. Diacetylactis. J Bacteriol 174:4838–4841

Song Huanlu (2002) The primary study on Diacetyl biosynthesis by lactic acid Bacteria. Food Ferment Ind 28:47–50

Song Wei, Ma Xia, Zhang Bailin (2008) Physiological benefits and fortifications of γ-Aminobutyric Acid in dairy products. J Dairy Sci Technol 31:297–302

Speckman RA, Collins EB (1968) Separation of diacetyl, acetoin, and 2, 3-butylene glycol by salting-out chromatography. Anal Biochem 22:154–160

Takahashi H, Tiba M, Yamazaki T et al (1959) On the site of action of gamma-aminobutyric acid on blood pressure. Jpn J Physiol 8:378–390

Udenfriend S (1950) Identification of gamma-aminobutyric acid in brain by the isotope derivative method. J Biol Chem 187:65–69

Usuki S, Ito Y, Morikawa K et al (2007) Effect of pre-germinated brown rice intake on diabetic neuropathy in streptozotocin- induced diabetic rats. Nutr Metab 4:25

Wang Zhen (1992) Dictionary of chemical technology. Chemical Industry Press, Bei Jing

Xia Jiang (2006) Breeding of γ-aminobutyric acid-producing lactobacillus and optimization of fermentation conditions. Zhejiang University, Hang Zhou

Xia Jiang, Mei Lehe, Huang Jun et al (2006) Screening and mutagenesis of Lactobacillus brevis for biosynthesis of γ-aminobutyric acid. J Nucl Agric Sci 20:379–382

Xian Qianlong (2013) Selection of γ-aminobutyric acid-producing lactic acid Bacteria and the development of functional yoghurt. Guangxi University of Technology, Liu Zhou

Xie Haiyan, Yin Dulin (2000) Catalytic oxidation of Butan-2-one to Diacetyl. Hunan Chem Ind 30:22–23

Xu Jianjun, Jiang Bo, Xu Shiying (2002) Screening of lactic acid Bacteria for biosynthesis of γ-amino butyric acid. Food Sci Technol:7–8

Yang Jiebin, Guo Xinghua, Zhang Chi et al (1996) Lactic acid Bacteria: biological basis and application. China Light Industry Press, Bei Jing

Yang Lijie, Wang Junhu (2004) Genetic manipulation of the pathway for diacetyl metabolism in Lactococcus lactis. Zhonggue Rupin Gongye 32:24–29

Yang LJ, Wang JH (1996) Genetic manipulation of the pathway for diacetyl metabolism in Lactococcus lactis. Appl Environ Microbiol 62:2641–2643

Yu Peng, Zhang Lanwei, Xu Qian et al (2006) Screening mutagenized Lactococcus Lactis subsp. lactis Biovar Diacetyl strains overproducing Diacetyl. J Dairy Sci Technol 29:218–220

Zheng Yingfu, Han Zhenrong, Zhao Chunhai (2005) A review on improving diacetyl formation in Lactococcus lactis. China Biotechnol 25:186–189

Chapter 2
Lactic Acid Bacteria and Conjugated Fatty Acids

Wei Chen, Bo Yang, and Jianxin Zhao

2.1 Introduction

Conjugated fatty acid (CFA) refers to a group of positional and geometric isomers of polyunsaturated fatty acid possessing conjugated double bonds. Conjugated double bonds, conjugated triple bonds, and conjugated quadruple bonds are the typical conjugated fatty acid forms, in which conjugated octadecadienoic acid and conjugated octadecatrienoic acid are most common isomers, such as conjugated linoleic acid (CLA), conjugated linolenic acid (CLNA), and conjugated steariconic acid CSA (Yang et al. 2015).

2.1.1 Conjugated Linoleic Acid

Conjugated linoleic acid (CLA) is a generic term of octadecadienoic acid with conjugated double bonds, referring to a group positional and geometric isomer of linoleic acid (LA), in which each conjugated double bond exists in two types, *cis* (*c*) and *trans* (*t*). In theory, according to the position of double bonds, 54 isomers of CLA could be synthesized; however, until now, only 28 isomers have been identified, including conjugated double bond on C_7,C_9, C_8,C_{10}, C_9, C_{11}, C_{10},C_{12}, C_{11}, and C_{13}. *c*9, *t*11-CLA (rumenic acid) was the most abundant CLA isomer, followed by *t*10,*c*12-CLA (Andrade et al. 2012).

CLA has attracted much attention due to its physiological effects, such as anti-inflammation, anticancer, reduction of atherosclerosis, anti-obesity, amelioration of

W. Chen (✉) · B. Yang · J. Zhao
Jiangnan University, Wuxi, China
e-mail: chenwei66@jiangnan.edu.cn; jxzhao@jiangnan.edu.cn

© Springer Nature Singapore Pte Ltd. and Science Press 2019
W. Chen (ed.), *Lactic Acid Bacteria*,
https://doi.org/10.1007/978-981-13-7283-4_2

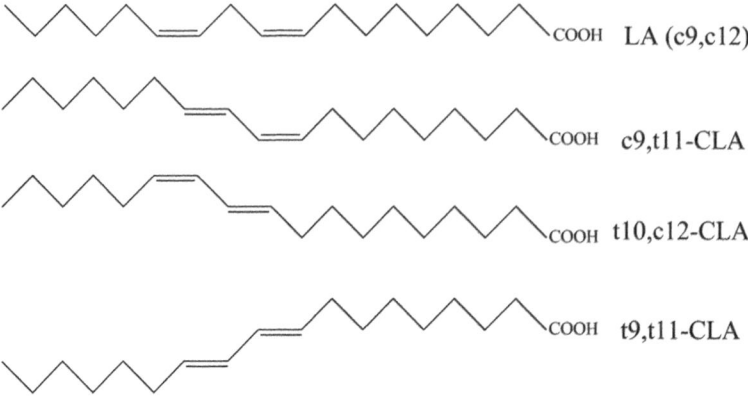

Fig. 2.1 Structure of linoleic acid and major CLA isomers

diabetes, promotion of bone growth, and immune regulation. As reported, the biological function was isomer-dependent, in which $c9,t11$-CLA and $t10,c12$-CLA were recognized as the CLA isomers with best physiological effects (Fig. 2.1). The major physiological functions of $c9,t11$-CLA were anti-cancer, anti-inflammation, and immune regulation, whereas $t10,c12$-CLA has significant benefits on anti-obesity and regulation of lipid metabolism. Additionally, $t9,t11$-CLA was reported with anti-inflammation function (Yang et al. 2015).

CLA naturally occurs in ruminant milk and issues; therefore, ruminant dairy and meat products are the main source of CLA in the daily diet, in which $c9, t11$-CLA comprised of 80–90% fatty acid of the total dairy lipids and $t10,c12$-CLA comprised of only 1% of total dairy lipids. Moreover, other CLA isomers, such as $t7,c9$-CLA, $c8,t10$-CLA, $t10,c12$-CLA, and $t11,c13$-CLA, could be detected in the milk (Jensen 2002). In ruminant animals, two major sources of CLA were reported: (1) CLA was mainly produced as one of the intermediates by some ruminant bacteria in the process of catalyzing LA into stearic acid (C18:0), and (2) numerous researches have reported that $c9,t11$-CLA could be generated by $\Delta9$-dehydrogenase in the mammary gland with vaccenic acid ($t11$-C18:1) as substrate. Many studies on the CLA-producing mechanism in ruminant bacteria have been carried out (Kepler et al. 1966, 1971; Kepler and Tove 1967; Polan et al. 1964; Rosenfeld and Tove 1971) that LA could be quickly transformed into CLA by linoleic acid isomerase in ruminant bacteria and then transferred into vaccenic acid at a slower rate. After vaccenic acid was accumulated to a certain level, it would be further transformed to stearic acid. Other studies have demonstrated that some vaccenic acid in ruminant animals could be absorbed and then transported to other tissues. Vaccenic acid in the mammary bland could be further transferred into CLA through catalyzing by $\Delta9$-dehydrogenase (Bauman et al. 2001). It has been identified that CLA generated from this process could comprise of 60–70% of total CLA in the milk (Corl et al. 2001).

2.1.2   Conjugated Linolenic Acid

Conjugated linolenic acid (CLNA) was one of the derivates from linolenic acid (LNA, C18:3) with conjugated double bonds, comprising of different isomers (Fig. 2.2). CLNA was firstly discovered in the nineteenth century; however, it was not attracted much attention due to rare awareness of its physiological effects. Till 1987, Nuteren and Christ-Hazelhof firstly identified the biological activity of CLNA when they studied the inhibitory effect of fatty acids derived from plant seeds on the synthesis of prostaglandin E2 (PGE2) (Nugteren and Christ 1987). Later, anti-cancer and anti-obesity activities of naturally CLNA from bifidobacteria were reported by other researchers (Coakley et al. 2009; Hennessy et al. 2012; Destaillats et al. 2005.) The bifidobacterial CLNA isomers analyzed included *c*9,*t*11,*c*15-CLNA, *t*9,*t*11,*c*15-CLNA, *c*9,*t*11,*c*13-CLNA, *c*9,*t*11,*t*13-CLNA, *c*6,*c*9,*t*11-CLNA, and *c*6,*t*9,*t*11-CLNA.

CLNA was widely distributed in nature, such as milk and ruminant meet. In addition, CLNA occurs in some plant seeds, for instance, pomegranate seeds, tung oil seeds, *momordica charantia* seeds, calendula seeds, etc. The CLNA isomers derived from plant seeds consist of many kinds of isomers, and thus proper

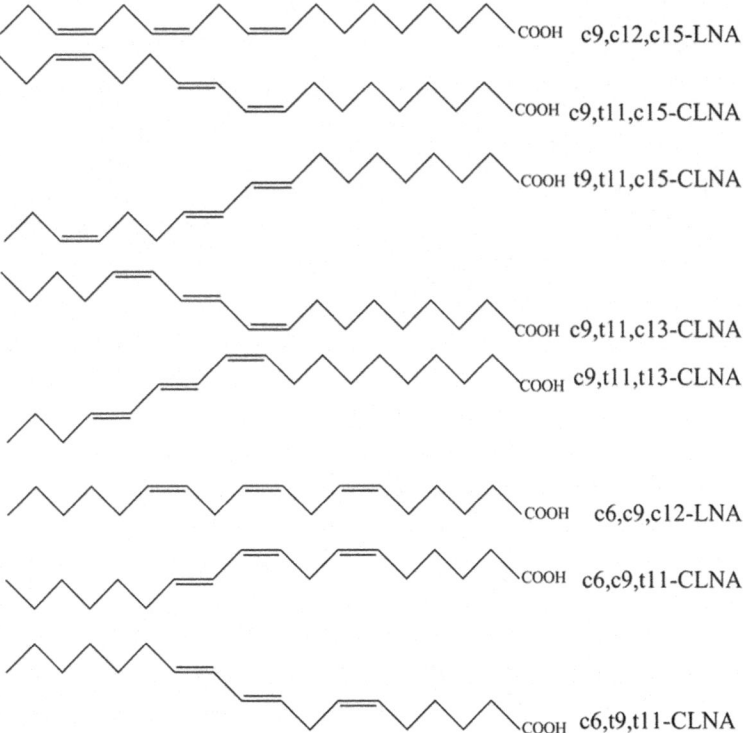

Fig. 2.2 Structure of α-linolenic acid, γ- linolenic acid, and conjugated linolenic acids

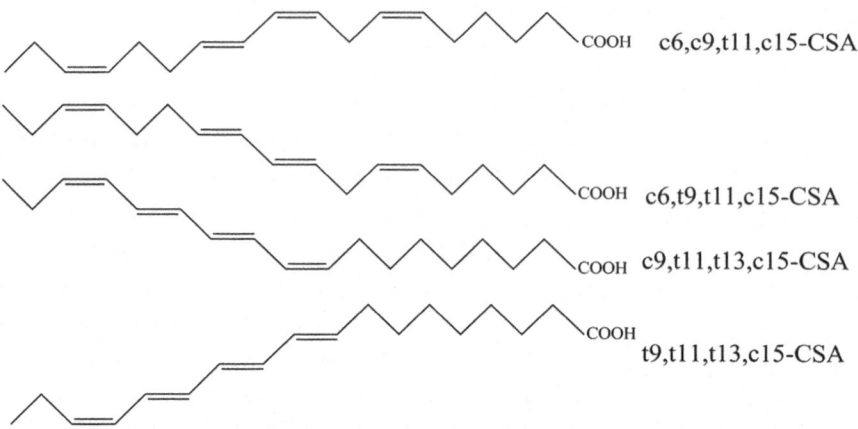

Fig. 2.3 Structure of conjugated stearidonic acids

separation methods would be key factors to obtain the pure isomers (Smith Jr. 1971). Recently, supercritical CO_2 fluid extraction and low-temperature-crystallization methods have been applied to separate CLNA. However, recent separation methods could not address the commercial requirements due to the limited amount of food grade plants.

2.1.3 Other Conjugated Fatty Acid

Despite CLA and CLNA, another conjugated fatty acid is conjugated stearidonic acid (CSA). CSA has been identified to possess many physiological activities, such as anti-tumor, antiatherosclerosis, and hypoglycemic activity. Besides, CSA could be applied in lipid peroxidation to evaluate the antioxidant agents. The identified isomers of CSA include c6,c9,t11,c15-CSA, c6,t9,t11,c15-CSA, c9,t11,t13,c15-CSA, and t9,t11,t13,c15-CSA (Fig. 2.3) (Hennessy et al. 2012).

2.2 Lactic Acid Bacteria with Conjugated Fatty Acid Production Ability

Microbial CFA producers have been studied for decades, which started in 1960s. Kepler et al. (1966, 1970, 1971), Kepler and Tove (1967) originally found that *Butyrivibrio fibrisolvens*, one of the ruminant bacteria, could convert LA to CLA. Then a variety of microbes showed the property of CLA production, especially that lactic acid bacteria could generate c9,t11-CLA and t9,t11-CLA. With the increase of research of CLA production in lactic acid bacteria, CLNA, CSA, and other CFA were found in lactic acid bacteria metabolites. Lactic acid bacteria with CFA production ability include *Lactobacillus* (*L. plantarum*, *L. acidophilus*, *L. casei*, *L. reuteri*, *L. fermentum*, *L. bulgaricus*, *L. rhamnosus*), *Bifidobacterium* (*B. breve*, *B. longum*, *B. animalis* subsp. *lactis*), *Lactococcus lactis*, *Streptococcus thermophiles*, etc.

2.2.1 Conjugated Fatty Acid Production in Lactobacilli

Lactobacillus was widely reported with CFA production ability, especially CLA production, which consisted of almost each species of lactobacilli (Table 2.1).

2.2.1.1 Lactobacillus plantarum

L. plantarum was the most widely studied strain among lactobacilli with CLA-production ability. In 2002, Kishino et al. screened many lactic acid bacteria strains with CLA-production ability, including *Lactobacillus*, *Enterococcus*, *Pediococcus*, and *Propionibacterium* (Kishino et al. 2002), in which *L. plantarum* AKU1009a was the strain with the best CLA-generation ability. Further study demonstrated that *L. plantarum* AKU1009a could even transform ricinoleic acid into CLA directly. Interestingly, the washed cells of this strain could catalyze α-linolenic acid into $c9,t11,c15$-CLNA and $t9,t11,c15$-CLNA with a 40% of total conversion rate. Comparatively, it has a better ability of CLNA from γ-linolenic acid with a conversion rate up to 68%. CSA could be produced from stearidonic acid into $c6,c9,t11,c15$-CSA and $c6,t9,t11,c15$-CSA by the strain (Kishino et al. 2010). The concentration of ricinoleic acid utilized by the washed cells of *L. plantarum* JCM1551 was up to 2400 mg/L with $c9,t11$-CLA and $t9,t11$-CLA as the main isomers (Andrade et al. 2012). *L. plantarum* NCUL005 has also been reported to produce CLA with final concentration of 623 mg/L growing in MRS medium in the presence of free LA (Andrade et al. 2012). Furthermore, growing and washed cells of *L. plantarum* ZS2058 could both transform free LA into CLA with conversion rate as 54.3% and 46.75%, respectively (Yang et al. 2017).

Except growing cells in the MRS medium or washed cells in the proper reaction solution, the strains added into the fermentis medium including sunflower oils or soymilk could also be used as the CLA producers (Li et al. 2012). VSL3# was the most widely used probiotics including eight strains, and studies have reported that all the eight strains could produce CLA, in which the conversion rate of *L. plantarum* strain was about 60% with $c9, t11$-CLA and $t9,t11$-CLA as the main isomers (Ewaschuk et al. 2006).

Furthermore, the CLA production mechanism by lactobacilli was also identified. Shimizu et al. firstly found that LA was firstly transformed into 10-hydroxy-*cis*-12 octadecenic acid and 10-hydroxy-*trans*-12-octadecenic acid, and then these two intermediates were both transferred into CLA (Ogawa et al. 2001). Further analysis

Table 2.1 High conjugated fatty acid producers in lactobacilli

Strains	Substrate and conditions	Products	Isomers	Reference
L. acidophilus L1	LA; growth cells	CLA (131 mg/L)	c9,t11-CLA; t9,t11-CLA	Alonso et al. (2003)
L. acidophilus La-5	LA; washed cells	CLA (388 mg/g)	c9,t11-CLA; t9,t11-CLA	Macouzet et al. (2009)
L. acidophilus F0221	LA; growth cells	CLA (161 mg/L)	c9,t11-CLA; t9,t11-CLA	Li et al. (2011)
L. acidophilus ADH	LA; growth cells	CLA (630 mg/L)	c9,t11-CLA; t9,t11-CLA	Xu et al. (2008)
L. acidophilus CCRC14079	LA; growth cells	CLA (105 mg/L)	c9,t11-CLA; t9,t11-CLA	Lin et al. (1999)
L. acidophilus AKU1137	LA; washed cells	CLA (4900 mg/L)	c9,t11-CLA; t9,t11-CLA	Ogawa et al. (2001)
L. acidophilus IAM10074	LA; washed cells	CLA (600 mg/L)	c9,t11-CLA; t9,t11-CLA	Ogawa et al. (2005)
L. acidophilus AKU1122	LA; washed cells	CLA (120 mg/L)	c9,t11-CLA; t9,t11-CLA	Ogawa et al. (2005)
L. acidophilus CRL730	LA; growth cells	CLA (23.8%)	c9,t11-CLA; t9,t11-CLA	van Nieuwenhove et al. (2007)
L. acidophilus Q42	LA; growth cells	CLA (20%)	c9,t11-CLA; t9,t11-CLA	
L. brevis IAM1082	LA; washed cells	CLA (550 mg/L)	c9,t11-CLA; t9,t11-CLA	Ogawa et al. (2005)
L. casei E5	LA; growth cells	CLA (111 mg/L)	c9,t11-CLA; t9,t11-CLA	Alonso et al. (2003)
L. casei E10	LA; growth cells	CLA (80 mg/L)	c9,t11-CLA; t9,t11-CLA	Alonso et al. (2003)
L. casei CRL431	LA; growth cells	CLA (175 mg/L)	c9,t11-CLA; t9,t11-CLA	van Nieuwenhove et al. (2007)
L. casei CRL87	LA; growth cells	CLA (17%)	c9,t11-CLA; t9,t11-CLA	
L. curvatus LMG13553	LA; growth cells	CLA (1.6%)	c9,t11-CLA; t9,t11-CLA	Gorissen et al. (2011)
	LNA; growth cells	CLNA (22.4%)	c9,t11,c15-CLNA; t9,t11,c15-CLNA	
L. paracasei IFO12004	LA; washed cells	CLA (200 mg/L)	c9,t11-CLA; t9,t11-CLA	Ogawa et al. (2005)
L. bulgaricus CCRC14009	LA; washed cells	CLA (16%)	c9,t11-CLA; t9,t11-CLA	Lin et al. (2006)
L. pentosus IFO12011	LA; washed cells	CLA (30 mg/L)	c9,t11-CLA; t9,t11-CLA	Ogawa et al. (2005)
L. plantarum AKU1009a	LA; washed cells	CLA (40,000 mg/L)	c9,t11-CLA; t9,t11-CLA	Andrade et al. (2012)
	Ricinoleic acid; washed cells	CLA (1650 mg/L)	c9,t11-CLA; t9,t11-CLA	

(continued)

Table 2.1 (continued)

Strains	Substrate and conditions	Products	Isomers	Reference
L. plantarum JCM1551	Ricinoleic acid; washed cells	CLA (2400 mg/L)	c9,t11-CLA; t9,t11-CLA	Andrade et al. (2012)
L. plantarum Ip5	LA; growth cells	CLA (142 mg/L)	c9,t11-CLA; t9,t11-CLA	
L. plantarum NCUL005	LA; growth cells	CLA (623 mg/L)	c9,t11-CLA; t9,t11-CLA	
L. plantarum ATCC8014	LA; growth cells	CLA (4.6%)	c9,t11-CLA; t9,t11-CLA	Gorissen et al. (2011)
	LNA; growth cells	CLNA (26.8%)	c9,t11,c15-CLNA; t9,t11,c15-CLNA	
L. plantarum ZS2058	LA; growth cells	CLA (54.3%)	c9,t11-CLA; t9,t11-CLA	Yang et al. (20140
	LA; washed cells	CLA (46.75%)	c9,t11-CLA; t9,t11-CLA	Xu et al. (2008)
L. plantarum DSM20179	LA; growth cells	CLA (240 mg/L)	c9,t11-CLA	Khosravi et al. (2015)
L. plantarum AKU1138	LA; washed cells	CLA (450 mg/ml)	c9,t11-CLA; t9,t11-CLA	Kishino et al. (2002)
L. plantarum LT2–6	LA; washed cells	CLA (52.4%)	c9,t11-CLA; t9,t11-CLA	Zhang (2004)
L. plantarum A6-1F	LA; washed cells	CLA (276 mg/ml)	c9,t11-CLA; t9,t11-CLA	Zhao et al. (2011)
L. rhamnosus PL60	LA; washed cells	CLA (4438 mg/g)	c9,t11-CLA; t10,t12-CLA	Lee et al. (2006a)
L. rhamnosus C14	LA; growth cells	CLA (190 mg/L)	c9,t11-CLA; t9,t11-CLA	van Nieuwenhove et al. (2007)
L. reuteri ATCC55739	LA; growth cells	CLA (300 mg/L)	c9,t11-CLA; t9,t11-CLA	Lee et al. (2003)
L. sakei LMG13558	LA; growth cells	CLA (4.2%)	c9,t11-CLA; t9,t11-CLA	Gorissen et al. (2011)
	LNA; growth cells	CLA (60.1%)	c9,t11,c15-CLNA; t9,t11,c15-CLNA	
L. sakei CG1	LNA; growth cells	CLNA (28.3%)	c9,t11,c15-CLNA; t9,t11,c15-CLNA	

has identified multiple enzymes that were involved in CLA production, including hydrogenase, oxidoreductase, and isomerase (Kishino et al. 2013).

In China, Zhou et al. firstly separated one strain, named *L. plantarum* ZS2058, from pickled vegetables in Sichuan province possessing high CLA-producing ability (Zhou et al. 2004). The optimal condition of CLA production by *L. plantarum* ZS2058 has also been reported by Xu et al. (2008), and furthermore, the separation of linoleic acid isomerase from this strain has also been carried out (Gu et al. 2008). Yang et al. screened of the CLA generation by some lactobacilli strains, and results showed that *L. plantarum* ZS2058 possessed the highest conversion rate, and the enzymatic activity assay demonstrated that the process generating CLA by *L. plantarum* ZS2058 consisted of multiple reactions with 10-hydroxy-*cis*-12-octadecenic acid, 10-oxo-*cis*-12-octadecenic acid, 10-oxo-*trans*11-octadecenic acid, and 10-hydroxy-*trans*11-octadecenic acid as the substrates (Yang et al. 2014).

2.2.1.2 *Lactobacillus acidophilus*

The conversion of CLA by *L. acidophilus* CCRC14079 was firstly studied by Lin et al. (1999), in which results revealed that this strain could transform about 10% of free linoleic acid in the milk lipids into CLA. The highest conversion rate of CLA produced by this strain was obtained when the strain was cultivated in the medium in the presence of free LA for 24 h. *L. acidophilus* L1 and O16 were another two strains possessing CLA production ability with conversion rate more than 50% growing in the medium or dried skimmed milk system added with free linoleic acid. Washed cells of *L. acidophilus* AKU1137 could generate CLA with the final concentration of 4.9 g/L. In this study, large amount of hydroxyl fatty acids could also be detected, and both of the concentrations of CLA and hydroxyl fatty acids present positive linear relationship (Ogawa et al. 2001). Kim and Liu screened the CLA production ability by eight lactobacilli strains growing in the MRS medium and skimmed milk reaction system, and results demonstrated that *L. acidophilus* 96 could transform LA into CLA, while four strains could not generate CLA at this condition. Macouzet et al. reported that washed cells of *L. acidophilus* La-5 could accumulate CLA when the strain grew in the MRS medium added with free LA at the concentration of 0.4 g/L or 0.37% milk lipids. Other studies showed that limiting oxygen could reduce the ratio of the *c*9,*t*11-CLA and *t*9,*t*11-CLA without influencing the total amount of CLA (Kim and Liu 2010).

2.2.1.3 *Lactobacillus reuteri*

Rosson et al. (1999) reported that *L. reuteri* PYR8, separated from the rat intestine, could transform about 60% of free LA into *c*9,*t*11-CLA. Lee et al. optimized the reaction condition of producing CLA by *L. reuteri* ATCC55739 and found that the immobilized cells could produce 175 mg/L CLA with the concentration of free LA as 500 mg/L, possessing the production efficiency of 175 mg/(L.h), about 5.5 times than that produced by washed cells (Sun et al. 2003). Further study performed by Roman et al. showed that cholate could not influence the CLA production by *L. reuteri* ATCC55739 in vitro. Moreover, Hernandez et al. also investigated the effect of temperature, concentration of LA, oxygen, and pH on the CLA production by a *L. reuteri* strain (Hernandezmendoza et al. 2010). Results showed that the concentration of CLA produced by this strain at different conditions showed significant difference, and the highest CLA conversion rate was obtained when the strain was cultivated at 10 °C for 20 h in a microanaerobic environment with free LA as the substrate at the concentration of 20 mg/mL. CLA production would decrease with pH decreasing from 6.5 to 5.5.

Rosson et al. (1999) first tried to separate the putative linoleic acid isomerase and also cloned these genes exogenously. However, the proteins cloned from *L. reuteri* were identified to produce trace of hydroxyl fatty acid, other than CLA, though Rosson modulated some other factors, which might influence the enzymatic activity, such as expression system.

2.2.1.4   *Lactobacillus casei*

In 2003, Alonso et al. discovered two *L. casei* strains which could produce CLA in the MRS medium or skimmed milk with the ratio of *c*9,*t*11-CLA exceeding 80% (Alonso et al. 2003). The study of van Nieuwenhove showed that *L. casei* CRL431 showed the highest CLA conversion rate (35.9%) among the eight studied strains (Nieuwenhove et al. 2010). Results also demonstrated that the concentration of CLA produced by these eight strains growing in the buffalo milk in the presence of free LA (200 mg/L) was two to three times than that in the MRS medium. Interestingly, all the tested eight strains could produce CLA with LA concentration up to 1000 mg/mL. The *L. casei* strain in the probiotics VSL#3 possessed the CLA conversion rate exceeding 60% with *c*9, *t*11-CLA and *t*10, *c*12-CLA as the main isomers.

2.2.1.5   Other Lactobacilli Strains

Recently, *L. rhamnosus* PL60 was the only strain in this species identified to produce CLA with t10,c12-CLA as the predominant isomer. In 2006, Lee investigated the physiological effect of *L. rhamnosus* PL60 in vivo. Results showed that comparative to the negative groups, the weight of the mice feeded with this strain decreased significantly, as well as the white adipose tissue. Further analysis identified that this strain possessed perfect anti-obesity effect due to its production of *t*10,*c*12-CLA. And this strain could also colonize in the gut of volunteers (Lee and Lee 2009).

Additionally, Romero-Pérez also reported a *L. paracasei* strain could also convert 85% of free LA into CLA (Romero-Pérez et al. 2013). Florence also revealed that the combination of *B. lactis*, *S. thermophilus*, and *L. bulgaricus* could increase the CLA in the dairy products (Florence et al. 2012). Other studies also reported that *L. sake* and *L. curvatus* could also transform ALA into CLNA with the conversion rate of 22.4% and 60.1%, respectively (Gorissen et al. 2011). Ewaschuk et al. also reported that the CLA conversion rate of *L. bulgaricus* and *S. thermophilus* tested in this study ranged 60–70% (Ewaschuk et al. 2006).

2.2.2 Conjugated Fatty Acid Production in Bifidobacteria

The first *Bifidobacterium* with CLA production was reported by Coakley (Coakley et al. 2003). In their study, 15 *Bifidobacterium* strains were screened for CLA generation in the medium in the presence of free LA, in which 9 strains showed perfect CLA-producing ability with *c*9, *t*11-CLA as the main isomer. Among all the tested strains, *B. breve* and *B. dentium* possessed higher CLA conversion rate, in which *B. breve* NCFB2258 could convert about 66% of free LA to *c*9,*t*11-CLA and 6.2% of LA to *t*9,*t*11-CLA. Moreover, nearly all the produced CLA existed in the supernatants of the medium. As bifidobacteria was one of the pioneer colonized species in neonates and infants fed with breast milk, a number of researchers isolated bifidobacteria from infants and analyzed their CLA production abilities. Chung (Chung et al. 2008) evaluated 150 bifidobacterial strains for their CLA-generation ability, and 4 strains among them could produce CLA with conversion rate exceeding 80%, especially the conversion rate of LA in one strain exceeded 90%. Additionally, 30 bifidobacteria were investigated for their CLA and CLNA production ability, and results demonstrated that the highest CLA conversion rate was 53%, which was 78% of CLNA conversion rate (Gorrisen et al. 2010). Major bifidobacterial CLA producers were listed in Table 2.2.

2.2.2.1 Bifidobacterium breve

Coakley and colleagues investigated many bifidobacteria strains for CLA production and found *B. breve* NCFB2257 and *B. breve* NCFB 2258 showed the highest CLA conversion rate of LA conversion up to 65% (Coakley et al. 2003). Rosberg-Cody and colleagues isolated and screened CLA producers from neonates' gut intestines and showed significant difference in CLA generation among different species, even different strains which belong to the same species, in which *B. breve* exhibited much higher conversion rate than all the other bifidobacterial species (Rosberg-Cody et al. 2004). Barrett et al. isolated from neonates healthy adults and elderly subjects suffered with *Clostridium difficile* infection and developed a rapid method for CLA producer screening. In their results, five strains could transfer free LA to CLA with conversion rate exceeding 20%, in which the highest conversion was 75% for a *B. breve* (Barrett et al. 2007). Chung et al. screened 100 and 50 bifidobacteria for CLA production, and only 4 strains could produce CLA with a conversion rate over 80%, in which *B. breve* LMC017 exhibited the highest conversion. This strain could convert 91.1% of free LA or 78.8% of LA monoglyceride into CLA. Another study revealed that the CLA conversion rate was substrate-dependent. For example, when different forms of LA (monolinolein, dilinolein, 50% safflower oil monolinolein, 90% safflower oil monolinolein) were added in the skim milk to serve as substrate for CLA production, the highest conversion rate of CLA by *B. breve* LMC520 was obtained when LA monolinolein or 90% safflower oil monolinolein as substrates (Choi et al. 2008)

Table 2.2 High conjugated fatty acids producers in bifidobacteria

Strains	Substrate and conditions	Products	Isomers	Reference
B. breve LMC017	LA; growth cells	CLA (474 mg/L)	c9,t11-CLA; t9,t11-CLA	Chung et al. (2008)
B. breve LMG11040	LA; growth cells	CLA (44%)	c9,t11-CLA; t9,t11-CLA	Gorrisen et al. (2010)
	LNA; growth cells	CLNA (65.5%)	c9,t11,c15-CLNA; t9,t11,c15-CLNA	
B. breve LMG11084	LA; growth cells	CLA (53.5%)	c9,t11-CLA; t9,t11-CLA	
	LNA; growth cells	CLNA (72%)	c9,t11,c15-CLNA; t9,t11,c15-CLNA	
B. breve LMG11613	LA; growth cells	CLA (19.5%)	c9,t11-CLA; t9,t11-CLA	
	LNA; growth cells	CLNA (55.6%)	c9,t11,c15-CLNA; t9,t11,c15-CLNA	
B. breve LMG13194	LA; growth cells	CLA (24.1%)	c9,t11-CLA; t9,t11-CLA	
	LNA; growth cells	CLNA (63.3%)	c9,t11,c15-CLNA; t9,t11,c15-CLNA	
B. breve NCFB2257	LA; growth cells	CLA (231 mg/L)	c9,t11-CLA; t9,t11-CLA	Coakley et al. (2003)
B. breve NCTC11815	LA; growth cells	CLA (215 mg/L)	c9,t11-CLA; t9,t11-CLA	
B. breve NCIMB8815	LA; growth cells	CLA (242 mg/L)	c9,t11-CLA; t9,t11-CLA	
B. breve NCIMB8807	LA; growth cells	CLA (128 mg/L)	c9,t11-CLA; t9,t11-CLA	
B. breve pattern A	LA; growth cells	CLA (76.6%)	c9,t11-CLA; t9,t11-CLA	Barrett et al. (2007)
B. breve NCFB2258	LA; growth cells	CLA (398 mg/L)	c9,t11-CLA; t9,t11-CLA	Coakley et al. 2003
B. breve NCIMB702258	ALA; growth cells	CALA (199 mg/L)	c9,t11,c15-CLNA; t9,t11,c15-CLNA	Hennessy et al. (2012)
	GLA; growth cells	CGLA (149 mg/L)	c6,c9,t11-CLNA; c6,t9,t11-CLNA	
	SA; growth cells	CSA (38 mg/L)	c6,c9,t11,c15-CSA; c6,t9,t11,c15-CSA	
B. breve NCIMB8807	LA; growth cells	CLA (277 mg/L)	c9,t11-CLA; t9,t11-CLA	
	ALA; growth cells	CALA (272 mg/L)	c9,t11,c15-CLNA; t9,t11,c15-CLNA	
B. breve DPC6330	LA; growth cells	CLA (300 mg/L)	c9,t11-CLA; t9,t11-CLA	
	ALA; growth cells	CALA (331 mg/L)	c9,t11,c15-CLNA; t9,t11,c15-CLNA	
	GLA; growth cells	CGLA (81 mg/L)	c6,c9,t11-CLNA; c6,t9,t11-CLNA	
	SA; growth cells	CSA (41 mg/L)	c6,c9,t11,c15-CSA; c6,t9,t11,c15-CSA	

(continued)

Table 2.2 (continued)

Strains	Substrate and conditions	Products	Isomers	Reference
B. breve KCTC10462	LA; growth cells	CLA (160 mg/L)	c9,t11-CLA; t9,t11-CLA	Oh et al. (2003)
B. breve KCTC3461	LA; growth cells	CLA (350 mg/L)	c9,t11-CLA; t9,t11-CLA	Song et al. (2005)
B. breve LMC520	LA; growth cells	CLA (280 mg/L)	c9,t11-CLA; t9,t11-CLA	Choi et al. (2008)
	LNA; growth cells	CLNA (90%)	c9,t11,c15-CLNA; t9,t11,c15-CLNA	Park et al. (2011)
B. dentium	LA; growth cells	CLA (12.5%)	c9,t11-CLA; t9,t11-CLA	Barrett et al. (2007)
B. bifidum LMG10645	LA; growth cells	CLA (40.7%)	c9,t11-CLA; t9,t11-CLA	Gorrisen et al. (2010)
	LNA; growth cells	CLNA (78.4%)	c9,t11,c15-CLNA; t9,t11,c15-CLNA	
B. bifidum CRL1399	LA; growth cells	CLA (24.8%)	c9,t11-CLA; t9,t11-CLA	van Nieuwenhove et al. (2007)
B. animalis subsp. lactis Bb12	LA; growth cells	CLA (170 mg/L)	c9,t11-CLA; t9,t11-CLA	Coakley et al. (2003)
B. longum DPC6320	LA; growth cells	CLA (205 mg/L)	c9,t11-CLA; t9,t11-CLA	Hennessy et al. (2012)
B. longum	LA; growth cells	CLA (60.1%)	c9,t11-CLA; t9,t11-CLA	Barrett et al. (2007)
B. infantis	LA; growth cells	CLA (18.1%)	c9,t11-CLA; t9,t11-CLA	
B. pseudocatenulatum KCTC10208	LA; growth cells	CLA (135 mg/L)	c9,t11-CLA; t9,t11-CLA	Oh et al. (2003)
B. pseudolongum LMG11595	LA; growth cells	CLA (42.2%)	c9,t11-CLA; t9,t11-CLA	Gorrisen et al. (2010)
	LA; growth cells	CLNA (62.7%)	c9,t11,c15-CLNA; t9,t11,c15-CLNA	

Gorissen et al. assessed the ability of producing CLA and CLNA by 36 bifidobacteria with free LA and ALA as the substrate, respectively, and found that six strains could transfer LA into CLA, but the conversion ratio differed significantly, in which the CLA conversion rate by *B. breve* LMG 11613 was 19.5% while 53.5% by *B. breve* LMG11084 (Gorissen et al. 2011). Furthermore, c9,t11-CLA was the predominant isomer in those bifidobacteria strains comprising 51.3–82.2% of total CLA. Due to the high CLA production, bifidobacteria was chosen by some researches to be used as the starter cultures to increase the CLA content in dairy products. The study performed by Hennessy et al. showed that the conversion rate of *B. breve* NCIMB702258 in milk system consisting of different additives presented significant difference and finally the content of c9,t11-CLA produced by the strain with additives comprising of yeast extract, casein hydrolysate, peptone, acetate, butyrate, and propionate was comparable to that when the strain grew in expensive MRS medium (Hennessy et al. 2010). Hennessy et al. also studied the ability of bifidobacteria to transfer different polyunsaturated fatty acids into their corresponding conjugated forms. Results clearly showed that the conversion rate was substrate-dependent, in which the conversion rate of CLA ranged from 12% to 97% with $c9,t11$-CLA and $t9,t11$-CLA as the major isomers, while for GLA and ALA, the conversion ratio ranged from 0–83% to 3.8–27%, respectively. In those strains they assessed, *B. breve* DPC 6330 showed the highest ability of producing conjugated fatty acids, which could convert 70% of LA into CLA, 90% of α-LNA into CLNA, 17% of γ-LNA into CLNA, and 28% of stearidonic acid into CSA, respectively. Ewaschuk et al. analyzed the commercial VSL#3 probiotics, which consisted of eight different strains, and found that both VSL#3 and each strain in it could generate CLA at a different level, in which *B. breve* showed the highest CLA-producing ability with 70% LA transferred to CLA (Ewaschuk et al. 2006).

2.2.2.2 *Bifidobacterium animalis*

Coakley and colleagues (2003) found that *B. animalis* Bb-12 could convert approximately 27% of LA into $c9,t11$-CLA when it grew in the MRS medium plus free LA. Rodriguez-Alcala et al. studied on the possible utilization of 22 probiotics including five bifidobacteria strains and selected two strains which could generate CLA in the skimmed milk with free linoleic acid or safflower oil. And the major isomers were c9,t11-CLA and t10,c12-CLA (Rodríguez-Alcalá et al. 2011). With the optimal condition, the conversion rate of CLA generated by *B. animalis* BLC was highest with free LA as substrate. *B. animalis* Bb12-1 could produce more CLA when ricinoleic acid served as substrate. These studies suggest that it's possible to utilize bifidobacteria strains as the starter cultures in the milk to increase the content of CLA with LA in different types as the substrate.

2.2.2.3 *Bifidobacterium longum*

With a rapid screening method for CLA production, Barrett et al. isolated a number of bifidobacteria strains from the feces of infant, health adults, and elder people infected with *C. difficile* which could generate CLA, and results revealed that four strains belonging to *B. longum* could produce CLA with a conversion ratio exceeding 20% (Barrett et al. 2007). Similar to *B. breve*, the main CLA isomers produced by *B. longum* was also $c9,t11$-CLA. Roberg-Cody and colleagues isolated a few strains belonging to *Bifidobacterium* genus with high CLA production ability and discovered that the CLA-producing ability among different bifidobacteria species present significant difference, in which *B. longum* strain could generate $c9,t11$-CLA and $t9,t11$-CLA in high conversion rate (Rosberg-Cody et al. 2011). Ewaschuk et al. studied the CLA production ability of each strain in VSL3# probiotics and found *B. longum* could transform ~70% of LA to CLA (Ewaschuk et al. 2006).

2.2.2.4   Other Bifidobacteria

B. breve, *B. longum*, and *B. animalis* were the widely studied *Bifidobacterium* species with CLA production. Other species have also been reported with CLA production ability. For example, Oh et al. isolated a number of *B. pseudocatenulatum* which could convert free LA to CLA with a high conversion rate (Oh et al. 2003). Rosberg-Cody et al. also reported that one *B. bifidum* strain possessed CLA-producing ability with conversion rate of 17.9% (Rosberg-Cody et al. 2011). Gorissen et al. found that *B. bifidum* LMG 10645 and *B. pseudocatenulatum* LMG11595 exhibited the capability of generating CLA and CLNA, and for *B. bifidum* LMG 10645, the *c*9,*t*11-CLA was up to 82% of total CLA it produced, while for *B. pseudocatenulatum* LMG11595, it was only 35.1% of total CLA. Furthermore, one *B. bifidum* strain could accumulate *c*9,*t*9-CLA in the skimmed milk in the presence of hydrolyzed soybean oil (Xu et al. 2004). Ewaschuk et al. reported that one *B. infantis* strain could produce *c*9,*t*11-CLA and *t*10,*c*12-CLA with a total conversion rate exceeding 70% (Ewaschuk et al. 2006).

2.2.3   Conjugated Fatty Acid Production by Other Lactic Acid Bacteria

Numerous of other lactic acid bacteria were reported to produce CLA, especially food fermentation involving lactococci and streptococci (Table 2.3).

    Lc. lactis subsp. *cremoris* CCRC12586, *Lc. lactis* subsp. *lactis*, and *S. thermophilus* CCRC12257 could produce CLA in the skimmed milk in the presence of free LA (Lin et al. 1999). Kim and Liu found five *Lc. lactis* strains could transform LA into CLA in the skimmed milk, among which three strains could also generate CLA when they were cultivated in MRS medium (Kim and Liu 2010). Among all the strains analyzed, *Lc. lactis* I-01 presented the highest CLA-producing ratio when it grew in the MRS medium or skimmed milk at the concentration of 0.1 mg/mL (Kim and Liu 2010). Kishino et al. analyzed 250 lactic acid bacteria for CLA production, which were belonged to lactobacilli, streptococci, pediococci, leuconostoc, propionibacteria, bifidobacteria, and enterococci. They found that the strains tested of lactobacilli, propionibacteria, pediococci, and lactococci could produce a large amount of CLA in MRS medium and the predominant CLA was *c*9, *t*11-CLA isomer. Moreover, 10-hydroxy-*cis*-12-octadecenic acid (10-HOE) was detected during CLA production in the research, which was considered as an intermediate during LA conversion to CLA (Ando et al. 2003). Xu and colleagues (2004) reported that CLA production by *E. faecium* and *P. acidilactici* was substrate-dependent, in which some strains could produce CLA in the skimmed milk with hydrolyzed soybean oil as the substrate, rather than unhydrolyzed soybean oil. El-Salam and colleagues

Table 2.3 Other high CFA producers in lactic acid bacteria

Strains	Substrate and conditions	Products	Isomers	Reference
E. faecium M74	LA; washed cells	CLA (1 mg/g)		Xu et al. (2004)
E. faecium	Enzymatic sesame oil; washed cells	CLA (104 mg/ml)	c9,t11-CLA; t9,t11-CLA	El-Salam et al. (2010)
E. faecium AKU1021	LA; washed cells	CLA (100 mg/ml)	c9,t11-CLA; t9,t11-CLA	Kishino et al. (2002)
P. acidilactici AKU1059	LA; washed cells	CLA (1400 mg/ml)	c9,t11-CLA; t9,t11-CLA	
P. acidilactici	LA; washed cells	CLA (1 mg/g fat)		Xu et al. (2004)
Lc. lactis 210	LA; growth cells	CLA (2 mg/g fat)		Kim and Liu (2002)
Lc. lactis IO-1	LA; growth cells	CLA (4 mg/g fat)		
Lc. lactis LMG S19870	LA; growth cells	CLA (46 mg/L)		Rodríguez-Alcalá et al. (2011)
Lc. lactis subsp. lactis CCRC12586	LA; growth cells	CLA (63 mg/L)		Lin et al. (1999)
Lc. lactis subsp. lactis CCRC10791	LA; growth cells	CLA (78 mg/L)		
Lc. lactis subsp. lactis	Enzymatic sesame oil; washed cells	CLA (21.6 mg/ml)	c9,t11-CLA; t9,t11-CLA	El-Salam et al. (2010)
Leu. mesenteroides subsp. mesenteroides	Enzymatic sesame oil; washed cells	CLA (198 mg/ml)	c9,t11-CLA; t9,t11-CLA	
S. thermophilus CRL728	LA; growth cells	CLA (33.9%)		van Nieuwenhove et al. (2007)
S. thermophilus CCRC12257	LA; growth cells	CLA (74 mg/L)		Lin et al. (1999)

demonstrated that some strains of lactobacilli, propionibacteria, lactococci, enterococci, and pediococci could grow well in the reconstituted milk plus 0.2% of enzymatic sesame oils and synthesize CLA with those hydrolyzed oil as substrate. Interestingly, they found the best CLA producers were *Lc. lactis* subsp. *lactis* strain and *Leu. mesenteroides* subsp. *mesenteroides*. Another study carried out by Rodriguez-Alcala showed that free LA and safflower oil could be utilized by *Lc. lactis* LMG 19870 as the substrate to generate CLA. CLA produced by the strain with free LA as substrate in the skimmed milk was up to 45.51 mg/L, whereas safflower oil served as substrate, and the concentration of CLA was 23.1 mg/L, nearly a half of that from free LA as substrate (Rodríguez-Alcalá et al. 2011).

2.3 The Mechanism for Conjugated Fatty Acid Production in Lactic Acid Bacteria

2.3.1 Conjugated Fatty Acid Production Mechanism in Lactobacilli

The reason why bacteria produced CLA was still unclear. The most accepted reason was biological detoxification, as those strains eliminate the toxic effect of free LA on the cells. However, this assumption was only identified by a few bacteria (Maia et al. 2007, 2010). In fact, LA was necessary for the growth of bacteria; however, the growth-promoting effect would be replaced by the stress effect when its concentration increased to some specific concentration. The relation between myosin-cross-reactive antigen (MCRA) and the anti-environmental stress has been already identified in bacteria. Recent studies have showed that most bacteria could produce 10-HOE from LA and then further conversion to CLA, suggesting that these transformations of LA into other substrates could decrease the toxic effect of free LA on the cells.

Compared with rumen bacteria, linoleic acid isomerase was believed as key factor in the mechanism of CLA production by lactic acid bacteria. Unfortunately, it remains unclear. Rosson et al. (1999) firstly tried to separate the putative linoleic acid isomerase and identified its function through overexpression and activity confirmation. Finally, a protein, with molecular weight as 67 kDa, was obtained and the optimal pH was 6.8~7.5. This protein was identified to be homology to myosin-cross-reactive antigen, widely present in the bacteria and predicted as fatty acid isomerase (Kil et al. 1994). However, when LA was catalyzed by the recombinant protein, only a trace of hydroxyl fatty acid was produced rather than CLA. Due to the instability structure of hydroxyl fatty acid, it would be degraded by heat, strong base (or acid), or even some methyl method, which would be the possible reasons why few studies reported the production of hydroxy fatty acids. Ogawa et al. found that LA would be firstly transformed into hydroxyl fatty acid by L. acidophilus AKU1137 and hydroxyl fatty acids was then quickly converted into CLA when its concentrated at a certain extent (Ogawa et al. 2001). With further GC-MS and NMR analysis, the hydroxyl fatty acids were confirmed as 10-hydroxy-cis-12-C18:1 (10-HOE) and 10-hydroxy-trans-11-C18:1. This was the first report to identify 10-HOE as the intermediate during CLA production in lactic acid bacteria. And they presumed that the pathway for CLA production involved hydration, dehydration, and isomerization.

Even though the 10-HOE was identified as the intermediate during CLA generation, the key enzymes involved in the following reactions were unclear. As high homologous to the putative linoleate isomerase purified by Rosson et al. (1999), myosin-cross-reactive antigen (MCRA) received more research of interests and firstly confirmed as fatty acid hydratase in S. pyogenes, where it was original found. Volkov and colleagues cloned the MCRA-encoding gene and expressed it in E. coli. With LA as substrate, it revealed that MCRA was fatty acid hydratase with FAD as the cofactor. This enzyme could transform LA into 10-HOE and 10,13-dihydroxyl-stearic acid (10,13-diHOA), which was also identified to be related with the pathogenicity of this strain (Volkov et al. 2010).

Kishino et al. successfully separated a triple-component linoleic acid isomerase from *L. plantarum* AKU1009a through differential centrifugation. The first protein was identified to be membrane protein and involved in transforming LA into 10-HOE (Kishino et al. 2011). In combining the latter two separates, *t*9,*t*11-CLA could be produced, suggesting that the linoleic acid isomerase in lactobacilli was not a single enzyme; instead, three proteins were demanded for CLA production. Later on, the first protein was finally approved to be MCRA through the N-terminal sequence, recombinant technology, and enzymatic activity confirmation (Kishino et al. 2011). Though no other intermediates, except 10-HOE, could be detected in the enzymatic reaction, the author still presented the possible pathway for CLA production, including hydrogenation, dehydration, double bond migration, hydrogenation, and dehydration. These observations provided a novel direction in this field. In 2013, the detailed pathway of generating CLA by *L. plantarum* was elucidated (Kishino et al. 2013).

At the same time, Yang et al. (2013, 2014) showed that MCRA-encoding genes cloned from different lactic acid bacteria were approved to be fatty acid hydratase, neither linoleic acid isomerase. They found that *L. plantarum* ZS2058 could accumulate several intermediates during CLA production. Further analysis showed that those intermediates were 10-HOE, 10-oxo-*cis*-12-octadecenic acid, and 10-oxo-*trans*-11-ocradecenic acid. Through bioinformatics analysis and comparison with Kishino et al., the genetic determinates for CLA production in *L. plantarum* ZS2058 were fully confirmed. Their results showed that three proteins were involved, which were myosin-cross-reactive antigen, short chain dehydrogenase/oxidoreductase, and acetoacetate decarboxylase (Fig. 2.4) (Yang et al. 2014). Furthermore, the determinants for generating CLA in *L. plantarum* ZS2058 were knocked out based on the *cre-lox*-based system. Neither intermediate could be detected in the corresponding gene deletion mutant. Meanwhile all those mutants could recover the ability to convert LA into CLA when the corresponding gene was complemented, which indicated that the triple-component linoleic acid isomerase system was the unique pathway for CLA production in *L. plantarum* (Yang et al. 2017).

2.3.2 Conjugated Fatty Acid Production Mechanism in Bifidobacteria

To date, no detailed characterization of bifidobacterial production mechanism has been developed. Rosberg-Cody firstly cloned MCRA-encoding gene from *B. breve* NCFB2258, and then the gene was inserted into the vector (Rosberg-Cody et al. 2011). Recombinant *E. coli* strains with the vector inserted with *mcra* gene could only transform LA into 10-HOE, suggesting that *mcra* gene was linoleic acid

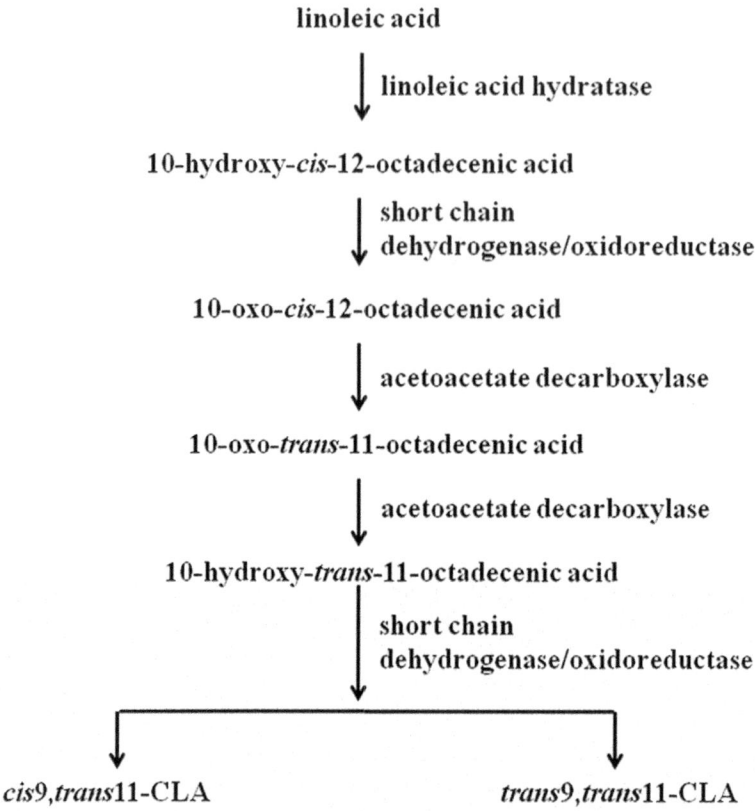

Fig. 2.4 CLA production pathway in *L. plantarum* ZS2058

hydratase, neither linoleic acid isomerase. The study demonstrated that MCRA from *B. breve* also utilized FAD as the cofactor, which was highly homologous to that from *L. reuteri* ATCC55739. Further study showed that deletion of *mcra* gene in *B. breve* NCFB2258 had no influence on CLA production (O'Connell et al. 2013) which indicated that MCRA was not involved in the converting LA in CLA by *Bifidobacterium*. Thus, identification of the bifidobacterial CLA production needs more investigation.

References

Alonso L, Cuesta EP, Gilliland SE (2003) Production of free conjugated linoleic acid by *Lactobacillus acidophilus* and *Lactobacillus casei* of human intestinal origin. J Dairy Sci 86(6):1941–1946

Ando A, Ogawa J, Kishino S, Shimizu S (2003) CLA production from ricinoleic acid by lactic acid bacteria. J Am Oil Chem Soc 80(9):889–894

Andrade JC, Ascenção K, Gullón P, Henriques SMS, Pinto JMS, Rocha-Santos TAP, Gomes AM (2012) Production of conjugated linoleic acid by food-grade bacteria: a review. Int J Dairy Technol 65(4):467–481

Barrett E, Ross RP, Fitzgerald GF, Stanton C (2007) Rapid screening method for analyzing the conjugated linoleic acid production capabilities of bacterial cultures. Appl Environ Microbiol 73(7):2333–2337

Bauman DE, Corl BA, Baumgard LH (2001) Conjugated linoleic acid (CLA) and the dairy cow. In: Garnsworthy PC, Wiseman J (eds) Recent advances in animal Nutrition. Nottingham University Press, Nottingham, pp 221–250

Choi NJ, Park HG, Kim YJ, Kim IH, Kang HS, Yoon CS, Yoon HG, Park SI, Lee JW, Chung SH (2008) Utilization of monolinolein as a substrate for conjugated linoleic acid production by *Bifidobacterium breve* LMC 520 of human neonatal origin. J Agric Food Chem 56:10908–10912

Chung SH, Kim IH, Park HG, Kang HS, Yoon CS, Jeong HY, Choi NJ, Kwon EG, Kim YJ (2008) Synthesis of conjugated linoleic acid by human-derived *Bifidobacterium breve* LMC 017: utilization as a functional starter culture for milk fermentation. J Agric Food Chem 56(9):3311–3316

Coakley M, Ross RP, Nordgren M, Fitzgerald G, Devery R, Stanton C (2003) Conjugated linoleic acid biosynthesis by human-derived *Bifidobacterium* species. J Appl Microbiol 94:138–145

Coakley M, Banni S, Johnson MC, Mills S, Devery R, Fitzgerald G, Stanton C (2009) Inhibitory effect of conjugated α-linolenic acid from bifidobacteria of intestinal origin on SW480 cancer cells. Lipids 44(3):249–256

Corl BA, Baumgard LH, Dwyer DA, Griinari JM, Phillips BS, Bauman DE (2001) The role of delta(9)-desaturase in the production of *cis*-9, *trans*-11 CLA. J Nutr Biochem 12(11):622–630

Destaillats F, Trottier JP, Galvez JMG, Angers P (2005) Analysis of α-linolenic acid biohydrogenation intermediates in milk fat with emphasis on conjugated linolenic acids. J Dairy Sci 88(9):3231–3239

Ewaschuk JB, Walker JW, Diaz H, Madsen KL (2006) Bioproduction of conjugated linoleic acid by probiotic bacteria occurs in vitro and in vivo in mice. J Nutr 136(6):1483

Florence ACR, Béal C, Silva RC, Bogsan CSB, Pilleggi ALOS, Gioielli LA, Oliveiraa MN (2012) Fatty acid profile, trans-octadecenoic, α-linolenic and conjugated linoleic acid contents differing in certified organic and conventional probiotic fermented milks. Food Chem 135(4):2207–2214

Gorissen L, Weckx S, Vlaeminck B, Raes K, De VL, De SS, Leroy F (2011) Linoleate isomerase activity occurs in lactic acid bacteria strains and is affected by pH and temperature. J Appl Microbiol 111(3):593–606

Gorrisen L, Raes K, Weckx S, Dannenberger D, Leroy F, De Vuyst L, De Smet S (2010) Production of conjugated linoleic acid and conjugated linolenic acid isomers by Bifidobacterium species. Appl Microbiol Biotechnol 87(6):2257–2266

Gu ST, Chen HQ, Ye Q, Tian FW, Chen W, Zhang H (2008) Study on location of linoleate isomerase from *Lactobacillus plantarum* ZS2058. Sci Technol Food Ind 29(12):57–60

Hennessy AA, Ross RP, Devery R, Stanton C (2010) Optimization of a reconstituted skim milk based medium for enhanced CLA production by bifidobacteria. J Appl Microbiol 106(4):1315–1327

Hennessy AA, Barrett E, Ross RP, Fitzgerald GF, Devery R, Stanton C (2012) The production of conjugated α-linolenic, γ-linolenic and stearidonic acids by strains of bifidobacteria and propionibacteria. Lipids 47(3):313

Hernandezmendoza A, Lopezhernandez A, Hill CG, Garcia HS (2010) Bioconversion of linoleic acid to conjugated linoleic acid by *Lactobacillus reuteri* under different growth conditions. J Chem Technol Biotechnol Biotechnol 84(2):180–185

Jensen RG (2002) The composition of bovine milk lipids: January 1995 to December 2000. J Dairy Sci 85(2):295–350

Kepler CR, Tove SB (1967) Biohydrogenation of unsaturated fatty acids. III. Purification and properties of a linoleate delta-12-*cis*, delta-11-*trans*-isomerase from *Butyrivibrio fibrisolvens*. J Biol Chem 246(16):5686–5692

Kepler CR, Hirons KP, Mcneill JJ, Tove SB (1966) Intermediates and products of the biohydrogenation of linoleic acid by *Butyrinvibrio fibrisolvens*. J Biol Chem 241(6):1350–1354

Kepler CR, Tucker WP, Tove SB (1970) Biohydrogenation of unsaturated fatty acids. IV. Substrate
specificity and inhibition of linoleate delta-12-cis, delta-11-trans-isomerase from Butyrivibrio
fibrisolvens. J Biol Chem 245(14):3612–3620

Kepler CR, Tucker WP, Tove SB (1971) Biohydrogenation of unsaturated fatty acids.
V. Stereospecificity of proton addition and mechanism of action of linoleic acid delta 12-*cis*,
delta 11-*trans*-isomerase from *Butyrivibrio fibrisolvens*. J Biol Chem 246(9):2765–2771

Khosravi A, Safari M, Khodaiyan F, Gharibzahedi SM (2015) Bioconversion enhancement of conjugated linoleic acid by *Lactobacillus plantarum* using the culture media manipulation and
numerical optimization. J Food Sci Technol 52(9):5781–5789

Kil KS, Cunningham MW, Barnett LA (1994) Cloning and sequence analysis of a gene encoding
a 67-kilodalton myosin-cross-reactive antigen of *Streptococcus pyogenes* reveals its similarity
with class II major histocompatibility antigens. Infect Immun 62(6):2440–2449

Kim YJ, Liu RH (2010) Increase of conjugated linoleic acid content in milk by fermentation with
lactic acid bacteria. J Food Sci 67(5):1731–1737

Kishino S, Ogawa J, Omura Y, Matsumura K, Shimizu S (2002) Conjugated linoleic acid production from linoleic acid by lactic acid bacteria. J Am Oil Chem Soc 79(2):159–163

Kishino S, Ogawa J, Ando A, Yokozeki K, Shimizu S (2010) Microbial production of conjugated γ-linolenic acid from γ-linolenic acid by *Lactobacillus plantarum* AKU 1009a. J Appl
Microbiol 108(6):2012–2018

Kishino S, Park SB, Takeuchi M, Yokozeki K, Shimizu S, Ogawa J (2011) Novel multi-component
enzyme machinery in lactic acid bacteria catalyzing C=C double bond migration useful for
conjugated fatty acid synthesis. Biochem Biophys Res Commun 416(1):188–193

Kishino S, Takeuchi M, Park S-B, Hirata A, Kitamura N, Kunisawa J, Ogawa J (2013)
Polyunsaturated fatty acid saturation by gut lactic acid bacteria affecting host lipid composition. Proc Natl Acad Sci U S A 110(44):17808–17813

Lee K, Lee Y (2009) Production of *c9,t11*- and *t10,c12*-conjugated linoleic acids in humans by
Lactobacillus rhamnosus PL60. J Microbiol Biotechnol 19(12):1617

Li H, Liu Y, Bao Y, Liu X, Zhang H (2012) Conjugated linoleic acid conversion by six *Lactobacillus
plantarum* strains cultured in MRS broth supplemented with sunflower oil and soymilk. J Food
Sci 77(6):M330–M336

Lin TY, Lin CW, Lee CH (1999) Conjugated linoleic acid concentration as affected by lactic cultures and added linoleic acid. Food Chem 69(1):27–31

Maia MRG, Chaudhary LC, Figueres L, Wallace RJ (2007) Metabolism of polyunsaturated
fatty acids and their toxicity to the microflora of the rumen. Antonie Van Leeuwenhoek
91(4):303–314

Maia MR, Chaudhary LC, Bestwick CS, Richardson AJ, Mckain N, Larson TR, Wallace RJ (2010)
Toxicity of unsaturated fatty acids to the biohydrogenating ruminal bacterium, *Butyrivibrio
fibrisolvens*. BMC Microbiol 10(1):52

Nieuwenhove CPV, Oliszewski R, González SN, Chaia ABP (2010) Conjugated linoleic acid
conversion by dairy bacteria cultured in MRS broth and buffalo milk. Lett Appl Microbiol
44(5):467–474

Nugteren DH, Christ E (1987) Naturally occurring conjugated octadecatrienoic acids are strong
inhibitors of prostaglandin biosynthesis. Prostaglandins 33(3):403–417

O'Connell KJ, Motherway MO, Hennessey AA, Brodhun F, Ross RP, Feussner I, Stanton C,
Fitzferald GF, van Sinderen D (2013) Identification and characterization of an oleate hydratase-
encoding gene from *Bifidobacterium breve*. Bioengineered 4(5):313–321

Ogawa J, Matsumura K, Kishino S, Omura Y, Shimizu S (2001) Conjugated linoleic acid accumulation via 10-hydroxy-12 octadecaenoic acid during microaerobic transformation of linoleic
acid by *Lactobacillus acidophilus*. Appl Environ Microbiol 67(3):1246–1252

Ogawa J, Kishino S, Ando A, Sugimoto S, Mihara K, Shimizu S (2005) Production of conjugated
fatty acids by lactic acid bacteria. J Biosci Bioeng 100(4):355–364

Oh DK, Hong GH, Lee Y, Min S, Sin HS, Cho SK (2003) Production of conjugated linoleic acid by isolated *Bifidobacterium* strains. World J Microbiol Biotechnol 19(9):907–912

Park HG, Heo W, Kim SB, Kim HS, Bae GS, Chung SH, Kim YJ (2011) Production of conjugated linoleic acid (CLA) by *Bifidobacterium breve* LMC520 and its compatibility with CLA-producing rumen bacteria. J Agric Food Chem 59(3):984–988

Polan CE, Mcneill JJ, Tove SB (1964) Biohydrogenation of unsaturated fatty acids by rumen bacteria. J Bacteriol 88(4):1056–1064

Rodríguez-Alcalá LM, Braga T, Malcata FX, Gomes A, Fontecha J (2011) Quantitative and qualitative determination of CLA produced by *Bifidobacterium* and lactic acid bacteria by combining spectrophotometric and Ag+ HPLC techniques. Food Chem 125(4):1373–1378

Romero-Pérez GA, Inoue R, Ushida K, Yajima T (2013) A rapid method of screening lactic acid bacterial strains for conjugated linoleic acid production. Biosci Biotechnol Biochem 77(3):648–650

Rosberg-Cody E, Ross RP, Hussey S, Ryan CA, Murphy BP, Fitzgerald GF, Devery R, Stanton C (2004) Mining the microbiota of the neonatal gastrointestinal tract for conjugated linoleic acid-producing bifidobacteria. Appl Environ Microbiol 70(8):4635–4641

Rosberg-Cody E, Liavonchanka A, Gobel C, Ross RP, O'Sullivan O, Fitzgerald GF, Stanton C (2011) Myosin-cross-reactive antigen (MCRA) protein from *Bifidobacterium breve* is a FAD-dependent fatty acid hydratase which has a function in stress protection. BMC Biochem 12:9

Rosenfeld IS, Tove SB (1971) Biohydrogenation of unsaturated fatty acids. VI. Source of hydrogen and stereospecificity of reduction. J Biol Chem 246(16):5025

Rosson RA, Grund AD, Deng MD, Sanchez-Riera F (1999) Linoleate isomerase. World Patent, WO-99/32604 A1

Smith CR Jr (1971) Occurrence of unusual fatty acids in plants. Prog Chem Fats Other Lipids 11:137,139–137,177

Sun OL, Chang SK, Cho SK, Choi HJ, Ji GE, Oh DK (2003) Bioconversion of linoleic acid into conjugated linoleic acid during fermentation and by washed cells of *Lactobacillus reuteri*. Biotechnol Lett 25(12):935–938

Volkov A, Liavonchanka A, Kamneva O, Fiedler T, Goebel C, Kreikemeyer B, Feussner I (2010) Myosin cross-reactive antigen of *Streptococcus pyogenes* M49 encodes a fatty acid double bond hydratase that plays a role in oleic acid detoxification and bacterial virulence. J Biol Chem 285(14):10353–10361

Xu S, Boylston TD, Glatz BA (2004) Effect of lipid source on probiotic bacteria and conjugated linoleic acid formation in milk model systems. J Am Oil Chem Soc 81(6):589–595

Xu QY, Chen HQ, Tian FW, Zhao JX, Zhang H, Chen W (2008) Analysis of conjugated linoleic acid in the bioconversion process of *Lactobacillus plantarum* ZS2058. Food Ferment Ind 1:110–115

Yang B, Chen H, Song Y, Chen YQ, Zhang H, Chen W (2013) Myosin-cross-reactive antigens from four different lactic acid bacteria are fatty acid hydratases. Biotechnol Lett 35(1):75–81

Yang B, Chen H, Gu Z, Tian F, Ross RP, Stanton C, Zhang H (2014) Synthesis of conjugated linoleic acid by the linoleate isomerase complex in food-derived lactobacilli. J Appl Microbiol 117(2):430–439

Yang B, Chen H, Stanton C, Ross RP, Zhang H, Chen YQ, Chen W (2015) Review of the roles of conjugated linoleic acid in health and disease. J Funct Foods 15:314–325

Yang B, Gao H, Stanton C, Ross RP, Zhang H, Chen YQ, Chen W (2017) Bacterial conjugated linoleic acid production and their applications. Prog Lipid Res 68:26–36

Zhou LH, Zhang H, Chen W, Tian FW (2004) Screening and identification of lactic acid bacteria for biosynthesis of conjugated linoleic acid. J Wuxi Univ Light Ind 23(5):53–57

Chapter 3
Lactic Acid Bacteria and B Vitamins

Wanqiang Wu and Baixi Zhang

Vitamin is one of the most significant micronutrients in all the biological metabolism progress. The 13 vitamins which are necessary to human bodies can be divided into lipid-soluble (vitamins A, D, E, K) and water-soluble (vitamin C and eight kinds of B vitamins). B vitamins contain thiamine (vitamin B1), riboflavin (vitamin B2), niacin (vitamin B3), pantothenic acid (vitamin B5), vitamin B6, biotin (vitamin B7 or vitamin H), folic acid (vitamin B9), and cobalamin (vitamin B12). Each B-type vitamin has different chemical properties, and its derivatives often participate in metabolism (such as energy production, red blood cell synthesis, etc.) as specific co-enzymes in physiological activities and play an important role in maintaining homeostasis (Table 3.1). B vitamins exist in various foods (Table 3.2) and easy to be destroyed by cooking and processing. Thus, the deficiency of B vitamins is the common problem influencing human health.

Lactic acid bacteria (LAB) are a kind of microorganism widely used in fermented foods, which are able to extend the shelf life of foods, improve the nutritional value, and develop the flavor. Lactic acid bacteria can synthesize and release some beneficial substances in food. The vitamin is one of the functional components of LAB synthesize. With the rapid development of genomics, humans can not only identify potential vitamin synthesis strains through genetic information but also improve their vitamin production by investigating their genetically anabolic networks. Therefore, as food microorganisms, the LAB is a kind of ideal microorganism for solving the problem of insufficient vitamin intake in humans in the future.

W. Wu (✉) · B. Zhang
Jiangnan University, Wuxi, China
e-mail: wuwanqiang@jiangnan.edu.cn; zbx@jiangnan.edu.cn

© Springer Nature Singapore Pte Ltd. and Science Press 2019
W. Chen (ed.), *Lactic Acid Bacteria*,
https://doi.org/10.1007/978-981-13-7283-4_3

Table 3.1 Coenzyme form and function of vitamin B

Vitamin B	Coenzyme forms	Function
B_1	Thiamine phosphate (TPP)	Oxidative decarboxylation, aldehyde group transfer
B_2	Flavin-5-phosphoric acid (FMN), flavin-5'-adenosine diphosphate (FAD)	Transfer H^+/electron
B_6	Phosphopyridoxal	Transamination, decarboxylation
B_9	Tetrahydrofolic acid	Carbon carrier
B_{12}	Cobamide	Transmethylase

Table 3.2 Food sources of vitamins B

Vitamin B	Food sources
B_1	Seed epidermis, animal viscera and lean meat, vegetables, and fruits are not abundant
B_2	Animal liver, milk, eggs, beans, and green leafy vegetables
B_6	Widely available, especially in liver, milk, egg yolks, vegetables, fish, whole grains, beans
B_9	Animal livers, kidneys, eggs, beans, yeast, nuts, green leafy vegetables, and fruits
B_{12}	Animal food, liver, egg yolk, meat, shellfish; milk and dairy products contain small amounts

3.1 Folic Acid

3.1.1 Chemical Structure and Demand

Folate, also known as folic acid, is the vital nutrient in the human body, participating in important cell metabolism, such as the copying, repairing, and methylation of DNA and the formation of nucleotide, other vitamins and amino acids, and so on. In addition to the essential role in repairing and copying of DNA, it also has antioxidant effects and protects DNA from free radicals (Duthie et al. 2002).

The folate deficiency presents in plenty of diseases, like Alzheimer's disease (Hugenholtz et al. 2002; Luchsinger et al. 2007), coronary heart disease (Danesh and Lewington 1998; Daniel and Bobik 1998), osteoporosis (Baines et al. 2007), neural tube defects (Group 1991; Czeizel and Dudas 1992), and so forth. Due to the prevalence of folic acid deficiency, the researchers currently focus on developing the food with high folate content. In this section, the term "folate" includes all folate derivatives, such as polyglutamic acid commonly found in foods and synthetic folic acid, often used as a food fortifier.

Folic acid (or pteroyl-L-glutamic acid) is attached to the pteridine and L-glutamic acid by P-aminobenzoic acid (Fig. 3.1). The form of natural folate differs depending on the nature of the pteroyl and acridine substituents as well as the number of glutamic acid residues attached to the pteroyl group. Natural folic acid consists of

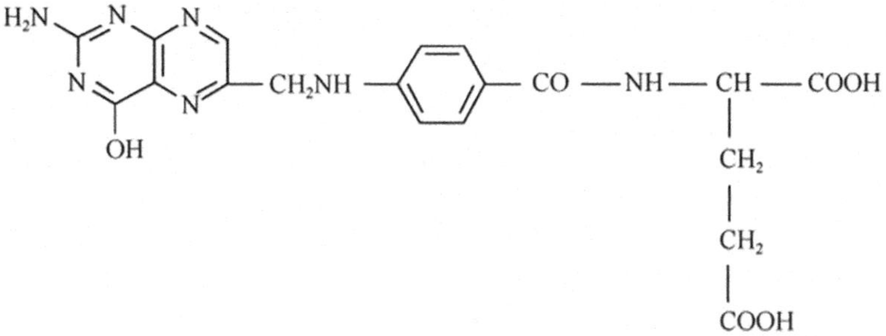

Fig. 3.1 Chemical structure of folic acid (pteroyl-L-glutamic acid)

5-methyltetrahydrofolate (5-MTHF), 5-formyltetrahydrofolate (5-formyl-THF), 10-formyltetrahydrofolate (10-formyl-THF), 5, 10-methylenetetrahydrofolate (5, 10-methylene-THF), 5-iminomethyltetrahydrofolate (5-formimino-THF), 5,6,7,8-tetrahydrofolate (THF), and dihydrofolate (DHF). Most of the natural folic acid is in the form of pteropolyglutamic acid, which is formed by the attachment of 2–7 glutamic acids on the amide (peptide) to γ-carboxyglutamic acid. The main folate in the cell is pteroglutamate, and the main extracellular folic acid is pteroyl monoglutamate. Natural pteropolyglutamic acid has up to 11 glutamic acid residues.

At present, the research on the nutrition of folic acid has been furthered deeply. The body cannot synthesize folic acid itself; therefore absorbing from food is necessary. Folic acid is found in many foods such as beans (soybeans, nuts, peas, etc.), leafy vegetables, citrus fruits, liver, and fermented or unfermented dairy products. Although soy and green leafy vegetables are good sources of folic acid, this does not meet the needs of the human body. The recommended daily intake (RDI) of folate from the US Food and Drug Administration (FDA) is 200–400 μg for adults and 400–600 μg for pregnant women. Despite the abundant sources of folic acid from food, lacking folic acid is still common, even in developed countries (Konings 2001; O'Brien et al. 2001). Based on these studies, some countries require mandatory folic acid supplementation. For example, in Canada and the United States, since 1998, folate has been mandatorily added in flour to reduce the incidence of neonatal neural tube defects.

3.1.2 Folate in Fermented Food

3.1.2.1 Folate in Fermented Dairy Products

Many important industrial LAB are able to synthesize folic acid. Commonly, there are *Streptococcus thermophilus*, *Bifidobacterium*, *Lactobacillus delbrueckii*, *Bulgarian subspecies*, etc. *Lactobacillus reuteri* CRL1098 is a well-known vitamin B12-producing strain that can also synthesize large amounts of folic acid (Santos et al. 2008). Therefore, the utilization of LAB to ferment the diary product can improve the level of folate and, to some extent, alleviate the deficiency of folic acid. However, the ability of microbial strains to produce or utilize folic acid is also related to the type and characteristics of the strain. Most researchers believe that *Streptococcus thermophilus* usually produces folic acid, while *Lactobacillus delbrueckii* subsp. *bulgaricus* consumes folic acid during growth, so proper combination of fermenting microbes is essential for the development of fermented foods and increasing vitamin content. The study shows that the combination of *Streptococcus thermophilus* and *Bifidobacterium animalis* increases the amount of folic acid in fermented dairy products by six times (Crittenden et al. 2003).

3.1.2.2   Folate in Other Fermented Foods

Studies also show that the use of certain LAB as a starter can produce large amounts of folic acid in vegetable fermentation. These starters are combinations of two or three following bacteria, *Lactobacillus plantarum*, *Lactococcus lactis*, *Leuconostoc*, *Lactobacillus flexeri*, *Lactobacillus pentosus*, *Streptococcus thermophilus*, *Lactobacillus delbrueckii* subsp. *bulgaricus*, *Lactobacillus acidophilus*, *Bifidobacterium animalis* subsp. *lactis*, and *Propionibacterium freudenreichii*. Finally, the highest yield of folic acid, usually 5-MTHF which can be directly utilized by the body, is twice that of the unfermented vegetables. Thus, with this research, new foods for vegetables can be developed to supplement the body's required folic acid (Jägerstad et al. 2004).

Another example of using LAB to increase the folic acid content in fermented foods is the fermentation of corn flour. After fermenting corn flour at 30 °C for 4 days with *Streptococcus faecalis*, the level of folic acid increases by nearly three times (Murdock and Fields 1984).

3.1.2.3   Folate-Producing LAB

Many studies have shown that industrially fermenting bacteria such as *Lactobacillus* and *Streptococcus thermophilus* have the ability to synthesize folic acid. Therefore, some fermented dairy products have higher levels of folic acid than unfermented dairy products. However, the ability of different industrially fermenting bacteria to produce or utilize folic acid varies widely. Most LAB cannot synthesize this vitamin (Hugenholtz and Kleerebezem 1999; Crittenden et al. 2003), but experiments have shown that *Lactobacillus plantarum* can produce folic acid in chemical synthesis media without folic acid. The folic acid content in milk is 20–60 g/L, while the folate content in yogurt may increase to above 200 g/L, depending on the fermentation and storage conditions (Wouters et al. 2002). The folate content also depends on the type of strain of *Streptococcus thermophilus* and the *Lactobacillus delbrueckii* subsp. *bulgaricus* which has been proved to reduce folate production during growth. Recent studies demonstrate that some probiotics (such as *Bifidobacteria*) can also synthesize folic acid (Table 3.3).

Table 3.3 The LAB in fermented dairy products which produces folate

LAB	Output	Reference
S. thermophilus CSCC2000	36.5 ng/g	Crittenden et al. (2003)
S. thermophilus CSCC2002	36.2 ng/g	
S. thermophilus CSCC2010	36.5 ng/g	
S. thermophilus CSCC2012	36.3 ng/g	
S. thermophilus CSCC2013	28.6 ng/g	
S. thermophilus CSCC2016	30 ng/g	
S. thermophilus CSCC2018	32 ng/g	
B. animalis lactate subspecies CSCC5127	13.5 ng/g	
B. animalis lactate subspecies CSCC5123	13 ng/g	
B. animalis CSCC1941	8.5 ng/g	
B. Bifidobacterium CSCC5128	9 ng/g	
B. longum subspecies *infantis* CSCC5187	20.5 ng/g	
B. breve CSCC5187	33.5 ng/g	
B. animalis lactate subspecies *Lafti* B94CSCC5127	10.5 ng/g	
E. Faecium CSCC5140	11.5 ng/g	
L. acidophilus ATCC4356	13.3 ng/mL	Rao et al. (1984)
Lactococcus lactis cremoris CM22	12.5 µg/L	Gangadharan et al. (2010)
Lactococcus lactis cremoris CM28	14.2 µg/L	
L. acidophilus N1	63.9 ± 5.2 µg/L	Lin and Young (2000)
L. acidophilus 4356	53.9 ± 4.6 µg/L	
L. bulgaricus 449	62.8 ± 2.1 µg/L	
L. bulgaricus 448	46.7 ± 5.0 µg/L	
S. thermophilus 573	59.6 ± 2.3 µg/L	
S. thermophilus MC	75.8 ± 6.5 µg/L	
L. Bifidobacterium ATCC15708	99.2 ± 3.8 µg/L	
B. longum B6	397 ± 60 µg/L	Hugenschmidt et al. (2010)
L. plantarum SM39	131 ± 196 µg/L	
L. brevis SM34	125 ± 28 µg/L	
L. reuteri ATCC55730	84 ± 32 µg/L	Pompei et al. (2007)
L. fermentum SM81	44 µg/L	
B. adolescentis MB114	65 µg/L	
B. adolescentis MB115	54 µg/L	
B. adolescentis MB227	54 µg/L	
B. adolescentis MB239	27 µg/L	
B. longum subspecies *infantis* ATCC15697	82 µg/L	
B. pseudocatenulatum MB116	41 µg/L	
B. pseudocatenulatum MB237	12 µg/L	

(continued)

Table 3.3 (continued)

LAB	Output	Reference
B. pseudocatenulatum MB264	95 µg/L	Sybesma (2003)
L. lactis cremoris B42	92 µg/L	
L. lactis cremoris B64	116 µg/L	
L. lactis cremoris B697	90 µg/L	
L. lactis cremoris B628	116 µg/L	
L. lactis cremoris NZ9000(6)	291 µg/L	
L. lactis cremoris NZ9010(6) *Ldh negative*	91 µg/L	
L. lactis cremoris B26	69 µg/L	
L. lactis cremoris B27	62 µg/L	
L. lactis cremoris B1172	63 µg/L	
L. lactis cremoris B1173	57 µg/L	
L. lactis cremoris B621	98 µg/L	
L. lactis cremoris B86	100 µg/L	
L. lactis cremoris B87	79 µg/L	
L. lactis cremoris B103	29 µg/L	
S. thermophilus B108	202 µg/L	
S. thermophilus B119	120 µg/L	
S. thermophilus B911	45 µg/L	
Leuconostoc B629	44 µg/L	
Leuconostoc WCFS-1	90 µg/L	
L. helveticus B219	89 µg/L	
L. helveticus B230	1 µg/L	
L. bulgaricus B194	41 µg/L	

3.1.2.4 Biosynthesis and Regulation Mechanism of Folic Acid in LAB

In LAB, plants and fungi, the de novo synthesis of folic acid includes the pteridine metabolic branch and the para-aminobenzoic acid (pABA) metabolic branch, both of which are indispensable.

The pteridine metabolic branch of lactic acid bacteria is as follows. Firstly, an important intermediate-6-hydroxymethyl-7,8-dihydropteridine pyrophosphate (DHPPP) is gradually synthesized through the participation of guanosine triphosphate (GTP) cyclized hydrolase I (GCHY-I), neopterin aldolase (DHNA), dihydroneopterin triphosphate pyrophosphatase (DHNTP), and 6-hydroxymethyldihydropterin pyrophosphate kinase (DHPPK). DHPPP then binds to pABA with the action of dihydropteroate synthase (DHPS) to form dihydropteroate, which is a direct precursor of folic acid synthesis. Finally, it is catalyzed into polyglutamic acid tetrahydrofolate by dihydrofolate reductase (DHFR) and folate glutamate synthase (FDGS).

The pABA metabolic branch of lactic acid bacteria is the formation of pABA by the branched acid with the action of 4-amino-4-deoxylated acid synthase and lyase and then the same as the pteridine metabolic branch. pABA will bind to DHPPP with the participation of DHPS, forming the direct precursor dihydropteroic acid; thus folic acid is synthesized.

Genes for folate biosynthesis have been identified in *Lactococcus lactis* (Sybesma et al. 2003), *Lactobacillus plantarum* (Kleerebezem et al., 2003), and *Lactobacillus bulgaricus* (van de Guchte et al., 2006). These genes encode enzymes that regulate the de novo synthesis of folic acid, including the *follB* gene encoding DHNA, the *folK* gene encoding DHPPPK, the *folE* gene encoding GCHI, the *folP* gene encoding the DHPS and the *folC* gene, the *folQ* (ylgG) gene encoding DHNTPase, the *folA* gene encoding DHFR, and so on. Not every kind of LAB produces folic acid because some LAB lack genes involved in folate biosynthesis, such as *Lactobacillus gargle* (Wegkamp et al., 2004) and *Lactobacillus saliva* (Claesson et al., 2006).

Metabolic engineering can increase folate levels in *Lactococcus* and *L. reuteri* (Sybesma et al. 2003; Wegkamp et al., 2007). By controlling the overexpression of the *folKE* gene encoding 6-hydroxymethyl-dihydropyrophosphate kinase and GTP cyclohydrolase in *Lactococcus*, the production of extracellular folate can be increased by ten times, and the total folate content of the product is increased three times more than the former. At the same time, overexpression of the *folA* gene encoding dihydrofolate reductase reduces total folic acid production by about 50%. In addition, simultaneous overexpression of *folKE* and *folC* facilitates the accumulation of intracellular folate. Overexpression of GTP cyclase I is also beneficial for increasing the flux of folate biosynthesis. Therefore, the appropriate combination of overexpressed *folKE* with other genes that overexpress or inhibit expression can significantly increase the yield of folic acid. Transgenic LAB are as safe as natural LAB (Leblanc et al. 2010).

In fact, we can not only increase folic acid production yield by overexpressing genes involved in folate biosynthesis but also increase folate production by overexpressing other genes involved in related metabolism. For example, an excess of pABA does not result in an increase in the active pool of folate itself. However, overexpression of pABA and folate biosynthesis gene clusters yields high levels of folate (Wegkamp et al. 2007). This is a process that is not affected by pABA supplementation.

After transferring the plasmid of *Lactococcus* MG1363 containing the complete folate synthesis genes (*folA, folB, folKE, folP, ylgG*, and *folC*) to *Lactobacillus gasseri* ATCC33323, the recombinant strain is thus transformed into a folate-producing strain (Wegkamp et al. 2004).

3.2   Riboflavin

3.2.1   Chemical Structure and Demand

The term "riboflavin" refers to all biologically active vitamin B_2, including riboflavin, riboflavin-5-phosphate (FMN), and riboflavin-5′-adenosine diphosphate (FAD). The chemical name of riboflavin is 7,8-dimethyl-10-(1′-D-ribosyl)-isorazine, and its molecular formula is $C_{17}H_2ON_4O_6$. The molecular structure

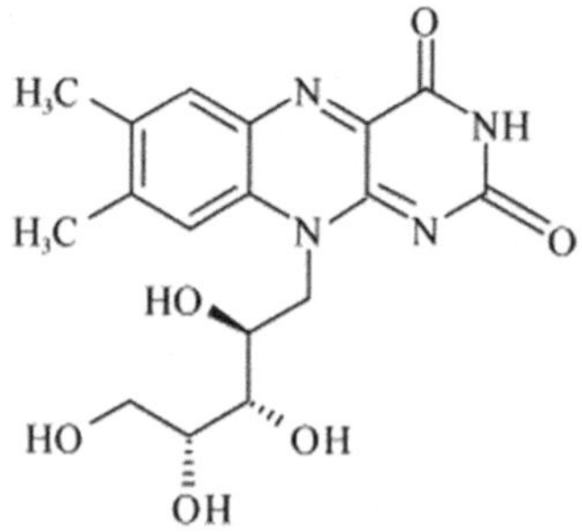

Fig. 3.2 Chemical
structure of riboflavin

includes a ribitol side-chain isorazine group (Fig. 3.2), which is the basis of all riboflavin derivatives. Riboflavin is an orange-yellow odorless needlelike crystal that is easily destroyed by sunlight and ultraviolet light, but it is stable to the thermal environment and is soluble in neutral liquid. Heating in the liquid at 120 ° C for 6 h was less damaged.

Riboflavin (vitamin B_2) was initially considered as a growth factor in rabbits. It plays a key role in cell metabolism as a precursor of riboflavin-5-phosphate (FMN) and 5′-adenosine diphosphate (FAD). Riboflavin is directly involved in the biological oxidation of carbohydrates, proteins, and fats in vivo and plays a variety of physiological functions.

The symptoms of riboflavin deficiency in the human body include throat congestion, sore throat, edema of mouth, cleft lip, glossitis, etc. (Wilson 1983). Severe cases are relatively rare. Human demand for riboflavin is closely related to gender, age, physiological status (pregnancy, lactation), and so on. Due to the inadequate storage of riboflavin, normal adults need to take 0.3–1.8 mg riboflavin daily to meet their metabolic needs. Although riboflavin is found in many foods such as dairy products, meat, eggs, and green vegetables, riboflavin deficiency is still widespread in both developed and developing countries (O'Brien et al. 2001; Blanck et al. 2002).

3.2.2   Riboflavin in Fermented Food

3.2.2.1   Riboflavin in Fermented Dairy Products

Dairy products contain riboflavin, but this is not the best source of such essential vitamins. Drinking milk contains 1.2 mg riboflavin per liter. In the case of low daily milk intake, increasing the content of riboflavin in milk has become an important method to prevent riboflavin deficiency. Clinical trial has shown intake of 200 g probiotics or traditional yogurt daily for 2 consecutive weeks can promote absorption of vitamin B_2 (Fabian et al. 2008). In addition, the content of riboflavin in dairy products varies with processing technology and microbial activity. For example, the contents of riboflavin and folic acid in cheese were 1.7 mg/L and 90 µg/L, respectively, and the contents of riboflavin and folic acid in yoghurt were 2.0 mg/L and

80 µg/L, respectively. The contents of riboflavin and folic acid in kefir were 1.3 mg/L and 50 µg/L, respectively, which were significantly higher than that in unfermented milk (1.2 mg/L) and folic acid (40 µg/L). The increase in riboflavin content in fermented dairy products is due to the use of a starter that can produce riboflavin during processing.

3.2.2.2 Riboflavin in Other Fermented Foods

Indonesian cardamom, also known as Tianbei, is a naturally fermented soybean product with very rich nutrition. It is known as "alternative food for meat" and is an indispensable food for Indonesians and a representative food for healthy food. Indonesian soybean sauce contains high concentration of B vitamins, including thiamine, riboflavin, vitamin B_6, folic acid, vitamin B_{12}, and so on. These vitamins are produced by a variety of microorganisms during the fermentation process. It was found that *Streptococcus*, *Lactobacillus casei*, *Lactobacillus plantarum*, and other LAB could be isolated from Indonesian soybean sauce and its soaking water (Keuth and Bisping 1993).

3.2.3   Riboflavin-Producing LAB

Recent studies have shown that many lactic acid bacteria can produce riboflavin (Table 3.4).

3.2.4   Biosynthesis and Regulation Mechanism of Riboflavin in LAB

Riboflavin biosynthesis begins with GTP and D-ribose-5-phosphoric acid and undergoes seven-step enzymatic reactions to synthesize riboflavin (Bacher et al. 2000). The imidazole ring of GTP is hydrolyzed to produce 4,5-diaminopyrimidine, which is converted to 5-amino-6-ribitol amino-2,4(1H, 3H)-pyrimidinedione by deamination, side-chain reduction, and dephosphorylation. Condensed 5-amino-6-ribitol amino-2,4(1H,3H)-pyrimidinedione from 6,7-dimethyl-8-ribitol group provided by ribulose-5-phosphate.The riboflavin and 5-amino-6-ribitol amino-2,4(1H, 3H)-pyrimidinedione were produced by the disproportionation with 5-amino-6-ribitol amino-2,4(1H, 3H)-pyrimidinedione,3,4-dihydroxy-2-butanone-4-phosphate and pyridine dioxide derivatives. This process requires two bifunctional enzymes, RibA and RibG, and riboflavin synthases, RibH and RibB. Among them, II/3,4-dihydroxybutyrone phosphate synthase hydrolyzed by GTP imidazole ring is a rate-limiting enzyme in riboflavin synthesis, which is encoded by *ribA* gene (Hu mbelin et al. 1999).

Table 3.4 Riboflavin-producing lactic acid bacteria

lactic acid bacteria	Yield	Reference
Lactobacillus fermentum MTCC8711	2.8 mg/L	Jayashree et al. (2010)
Lactobacillus fermentum GKJFE[a]	3.49 mg/L	
Lactobacillus plantarum NCDO1752	600 µg/L	Burgess et al. (2006)
Leuconostoc mesenteroides CB200	150 µg/L	
Leuconostoc mesenteroides CB201	160 µg/L	
Leuconostoc mesenteroides CB202	400 µg/L	
Leuconostoc mesenteroides CB203	300 µg/L	
Leuconostoc mesenteroides CB204	160 µg/L	
Leuconostoc mesenteroides CB205	150 µg/L	
Leuconostoc mesenteroides CB206	180 µg/L	
Leuconostoc mesenteroides CB207	500 µg/L	
Leuconostoc mesenteroides CB208	260 µg/L	
Leuconostoc mesenteroides CB209	120 µg/L	
Leuconostoc mesenteroides CB210	130 µg/L	
Lactococcus lactis subsp. *cremoris* NZ9000	700 µg/L	Burgess et al. (2004)
Lactococcus lactis subsp. *cremoris* NZ9000[a]	24 mg/L	Sybesma et al. (2004)
		Leblanc et.al. (2005)
Lactobacillus plantarum CRL725	(91 ± 11)ng/mL	Valle et al. (2014)
Lactobacillus plantarum CRL725 variant G[a]	(1100 ± 20)ng/L	

[a]Genetic engineering bacteria

Although LAB strains have the ability to synthesize riboflavin, most probiotic strains also consume riboflavin while synthesizing, thereby reducing riboflavin levels in the fermented product (Elmadfa et al. 2001). Therefore, how to improve the riboflavin production of LAB is our concern. Screening fermentation strains is a traditional method to improve riboflavin production of LAB. Screening of resistant strains with rosin is an effective method to obtain riboflavin-producing strains. Rose yellow pigment is a toxic riboflavin analogue that is often used to mutagenize food-grade riboflavin-producing microorganisms, such as *Lactococcus lactis* NZ9000 (riboflavin yield after mutation screening is 700 µg/L) (Burgess et al. 2004), *Lactobacillus plantarum* NCDO1752 (riboflavin yield after mutation screening is 600 µg/L), *Leuconostoc mesenteroides* spp. (riboflavin yield after mutation screening is 500 µg/L), and *Propionibacterium freudenreichi* spp. (riboflavin yield after mutation screening is 3000 µg/L) (Burgess et al. 2006). The high-yield riboflavin strain, *Lactococcus* CB010, screened by rosin resistance has similar bioavailability to pure riboflavin, can treat most riboflavin deficiency, and improve growth shrinkage, high activation coefficient values, and hepatomegaly in animal experiments (Leblanc et al. 2005). Other research proved that the strain of *Propionibacterium freudenreichii* B2336 screened by rosin resistance can produce riboflavin with high yield and is beneficial for the recovery of animal models of riboflavin deficiency. Foods fermented with *Propionibacterium freudenreichii* B2336 have higher levels of riboflavin and alleviate the clinical symptoms of most riboflavin deficiency (Leblanc et al. 2006).

In addition to traditional strain screening methods, another method to increase riboflavin production in fermented foods is genetic engineering, which can improve the production of riboflavin in lactic acid bacteria by genetic engineering. This method is currently a hot topic of research, and the food-grade fermentation strain after transformation has great potential in the future food field.

Hydrated ammonium oxalate-corynebacterium sulfate contains all the genes for riboflavin biosynthesis, and high-yield strains can be obtained by recombinant DNA technology. The recombinant strain produced and accumulated 17-fold more ribo-flavin than the original strain that was not genetically engineered (Koizumi et al. 2000). In the *Lactococcus* NZ9000 containing the pNZGBAH sequence, simultane-ous overexpression of four biosynthetic genes *ribG*, *ribH*, *ribB*, and *ribA* can greatly increase riboflavin production, eventually reaching 24 mg/L. Yogurt fermented with this strain as a feed for animals in animal experiments can alleviate the symptoms of riboflavin deficiency in mice (Leblanc et al. 2005).

Genetic engineering methods have significantly reduced production costs com-pared with current methods of vitamin production. Another advantage of this method is that the modified host microorganism can produce not only one vitamin. Sybesma et al. (2004) altered the biosynthesis pathway of folic acid and riboflavin in *Lactococcus* by site-directed mutagenesis and metabolic engineering and then screened for a mutant *Lactococcus lactis* NZ9000 by leucine-resistant resistance screening. The upstream regulatory region of the riboflavin biosynthesis gene of the mutant has changed, resulting in an increase in riboflavin production. At the same time, the GTP cyclohydrolase I of the mutant has overexpressed and increased the yield of folic acid (Sybesma et al. 2004).

3.3 Cobalamin

3.3.1 Chemical Structure and Demand

Vitamin B_{12}, also known as cobalamin, is a general term for a class of porphyrin compounds containing cobalt. Vitamin B_{12} is obtained in industrial production and is not naturally occurring (Rucker et al. 2001). The main forms of natural vitamin B_{12} are deoxyadenosylcobalamin (coenzyme B_{12}), methylcobalamin, and pseudoco-balamin. Structurally, the cobalamin molecule can be divided into three parts (Fig. 3.3): central glucolin ring, Coβ ligand attached to the adenosine methyl group, and a Co alpha ligand containing nucleotide rings (usually dimethylbenzimid-azoles). However, in some anaerobic bacteria, other forms of ligands such as ade-nine also exist, forming pseudocobalamin (pseudo B_{12}) and other active factors (Martens et al. 2002). The crystal structure of vitamin B_{12} is very complex. Hodgkin firstly analyzed it in 1956 by X-ray method (Hodgkin et al. 1956).

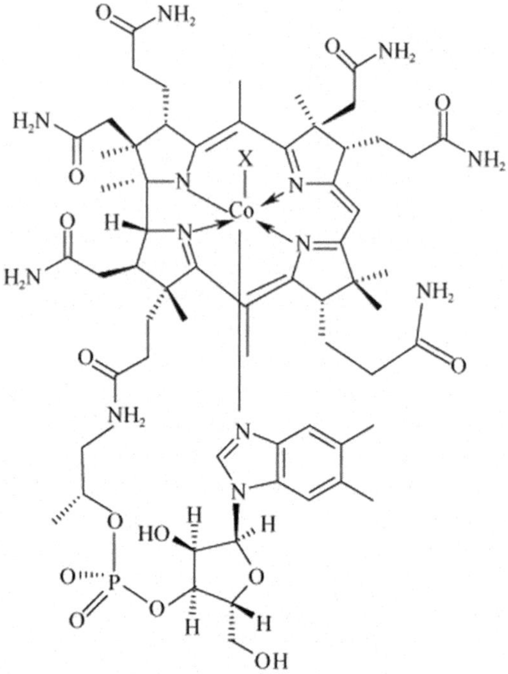

Fig. 3.3 Chemical
structure of vitamin B_{12}

Animals and plants cannot produce vitamin B_{12}, which is the only vitamin produced by microorganisms. Actinomycetes and bacteria are the main microorganisms that synthesize vitamin B_{12} (Roth et al. 1996; Smith et al. 2007). Among them, *Streptomyces* in actinomycetes (such as *Streptomyces antibiotics, Streptomyces faecalis, Streptomyces griseus, Streptomyces roseolus,* etc.) and *Flavobacterium, Bacillus megaterium, Bacillus subtilis, Lactobacillus,* and *Lactobacillus reuteri* in bacteria are the most commonly used fermentation strain in production. Bacteria in the rumen of adult ruminants and vegetarians can produce vitamin B_{12} in large quantities, which is the main source of the vitamin. There is no such microorganism in the small intestine of humans. Therefore, in order to obtain vitamin B_{12}, humans must take it from the outside. Good sources of vitamin B_{12} are animal meat (especially liver and kidney), eggs, dairy products, and functional foods. The daily demand for vitamin B_{12} in adults is 2–3 μg. The deficiency of vitamin B_{12} can affect hematopoietic function and damage the nervous system and cardiovascular system to varying degrees. Severe patients may suffer from malignant anemia.

3.3.2 Cobalamin in Fermented Food

In fermented food, cobalamin content in fermented dairy products is higher. After fermentation, the cobalamin content of milk increased. Especially in cheese and yogurt, the cobalamin content can reach 1.5 times than before fermentation (Wouters et al.

2002). Sufu is a traditional folk food of the Han nationality, which has been circulating for thousands of years in China. It is very popular in China and the Southeast Asia because of its good taste, high nutritional value, and unique flavor. A study showed that the content of cobalamin in sufu increased by two times during fermentation.

3.3.3 Cobalamin-Producing LAB

Lactobacillus reuteri is a representative strain of LAB for the fermentation of vitamin B_{12}. The *cob* cluster (methyltransferase gene cluster) of *Lactobacillus reuteri* strains has a distinct feature. There is a *hem* gene in the center of *cob* cluster of *Lactobacillus reuteri*, which can only be found in some *Clostridium* strains (Rodionov et al. 2003). Bioassays have confirmed that *Lactobacillus reuteri* can produce 50 μg/L of vitamin B12, and the production of *Lactobacillus reuteri* can meet the daily needs of adults. However, the cobalamin produced by the anaerobic pathway of *Lactobacillus reuteri* is pseudocobalamin (Santos et al. 2007). Although pseudocobalamin is a coenzyme of many anaerobic and facultative anaerobes, it does not play a role in animals and humans (Rucker et al. 2001). Recent data indicate that *Lactobacillus reuteri* strain CRL1098 can produce not only pseudocobalamin but also a small amount of other Guerin-like compounds, including coenzyme B_{12} under microaerobic conditions (Santos et al. 2007).

3.3.4 Biosynthesis and Regulation Mechanism of Cobalamin in LAB

The first biological model for studying the biosynthesis of vitamin B12 is *Propionibacterium freudenreichii*, which is an industrial strain of vitamin B12 (Battersby 1994). Researchers at Battersby and Scott Labs elucidated the aerobic synthesis pathway of vitamin B12 by studying *Pseudomonas aeruginosa* (Blanche et al. 1995). Escalante-Semerena (2007) found that the anaerobic synthesis of vitamin B_{12} is carried out by *Propionibacterium freudenreichii*, *Salmonella*, and *Bacillus megaterium*. The conclusions of these studies indicate that the synthesis of cobalamin can be carried out under both aerobic and anaerobic conditions, and the two synthetic pathways are slightly different. In the anaerobic pathway, the central Co^{2+} will be inserted at an earlier step. The porphyrin ring produces an unstable, difficult-to-separate intermediate, whereas under aerobic conditions, Co^{2+} inserts a porphyrin ring at a later step, resulting in a relatively stable intermediate.

The synthesis of cobalamin can be divided into three steps. The first step is the condensation of the precursor 5-aminolevulinic acid (ALA) to form uroporphyrin III. The second step is to catalyze the synthesis of adenosylcobalamin acid from uroporphyrinogen III. The third step is to catalyze the synthesis of adenosylcobalamin from adenosylcobalaminic acid.

The common methods to increase vitamin B_{12} production are random mutation and genetic engineering technology (Martens et al. 2002; Burgess et al. 2009). At present, different metabolic engineering methods can increase the production of vitamin B_{12} of *Propionibacterium freudenreichii*. Through metabolic engineering, *Propionibacterium freudenreichii* containing *cobA*, *cbiLF*, and *cbiEGH* increased the production of vitamin B_{12} by 1.7, 1.9, and 1.5 times, respectively. After overexpression of *hemA* gene and homologous *hemB* AND *cobA* genes from *Rhodopseudomonas* spp., the vitamin B_{12} yield of *Propionibacterium freudenreichii* was 2.2 times higher than that of wild-type strains (Piao et al. 2004).

In order to effectively utilize vitamin B_{12} biosynthetic genes, researchers have established a genetic transformation system of *Propionibacterium freudenreichii*. pRGO1 is a cryptic plasmid of *Propionibacterium* E214; Kiatpapan et al. (2000) determined the complete nucleotide sequence of pRGO1, which has a length of 6868 bp and a GC content of 65% and contains six open reading frames (orf1 to orf6). Kiatpapan et al. (2000) also constructed a shuttle vector pPK705, which shuttles between *Escherichia coli* and *Propionibacterium* containing orf1, orf2, orf5, and orf6. Finally, the vector successfully transformed *Propionibacterium freudenreichii* IFO12426, while high conversion efficiencies are also found in other species of *Propionibacterium*, and the pPK705 vector is also stably present in *Propionibacterium freudenreichii*. The successful construction of this host-vector system facilitates genetic research and factory production of vitamin B_{12}.

3.4   Other B Vitamins

3.4.1   *Vitamin B1*

Vitamin B1, also known as thiamine or anti-neuritis, is a combination of a pyrimidine ring and a thiazole ring (Fig. 3.4). It is a white crystalline or crystalline powder and the first vitamin to be found. It can be synthesized by fungi, microorganisms, and plants, while animals and humans can only be obtained from food.

Vitamin B1 is widely found in foods (especially in the epidermis and germ of grains). But the peel during food processing leads to a large loss of VB_1, resulting in insufficient intake of VB_1. The study found that vitamin B1 daily intake below

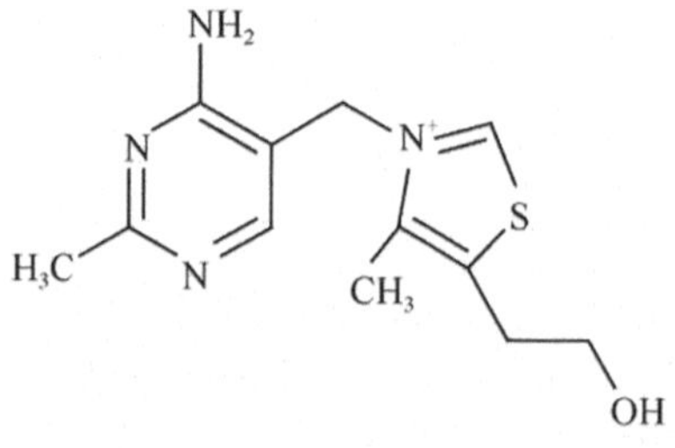

Fig. 3.4 Chemical structure of vitamin B1 and thiamine pyrophosphate

0.2 mg will result in vitamin deficiency. VB_1 deficiency leads to diabetic microangiopathy (Stirban et al. 2006), neurodegenerative diseases (Zhao et al. 2008), and so on.

Vitamin B1 is involved in glucose metabolism in the body (Haas 1988). After being phosphorylated in the body, it mainly exists in three forms, respectively. It is thiamine monophosphate (TMP), thiamine pyrophosphate (TPP), and thiamine triphosphate (TTP), the most important biologically active form of which is TPP. TPP is an important cofactor in the transketolase (TK) reaction of pyruvate dehydrogenase complex (PDHC), α-ketoglutarate dehydrogenase complex (KGDHC), and pentose phosphate pathway, among which PDHC and KGDHC are the important components of the cell's use of glucose to produce ATP pathways; TK is involved in gluconeogenesis (Gibson and Blass 2007). Therefore, VB_1 plays an important role in glucose metabolism. It also participates in numerous redox reactions in the body and improves energy metabolism in cells.

Owing to the rich sources of VB_1, there are few studies using biosynthesis for VB_1 production. Research have shown that the concentration of VB_1 in milk increased by 11% after 48 h of fermentation with *B. longum* (Hou et al. 2000), indicating that LAB have the ability to produce VB_1.

3.4.2 Vitamin B6

The activity of vitamin B_6 is reflected by three kinds of vitamin B_6 and their 5′-phosphate esters. The active coenzyme form of vitamin B_6 is pyridine 5′-phosphate (PLP). Vitamin B_6 is a coenzyme of about 140 enzymes in the human body, including transaminase, decarboxylase, racemase, and dehydratase. PLP is also involved in the metabolism of carbon groups, folic acid, and vitamin B12 and is involved in more than 80 biochemical reactions. It has an irreplaceable role in human protein metabolism, glycogen decomposition into glucose and lipid metabolism, hormone regulation, niacin formation, nucleic acid, and immune metabolism. Vitamin B_6 is a general term for pyridoxine, pyridoxal, and pyridoxamine (Fig. 3.5).

Vitamin B_6 is widely found in various foods, including animal liver, eggs, brown rice, sunflower seeds, walnuts, soybeans, carrots, and bananas. In addition to the body's intake of vitamin B_6 from the food, the intestinal bacteria in the body can also synthesize a part of vitamin B_6, so the primary deficiency of vitamin B6 is not

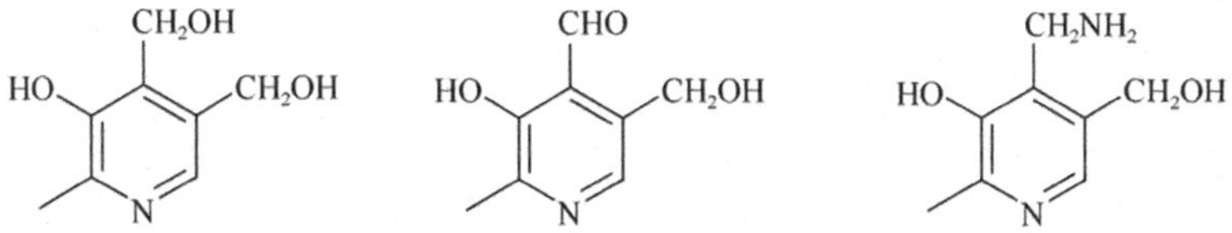

Fig. 3.5 Chemical structure of vitamin B_6 and its derivatives

widespread. The Chinese Nutrition Society recommends that vitamin B_6 can tolerate a maximum intake of 50 mg/day for children and 100 mg/day for adults.

The LAB starter of yoghurt, cheese, and other fermented foods can increase the production of vitamin B_6 (Shahani and Chandan 1979; Alm 1982). Recent studies have found that co-fermentation of *Streptococcus thermophilus* ST5, *Lactobacillus helveticus* R0052 ST5, or *Bifidobacterium longum* R0175 can increase the content of vitamins B1 and B6 in soybean products (Champagne et al. 2010).

References

Alm L (1982) Effect of fermentation on B-vitamin content of milk in Sweden. J Dairy Sci 65(3):353–359

Bacher A, Eberhardt S, Fischer M et al (2000) Biosynthesis of vitamin B2 (riboflavin). Annu Rev Nutr 20(1):153–167

Baines M, Kredan MB, Usher J et al (2007) The association of homocysteine and its determinants MTHFR genotype, folate, vitamin B12 and vitamin B6 with bone mineral density in postmenopausal British women. Bone 40(3):730–736

Battersby AR (1994) How nature builds the pigments of life: the conquest of vitamin B12. Science 264(5165):1551–1557

Blanche F, Cameron B, Crouzet J et al (1995) Vitamin B12: how the problem of its biosynthesis was solved. Angew Chem Int Ed 34(4):383–411

Blanck HM, Bowman BA, Serdula MK et al (2002) Angular stomatitis and riboflavin status among adolescent Bhutanese refugees living in southeastern Nepal. Am J Clin Nutr 76(2):430–435

Burgess C, O'Connellmotherway M, Sybesma W et al (2004) Riboflavin production in Lactococcus lactis: potential for in situ production of vitamin-enriched foods. Appl Environ Microbiol 70(10):5769–5777

Burgess CM, Smid EJ, Rutten G et al (2006) A general method for selection of riboflavin-overproducing food grade micro-organisms. Microb Cell Factories 5(2):24

Burgess CM, Smid EJ, van Sinderen D (2009) Bacterial vitamin B2, B11 and B12 overproduction: an overview. Int J Food Microbiol 133(1–2):1–7

Champagne CP, Tompkins TA, Buckley ND et al (2010) Effect of fermentation by pure and mixed cultures of Streptococcus hermophilus and Lactobacillus helveticus on isoflavone and B-vitamin content of a fermented soy beverage. Food Microbiology 7(7):968–972

Claesson MJ, Li Y, Leahy S et al (2006) Multireplicon genome architecture of *Lactobacillus salivarius*. Proc Natl Acad Sci U S A 103(17):6718–6723

Crittenden RG, Martinez NR, Playne MJ (2003) Synthesis and utilisation of folate by yoghurt starter cultures and probiotic bacteria. Int J Food Microbiol 80(3):217–222

Czeizel AE, Dudas I (1992) Prevention of the first occurrence of neural-tube defects by periconceptional vitamin supplementation. N Engl J Med 327(26):1832–1835

Danesh J, Lewington S (1998) Plasma homocysteine and coronary heart disease: systematic review of published epidemiological studies. J Cardiovasc Risk 5(4):229–232

Daniel R, Bobik TA (1998) Biochemistry of coenzyme B12-dependent glycerol and diol dehydratases and organization of the encoding genes. FEMS Microbiol Rev 22(5):553–566

Duthie SJ, Narayanan S, Brand GM et al (2002) Impact of folate deficiency on DNA stability. J Nutr 132(8):2444S–2449S

Elmadfa I, Heinzle C, Majchrzak D et al (2001) Influence of a probiotic yoghurt on the status of vitamins B(1), B(2) and B(6)in the healthy adult human. Ann Nutr Metab 45(1):13–18

Escalante-Semerena JC (2007) Conversion of cobinamide into adenosylcobamide in bacteria and archaea. J Bacteriol 189(13):4555–4560

Fabian E, Majchrzak D, Dieminger B et al (2008) Influence of probiotic and conventional yoghurt on the status of vitamins B1, B2and B6 in young healthy women. Ann Nutr Metab 52(1):29–36

Gangadharan D, Sivaramakrishnan S, Pandey A et al (2010) Folate-producing lactic acid bacteria from cow's milk with probiotic characteristics. Int J Dairy Technol 63(3):339–348

Gibson GE, Blass JP (2007) Thiamine-dependent processes and treatment strategies in neurodegeneration. Antioxid Redox Signal 9(10):1605–1619

Group M V S R (1991) Prevention of neural tube defects: results of the Medical Research Council Vitamin Study. Lancet 338(8760):131–137

Haas RH (1988) Thiamin and the brain. Annu Rev Nutr 8(1):483–515

Hodgkin DC, Kamper J, Mackay M et al (1956) Structure of vitamin B12. Nature 178(4524):64–66

Hou JW, Yu RC, Chou CC (2000) Changes in some components of soymilk during fermentation with bifidobacteria. Food Res Int 33(5):393–397

Hugenholtz J, Kleerebezem M (1999) Metabolic engineering of lactic acid bacteria: overview of the approaches and results of pathway rerouting involved in food fermentations. Curr Opin Biotechnol 10(5):492–497

Hugenholtz J, Hunik J, Santos H et al (2002) Nutraceutical production by propionibacteria. Lait 82(1):103–112

Hugenschmidt S, Schwenninger SM, Gnehm N et al (2010) Screening of a natural biodiversity of lactic and propionic acid bacteria for folate and vitamin B12 production in supplemented whey permeate. Int Dairy J 20(12):852–857

Hümbelin M, Griesser V, Keller T et al (1999) GTP cyclohydrolase II and 3, 4-dihydroxy-2-butanone 4-phosphate synthase arerate-limiting enzymes in riboflavin synthesis of an industrial Bacillus subtilis strain used for riboflavin production. J Ind Microbiol Biotechnol 22(1):1–7

Jägerstad M, Jastrebova J, Svensson U (2004) Folates in fermented vegetables-a pilot study. LWT-Food Sci Technol 37(6):603–611

Jayashree S, Jayaraman K, Kalaichelvan G (2010) Isolation, screening and characterization of riboflavin producing lactic acid bacteria from Katpadi, Vellore district. Recent Res Sci Technol 2(1):83–88. 430–435

Keuth S, Bisping B (1993) Formation of vitamins by pure cultures of tempe moulds and bacteria during the tempe solid substrate fermentation. J Appl Bacteriol 75(5):427–434

Kiatpapan P, Hashimoto Y, Nakamura H et al (2000) Characterization of pRGO1, a plasmid from Propionibacterium acidipropionici, and its use for development of a host-vector system in propionibacteria. Appl Environ Microbiol 66(11):4688–4695

Kleerebezem M, Boekhorst J, Kranenburg RV et al (2003) Complete genome sequence of Lactobacillus plantarum WCFS1. Proc Natl Acad Sci U S A 100(4):1990–1995

Koizumi S, Yonetani Y, Maruyama A et al (2000) Production of riboflavin by metabolically engineered Corynebacterium ammoniagenes. Appl Microbiol Biotechnol 53(6):674–679

Konings EJ (2001) Folate intake of the Dutch population according to newly established liquid chromatography data for foods. Am J Clin Nutr 73(4):765–776

Leblanc JG, Burgess C, Sesma F et al (2005) Ingestion of milk fermented by genetically modified Lactococcus lactis improves the riboflavin status of deficient rats. J Dairy Sci 88(10):3435–3442

Leblanc JG, Rutten G, Bruinenberg P et al (2006) A novel dairy product fermented with Propionibacterium freudenreichii improves the riboflavin status of deficient rats. Nutrition 22(6):645–651

Leblanc JG, Starrenburg M, Sybesma W et al (2010) Supplementation with engineered Lactococcus lactis improves the folate status in deficient rats. Nutrition 26(7–8):835–841

Lin M, Young C (2000) Folate levels in cultures of lactic acid bacteria. Int Dairy J 10(5):409–413

Luchsinger JA, Tang MX, Miller J et al (2007) Relation of higher folate intake to lower risk of Alzheimer disease in the elderly. Arch Neurol 64(1):86–92

Martens JH, Barg H, Warren MJ et al (2002) Microbial production of vitamin B12. Appl Microbiol Biotechnol 58(3):275–285

Murdock FA, Fields ML (1984) B-vitamin content of natural lactic acid fermented cornmeal. J Food Sci 49(2):373–375

O' Brien MM, Kiely M, Harrington KE et al (2001) The North/South Ireland food consumption survey: vitamin intakes in 18–64-year-old adults. Public Health Nutr 4(5A):1069–1079

Piao Y, Yamashita M, Kawaraichi N et al (2004) Production of vitamin B12 in genetically engineered Propionibacterium freudenreichii. J Biosci Bioeng 98(3):167–173

Pompei A, Cordisco L, Amaretti A et al (2007) Administration of folate-producing bifidobacteria enhances folate status in Wistar rats. J Nutr 137(12):2742–2746

Rao DR, Reddy AV, Pulusani SR et al (1984) Biosynthesis and utilization of folic acid and vitamin B12 by lactic cultures in skim milk. J Dairy Sci 67(6):1169–1174

Rodionov DA, Vitreschak AG, Mironov AA et al (2003) Comparative genomics of the vitamin B12 metabolism and regulation in prokaryotes. J Biol Chem 278(42):41148–41159

Roth JR, Lawrence JG, Bobik TA (1996) Cobalamin(coenzyme B12): synthesis and biological significance. Annu Rev Microbiol 50(2):137–181

Rucker RB, Suttie JW, Mccormick DB et al (2001) Handbook of vitamins, 3rd edn. CRC Press, Boca Raton

Santos F, Vera JL, Lamosa P et al (2007) Pseudovitamin B (12) is the corrinoid produced by *Lactobacillus reuteri* CRL1098 under anaerobic conditions. FEBS Lett 581(25):4865–4870

Santos F, Vera JL, Valdez G et al (2008) The complete coenzyme B12 biosynthesis gene cluster of *Lactobacillus reuteri* CRL1098. Microbiology 154(1):81–93

Shahani KM, Chandan RC (1979) Nutritional and healthful aspects of cultured and culture-containing dairy foods. J Dairy Sci 62(10):1685–1694

Smith AG, Croft MT, Moulin M et al (2007) Plants need their vitamins too. Curr Opin Plant Biol 10(3):266–275

Stirban A, Negrean M, Stratmann B et al (2006) Benfotiamine prevents macro- and microvascular endothelial dysfunction and oxidative stress following a meal rich in advanced glycation end products in individuals with type 2 diabetes. Diabetes Care 29(9):2064–2071

Sybesma WFH (2003) Metabolic engineering of folate production in lactic acid bacteria. Wageningen Universiteit, Wageningen

Sybesma W, Starrenburg M, Kleerebezem M et al (2003) Increased production of folate by metabolic engineering of Lactococcus lactis. Appl Environ Microbiol 69(6):3069–3076

Sybesma W, Burgess C, Starrenburg M et al (2004) Multivitamin production in Lactococcus lactis using metabolic engineering. Metab Eng 6(2):109–115

Valle MJD, Laiño JE, de Giori GS et al (2014) Riboflavin producing lactic acid bacteria as a biotechnological strategy to obtain bio-enriched soymilk. Food Res Int 62(8):1015–1019

Van de Guchte M, Penaud S, Grimaldi C et al (2006) The complete genome sequence of lactobacillus bulgaricus reveals extensive and ongoing reductive evolution. Proc Natl Acad Sci U S A 103:9274–9279

Wegkamp A, Starrenburg M, Hugenholtz J et al (2004) Transformation of folate-consuming Lactobacillus gasseri into a folate producer. Appl Environ Microbiol 70(5):3146–3148

Wegkamp A, Van OW, Smid EJ et al (2007) Characterization of the role of para-aminobenzoic acid biosynthesis in folate production by Lactococcus lactis. Appl Environ Microbiol 73(8):2673–2681

Wilson J (1983) Disorders of vitamins: deficiency, excess and errors of metabolism. In: Harrison's principles of internal medicine. McGraw-Hill Book Co, New York, pp 461–470

Wouters JTM, Ayad EHE, Hugenholtz J et al (2002) Microbes from raw milk for fermented dairy products. Int Dairy J 12(1):91–109

Zhao N, Zhong C, Wang Y et al (2008) Impaired hippocampal neurogenesis is involved in cognitive dysfunction induced by thiamine deficiency at early pre-pathological lesion stage. Neurobiol Dis 29(2):176–185

Chapter 4
Lactic Acid Bacteria and Bacteriocins

Qiuxiang Zhang

4.1 Bacteriocins: Introduction and Classification

Bacteriocin is a kind of antibiotic substance produced by bacteria in the metabolic process and has the function of resisting bacteria, fungi, or viruses. The intrinsic nature of bacteriocin is protein or polypeptide. For producing strain, bacteriocin is a biological weapon as it can inhibit or kill competitors in complex or harsh environments.

The history of bacteriocin can be traced back to the 1920s. In 1925, Gratia found that antagonism existed among *Escherichia coli* strains, which he believed was caused by a substance produced by the *E. coli* V strain. He further isolated the metabolite of the V strain. It seemed that it was a phage-like substance, which then was named as colicin. Later, it was found that not only gram-negative but also many gram-positive bacteria can produce similar substances. So these substances are known as bacteriocins altogether. Great differences exist between bacteriocins and antibiotics, another production generated by microorganisms. Bacteriocins are primary products synthesized by ribosome, encoded by genes, and can selectively inhibit or kill sensitive bacteria. Bacteriocins can be degraded into fragments by protease so that they do not change the normal intestinal flora. While antibiotics are secondary metabolites which are produced by certain microorganisms through enzymatic reactions. They are non-genetic encoding and kill all sensitive bacteria. In addition, the bacteriostatic spectrum of bacteriocins is relatively narrow and varied between each other, while the antibiotics are mostly broad spectrum.

To better distinguish bacteriocins from antibiotics, Konisky (1982) proposed the definition of bacteriocins as a class of bacteriostatic proteins or precursor polypeptides produced by certain bacteria through the ribosome mechanism. They can

Q. Zhang (✉)
Jiangnan University, Wuxi, China
e-mail: zhangqx@jiangnan.edu.cn

© Springer Nature Singapore Pte Ltd. and Science Press 2019   61
W. Chen (ed.), *Lactic Acid Bacteria*,
https://doi.org/10.1007/978-981-13-7283-4_4

inhibit not only homologous bacteria. The producing strain is autoimmune to their bacteriocins (Konisky 1982).

The naming of bacteriocins follows a specific rule, usually end with "cin" or "in". Before the suffix is the genus or species name of the bacterium. For example, the bacteriocin produced by *E. coli* is named colicin. Bacteriocin produced by *Bacillus subtilis* is named subtilin. Bacteriocin produced by *Leuconostoc gelidum* is named leucocin. And bacteriocin produced by *Pediococcus acidilactici* is named pediocin. Since some strains from the same species may produce different bacteriocins, they are distinguished by letters or numbers after cin or in. For example, lactacin F refers to the sixth bacteriocin reported in *Lactobacillus*, and bifilong Bb-42 is a bacteriocin produced by *Bifidobacterium longum* Bb-42 strain.

It is known that most bacteria in nature, including gram-positive and gram-negative bacteria, as well as certain archaea, can produce a variety of bacteriocins, such as lactic acid bacteria, *B. subtilis*, *Staphylococcus aureus*, *E. coli*, *Halobacterium*, and so on. Among them, the bacteriocins produced by lactic acid bacteria are nontoxic and diverse and has great application potential in the fields of food processing, agricultural production, and biomedicine. In particular, nisin, a bacteriocin produced by *Lactococcus lactis*, has been commercialized and is widely used as a natural food preservative in foods such as milk, meat, and canned foods.

Researches on bacteriocins of lactic acid bacteria have attracted great attention since the discovery of nisin. Bacteriocins from lactic acid bacteria are peptides produced during acid fermentation, which have relatively low molecular weight and can inhibit or kill gram-positive pathogens and spoilage in foods. Some of them even have obvious inhibitory effects on gram-negative bacteria. According to the relative molecular mass, chemical structure, thermal stability, antibacterial spectrum, and other characteristics of amino acids, Klaenhammer (1993) classified lactic acid bacteria bacteriocins into four categories: the first class is lantibiotics, the typical representative is nisin; the second class is a kind of small molecule (relative molecular mass < 10,000), thermostable, membrane-active peptide, such as Pediocin PA-1; the third type is macromolecule (relative molecular mass > 30,000), thermolabile protein such as helveticin; the fourth type is a composite bacteriocin containing lipids or binding to proteins. The first three categories have been recognized, and the view on the fourth type of bacteriocin is still controversial.

4.1.1   The First Group of Bacteriocins

The first group of bacteriocins is lantibiotics. Lantibiotics are heat-stable, with a relative molecular weight of less than 50,000 and a length of 19–38-amino acid residues, and are named after its special amino acids, such as lanthionine. The producing bacteria are generally gram-positive, such as *Lactococcus*, *Streptococcus*, *Micrococcus*, *Streptomyces*, *Lactobacillus*, *Bacillus*, *Staphylococcus*, *Enterococcus*, and so on. The active sites of these bacteriocin molecules contain a large number of rare amino acids, including lanthionine (Lan), beta-methyllanthionine (MeLan), dehydroalanine (Dha), and dehydrobutyrine (Dhb). The amino acid sequence of lantibiotics is varied, but it has certain regularity in its cyclization, synthesis, and mechanism of action. According to the difference of the topological structure and action mode of these bacteriocins, lantibiotics can be divided into three subgroups: Type A is slender, with positive charge and amphiphilic, which can form potential-dependent holes on the bacterial plasma membrane. Type B is spherical, without charge or negative charge, and plays its role mainly by destroying the function of enzymes. Type C is a two-component lantibiotics, which contains two polypeptide chains playing a synergistic antibacterial role.

4.1.1.1 Type A Lantibiotics

All the bacteriocins produced by lactic acid bacteria belong to type A. Type A lantibiotics can be further subdivided into slender AI and N-terminal linear and C-terminal spherical AII, represented by nisin and lacticin 481, respectively.

The modification of class AI is accomplished by dehydratase LanB and cyclase LanC, which are then transported out of the cells by LanT and excised by protease LanP. Nisin is a typical representative of this type. It is produced by some strains of *Lactococcus lactis* subsp. *lactis*. As a natural antimicrobial peptide, nisin can effectively kill or inhibit most gram-positive bacteria and has a strong inhibitory effect on pathogenic bacteria such as *Staphylococcus aureus*, *Clostridium botulinum*, and *Streptococcus hemolyticus*. The isoelectric point of nisin was about 9. The stability increased with the decrease of pH. At pH 2, the antimicrobial activity of nisin was not lost after being treated at 121 °C for 30 minutes.

Up to now, five natural variants of nisin have been found in nature, namely, nisin A (Gross and Morell 1971), nisin Z (Mulders et al. 1991), nisin Q (Zendo et al. 2003), nisin F (de Kwaadsteniet et al. 2008), and nisin U (Wirawan et al. 2006). Among them, nisin A and nisin Z are the earliest and most studied ones. Their differences lie only in the amino acid at position 27, where nisin A is His, while nisin Z is Asn. But there is no difference in their antimicrobial activities. The structural gene of nisin encodes 57 amino acids, and 23 amino acids at N-terminal are removed during secretion. The mature nisin contains 34 amino acid residues and consists of 5 thioether bridges. Nisin has amphiphilic amino acid distribution. Its N-terminal is hydrophobic, which can interact with the phospholipid end of the target cell membrane. The C-terminal contains more hydrophilic amino acid, which shows hydrophilicity. The structure of nisin in solution was determined by NMR. It was found that there are two domains in nisin molecule: N-terminal domain and C-terminal domain. The former contains the first three thioether rings (A–C ring), and the latter consists of D and E thioether rings. A–C ring and D–E ring are connected by a flexible hinge region, which follows the C-terminal domain. It is a flexible tail consisting of six amino acids (van den Hooven et al. 1996). Nisin U, a variant of nisin A secreted by *Streptococcus uberis*, is the only bacteriocin not produced by *Lactococcus lactis* in the nisin family. Nisin U is composed of 31 amino acids, which is close to the antimicrobial spectrum of nisin A. However, the antimicrobial activity of nisin U to *Streptococcus pyogenes* and *Lactococcus lactis* is not as good as that of nisin A. Nisin U and nisin A have the same bridging mode, and the strains producing these two bacteriocin are cross immune. If nisin U and nisin A were added to the cultures of *Lactococcus* and *Streptococcus*, self-induction or mutual induction would occur between the two peptides, which indicated the similarity of their functions (Wirawan et al. 2006).

The dehydration and cyclization of class AII lantibiotics are accomplished by a bifunctional modifying enzyme LanM. The precursor peptides of class AII lantibiotics have conserved common sequences. Translocation and processing of the precursor are also completed by only one enzyme LanT. Lactacin 481, a kind of class AII bacteriocins, is composed of 27 amino acids which contain two lanthionine, MeLan and Dhb. The expression of lactacin 481 is regulated by adjusting the transcriptional level of the upstream promoter *LctA* gene (Hindre et al. 2004). Paik et al. found that the bacteriocin sublancin 168 produced by *Bacillus subtilis* has both stable disulfide bond and lantibiotics structure, which is the first bacteriocin with two types of structure (Paik et al. 1998).

4.1.1.2 Type B Lantibiotics

Type B lantibiotics is a compact spherical structure with no charge or negative charge in neutral environment. Mersacidin produced by *Bacillus subtilis* and cinnamycin secreted by *Streptomyces cinnamomi* are the representatives of type B lantibiotics.

Mersacidin consists of 20 amino acids with a relative molecular weight of 1825 (Chatterjee et al. 1992). It is the only lantibiotics whose structure is determined by X-ray (Schneider et al. 2000). Mersacidin mainly inhibits gram-positive bacteria, including penicillin-resistant *Staphylococcus aureus*. That is because mersacidin inhibits the synthesis of peptidoglycan by impacting on the diphosphate group of lipid II and mersacidin (Brotz et al. 1997). Mersacidin acts in a manner similar to vancomycin, so mersacidin has great potential in the development of drugs to treat drug-resistant superbacterial infections and can be used to inhibit methicillin-resistant *Staphylococcus aureus* and vancomycin-resistant *Enterococcus*.

Cinnamycin contains 19 amino acids, with two beta-methylthionine residues linked to dehydrotyrosine via a nucleophilic cysteine at the N-terminal of the peptide chain. Its antibacterial activity can only be applied to a few strains, such as *Bacillus subtilis*. Cinnamycin destroys ATP-dependent calcium uptake and protein transport by increasing cell membrane permeability of sensitive bacteria. Moreover, the bacteriocin can bind to phosphatidylethanolamine on phospholipids, thereby inhibiting the competitive inhibition of phospholipase, preventing the conversion of phosphatidylethanolamine to lysophosphatidylethanolamine, and affecting the dissolution of red blood cells.

4.1.1.3   Type C Lantibiotics

Type C is a multicomponent lantibiotic. The two independent transcription-modified peptides have no or very low activity, but they have significant synergistic antibacterial effect. Lacticin 3147 produced by *Lactococcus lactis* subsp. DPC3147 could be attributed to this kind of bacteriocin. It contains two peptides, LtnA1 and LtnA2, with relative molecular weights of 3306 and 2847, respectively. It is speculated that LtnA1 binds to lipid II, and the compound of LtnA2 and LtnA1-lipid II together can insert into the cell membrane of sensitive bacteria more effectively to form pores. The combined action activity of LtnA1 and LtnA2 is 30 times that of LtnA1 alone (Morgan et al. 2005). Lacticin 3147 mainly inhibits gram-positive bacteria, including *Staphylococcus aureus*, *Enterococcus*, *Pneumococcus*, *Propionibacterium acnes*, and *Streptococcus mutans*, and foodborne pathogens such as *Listeria monocytogenes* and *Bacillus cereus*. It can also inhibit gram-negative bacteria when cooperated with polymyxin (Draper et al. 2013). Smb (Yonezawa and Kuramitsu 2005) produced by *Streptococcus mutans* GS5 and cytolysin secreted by *Enterococcus* are also composed of two peptides. Cytolysin, an exotoxin, also known as streptococcal hemolysin, is the only lantibiotic that inhibits both bacteria and eukaryotic cells (Tyne et al. 2013).

4.1.2   The Second Group of Bacteriocins

The second group of bacteriocins does not contain noncoding amino acid residues such as lanthionine. They can be divided into three subgroups, class IIa, class IIb, and class IIc.

4.1.2.1   Class IIa

Class II bacteriocin is the largest and most widely studied subclass of bacteriocin in second categories. So far, all class IIa bacteriocins have strong anti-*Listeria* activities. Some class IIa bacteriocins can also inhibit other spoilage bacteria in food, such as *Bacillus*, *Staphylococcus*, *Clostridium*, and so on. More than 30 kinds of class a lactic acid bacteria have been found. Typical representatives are pediocin PA-1, pediocin AcH, sakacin A, sakacin P, leucocin A, curvacin A, enterocin A, etc. *Lactobacillus*, *Pediococcus*, *Leuconostoc*, *Carnivora*, and *Enterococcus* are the main producers. Leucocin A produced by *Leuconostoc gelidum* UAL 187 is the first found class IIa bacteriocin (Hastings et al. 1991). But pediocin PA-1 produced by *Pediococcus lactis* is the most studied class IIa bacteriocin.

Class IIa bacteriocins generally contain 37 to 48 amino acid residues and a conserved YGNGVXaaC group at the N-terminus, which is generally considered to be a recognition sequence for membrane-bound protein receptors. As more and more new class IIa bacteriocins are discovered, their N-terminal groups can also be

expressed as YGNGVXaaCXaa(K/N)XaaXaaCXaaV(N/D)(W/K/R)Xaa(G/A/S)
(A/N) (in parenthesis are the conservative residues, and Xaa represents residues
with high frequency of variation). Another important feature of class IIa bacteriocins is that the N-terminal conserved region contains at least two cysteines forming
a disulfide bond. In general, peptides containing two disulfide bonds are more bacteriostatic than bacteriocins containing one disulfide bond (Eijsink et al. 1998).
Relative to the conservation of the N-terminus, the sequence similarity of the
C-terminus of class IIa bacteriocin is only 34% to 80% and forms an α-helix, which
acts as a transmembrane component when the cell membrane of the susceptible
bacteria forms a pore. Class IIa bacteriocins have no posttranslational modifications
other than the formation of disulfide bonds and are simpler in structure than other
bacteriocins. Although the minimum inhibitory concentrations of these bacteriocins
are different, there is a certain similarity in the inhibition spectrum (Fimland et al.
2005).

4.1.2.2   Class IIb

Class IIb bacteriocin is a two-component bacteriocin formed by two peptide oligo-
mers. The complete activity requires two peptides to interact with each other, mainly
including lactacin F, plantaricin S, plantaricin A, lactococcin C, and lactococcin
MN (Garneau et al. 2002). Class IIb bacteriocins can also be subdivided into two
categories, synergistic type (S type) and enhanced type (E type) (Marciset et al.
1997).

The antibacterial effect of S-type bacteriocin is formed by the interaction of two
peptides. The single peptide has no inhibitory effect. Typical S-type bacteriocins are
lactococcin G produced by *Lactococcus lactis*, lactacin F produced by *Lactobacillus
johnsonii*, and lactacin 705 produced by *Lactobacillus casei*. Lactococcin G consists of an alpha chain and a beta chain, which contains 39 and 35 amino acids,
respectively. The mode of action is to inhibit amino acid uptake, dissipate proton
momentum, and reduce intracellular ATP levels.

The two peptides of type E bacteriocins both have antibacterial effects. The bacteriostatic effect is significantly enhanced when the two peptides act together.
Thermophilin 13 produced by *Streptococcus thermophilus*, enterocin L50 (Cintas
et al. 2000) produced by *Enterococcus faecium*, and ABP-118 (Flynn et al. 2002)
produced by *Lactobacillus saliva* are typical representatives of E-type class IIb bacteriocins. The two polypeptides of plantaricin PlnJK have certain faint inhibitory
effects on *Escherichia coli* and *Listeria monocytogenes*. The antibacterial effect is
enhanced 1000 times when the two peptides are mixed with a concentration of 1:1
(Anderssen et al. 1998).

4.1.2.3 Class IIc

Class IIc bacteriocins include those non-lantibiotics that neither belong to class IIa nor class IIb and are generally considered to contain thiol-based activities and signal peptide encoding mechanisms (Cotter et al. 2005). Strains producing bacteriocin type IIc are various, and the diversity of bacteriocins is complex. Lactococcin B belongs to class IIc bacteriocins, which can be activated by mercaptan. The antimicrobial and immune proteins of sakacin P produced by *Lactobacillus sake* are coupled (Mathiesen et al. 2005). Class IIc bacteriocin also has some similar structures, such as GG sequence in leading sequence, ABC transporter, and some immune related-proteins. Some class IIc bacteriocins do not possess recognizable N-terminal signal peptides, such as enterocin Q produced by *Enterococcus faecium*, aureocin A53 produced by *Streptomyces aureus*, and BHT-B produced by *Streptococcus rattus*. Gassericin A is secreted by *Lactobacillus gasseri* LA39, which is isolated from human infant feces. It is encoded by chromosomes and has a very rare cyclic structure. It has inhibitory activity against many food pathogens, including *Listeria monocytogenes*, *Bacillus cereus*, and *Staphylococcus aureus*.

4.1.3 The Third Group of Bacteriocins

The third group of bacteriocins have a relatively high molecular weight (usually greater than 30,000). It is inactivated after heating at 100 °C for 30 minutes. These bacteriocins include helveticin J (Joerger and Klaenhammer 1986), lacticin A, lacticin B, caseicin 80 (Müller and Radler 1993), acidophilucin A, helveticin V-128, enterolysin A, etc. However, propionicin SM1 produced by *Propionibacterium acnes* is heat-resistant (Miescher et al. 2000). These bacteriocins can also be subdivided into two classes: lysozyme-like bacteriocins that inhibit bacteria through cell lysis and non-lysozyme antibacterial proteins. Only a few of the bacteriocins produced by *Lactobacillus* and *Bifidobacterium* have been reported. Only five of the third bacteriocins produced by lactic acid bacteria have been studied at the genetic level. They are lysostaphin produced by *Staphylococcus simulans*, helveticin J produced by *Lactobacillus helveticus* 481, zoocin A produced by *Streptococcus zooepidemicus* 4881, millericin B produced by *Streptococcus milleri* NMSCC 061, and enterolysin A produced by *Enterococcus faecalis* LMG 2333.

4.1.3.1 Lysozyme-Like Bacteriocins

Lysostaphin produced by *Staphylococcus simulans* is a representative of this kind of bacteriocin (King et al. 1980). Lysostaphin is a kind of metalloproteinase. Zn^{2+} is a necessary cofactor. It is composed of 246 amino acids, with the relative molecular weight of about 27,000. Lysostaphin is an extracellular enzyme. The producing strain first generates a 493-amino-acid proenzyme. The N-terminal of the enzyme has multiple repetitive sequences containing 13 amino acids. The N-terminal repeats are removed by cysteine protease after the proenzyme is secreted into the extracellular matrix, and then the mature lysostaphin is formed (Heinrich et al. 1987). Lysostaphin has many catalytic activity centers, of which endopeptidase, glycosidase, and amidase are related to the catalytic activity of hydrolyzing the cross-linked structure of bacterial cell wall peptidoglycan. Endopeptidase activity is the most important (Neumann et al. 1993). Endopeptidase can specifically hydrolyze Gly pentapeptide bridges in the cross-linked structure of bacterial cell wall peptidoglycans. Since this structure exists only in the cell wall of *Staphylococcus* and is most widely distributed in the cell wall of *Staphylococcus aureus*, lysostaphin can kill almost all staphylococci, but is invalid for other species of bacteria.

4.1.3.2   Non-lysozyme Antibacterial Proteins

These bacteriocins deplete ATP of target microorganisms through proton dynamics, leading to cell death. Dysgalacticin (relative molecular weight of 21,000) and streptococcin A-M57 (relative molecular weight of 17,000) are secreted by *Streptococcus dysgalactiae* subsp. *equisimilis* and *Streptococcus pyogenes*, respectively. The former has a narrow inhibitory spectrum, while the latter has a distinct inhibitory spectrum. It can inhibit non-streptococcal gram-positive bacteria such as *Micrococcus luteus* and *Lactococcus lactis*, most of *Listeria*, *Bacillus megagenes* and *Staphylococcus simulans*.

4.1.4   The Fourth Group of Bacteriocin

Some other proteins with bacteriostatic and bactericidal effects have been found during the research on bacteriocins. They not only inhibit gram-positive bacteria, but also inhibit gram-negative bacteria and fungi. These protein antagonists not fully conform to the definition of bacteriocins are called bacteriocin-like substances. Bacteriocin-like substances have wider application prospect than bacteriocins because of their stability in a wide pH range and broad-spectrum antibacterial properties.

Whitford et al. isolated a strain of *Streptococcus* LRC0255 from the rumen. Bacteriocin-like substance bovicin 255 produced by this strain was active at pH 1-12 and stable to heat. The activity of bovicin 255 was unchanged at 100 °C for 15 minutes. It inhibited gram-positive bacteria, but had no effect on gram-negative bacteria. It was sensitive to pronase and protease K but insensitive to pepsin and peptidase isolated from pig intestinal mucosa (Whitford et al. 2001).

Collado et al. isolated six strains of bifidobacteria from human feces, which not only inhibited gram-positive bacteria but also inhibited gram-negative bacteria and yeast. Bacteriocin-like substances produced by these bifidobacteria were stable at pH 3–10, and their activity remained unchanged at 100 °C for 10 min. They were resistant to alpha-amylase and phospholipase A but sensitive to protease. The relative molecular weight of the substances produced by BIR-0312 and BIR-0324 were 10,000-30,000, while those of the other four bacteriocins were all less than 10,000 (Collado et al. 2005a). Further studies found that these bacteriocins also inhibit *Helicobacter pylori* (Collado et al. 2005b).

There are relatively few reports on bacteriocin-like substances which can inhibit fungi. Magnusson and Schnürer from the Department of Microbiology of Swedish Agricultural University isolated a *Lactobacillus coryniformis* Si3 strain from silage. This strain produced a broad-spectrum inhibitory fungal protein. It has strong inhibitory effect on fungi including *Aspergillus, Penicillium, Mucor*, and *Fusarium* and also has weak inhibitory effect on yeast including *Kluyveromyces marxianus* and *Saccharomyces cerevisiae*. The activity was highest at pH 3.0–4.5 and was lost when pH was higher than 6, but it could be restored when pH was returned to 3.6. The activity was stable at 121 °C for 15 min. It was sensitive to protease K, trypsin, and pepsin. Relative molecular weight was 30,000 (Magnusson and Schnürer 2001).

4.2 Species and Structure of Bacteriocin of Lactic Acid Bacteria

4.2.1 Bacteriocins Produced by Lactobacillus

Many different *Lactobacillus* spp. were isolated from natural fermented dairy products, nondairy products, starters, animals, plants, and human intestines. The *Lactobacillus* spp. produce a wide range of bacteriocins, which differ greatly in molecular weight, biochemical characteristics, sensitive range, and mode of action. In addition to producing bacteriocins, *Lactobacillus* often produces acids, hydrogen peroxide, diacetyl, and other substances, which can antagonize certain microorganisms. Therefore, it is necessary to remove the above compounds from the samples before studying the bacteriostatic effect of *Lactobacillus*. Currently known *Lactobacillus* bacteriocins are more than 30 kinds, including plantaricin A, plantaricin B, plantaricin C, sakacin A, sakacin M, sakacin P, lactocin S, etc.

4.2.1.1 Plantaricin

Bacteriocin produced by *Lactobacillus plantarum* is collectively referred to as plantaricin. Most of the plantaricins have a broad spectrum of inhibition, and the inhibited bacteria include lactic acid bacteria and other gram-positive bacteria such as *Listeria, Staphylococcus aureus*, and *Listeria monocytogenes*. Plantaricin is complex in classification; plantaricin LR14 and plantaricin C belong to class I. Some belong to class II, such as plantaricin S and plantaricin NC8, due to their conserved YYGNGV/C region at the N-terminal, which conforms to the characteristics of class IIa bacteriocins. The yield of plantaricin mainly depends on the genetic characteristics and culture conditions of the producing bacteria. It is generally believed that plantaricin begins to synthesize and secrete in the middle of the exponential growth period and reaches the maximum yield at the stationary phase. Therefore, in the process of plantaricin production, conditions such as culture time, temperature, initial pH of the fermentation broth, and medium composition should be well controlled.

4.2.1.2 Gassericin A

Gassericin A is secreted by *Lactobacillus gasseri* LA 39. Isoleucine is connected to N-terminal, and alanine is connected to C terminal of Gassericin A. A total of 58 amino acids are linked to form a ring structure with a relative molecular weight of 5652 (Kawai et al. 2009). Gassericin A is thermostable and insensitive to pH. It is still active in the pH range of 2–12. It can inhibit some *Lactobacillus* and some foodborne spoilage bacteria such as *Listeria monocytogenes*, *Bacillus cereus*, and *Staphylococcus aureus* (Nakamura et al. 2013). Another strain of *Lactobacillus gasseri* LA158 isolated from infant feces can synthesize bacteriocin Gassericin T, which is also thermostable and has a bacteriostatic spectrum similar to that of Gassericin A. However, Gassericin T is a two-component bacteriocin and belongs to lacticin F group bacteriocins (Arakawa et al. 2009a, 2009b).

4.2.1.3 Sakacin

Sakacin A is secreted by *Lactobacillus sake* 706 and inhibits many *Lactobacillus* and gram-positive bacteria. The synthesized gene cluster is located on the plasmids. Holck et al. determined the protein sequence of sakacin A by Edman degradation reaction. The bacteriocin contained 41 amino acids, and its relative molecular weight was about 4308.7 (Mathiesen et al. 2005).

4.2.1.4 Lactacin

Lactacin B is produced by *Lactobacillus acidophilus* N2. It is a kind of thermostable, catalase-insensitive bacteriocin, the relative molecular weight of about 65,000. It can inhibit *Enterococcus faecalis* and homologous *Lactobacillus delbrueckii*, *Lactobacillus bulgaricus*, and *Lactobacillus helveticus* (Dobson et al. 2007). Lactacin F, secreted by *Lactobacillus acidophilus* 11088, is a hydrophobic peptide with 54–57 amino acids and a relative molecular weight of about 63,000. Lactacin F is the first nonlantibiotic bacteriocin produced by lactic acid bacteria with known DNA and protein sequence information. Both lactacin B and F are heat-stable and maintained antimicrobial activity at 121 °C for 15 min. Their solubility was similar, and both belonged to class II bacteriocins (Muriana and Klaenhammer 1991).

Although the similarity between the two bacteriocins is very high, there is a certain difference in their antibacterial spectrum. It is shown that the antibacterial spectrum of lactacin B is obviously narrower than that of lactacin F.

4.2.1.5 Lactocin

Lactocin S is a thermostable bacteriocin produced by *Lactobacillus sakei*. Its activity decreased by only half when it was treated at 100 °C for 1 h (Mortvedt et al. 1991). Lactocin 27 is a thermostable class II bacteriocin produced by *Lactobacillus helveticus*. It contains large amounts of glycine and alanine. It is sensitive to trypsin and streptomycin and insensitive to fig protease (Upreti and Hinsdill 1975).

4.2.1.6 Helveticin

Helveticin J is synthesized and secreted by *Lactobacillus helveticus* 481. Its regulatory gene cluster is located on chromosome and has a narrow inhibitory range. It belongs to class III bacteriocins. Its molecular weight is about 37,511. It is unstable and inactivated after 30 minutes at 100 °C (Joerger and Klaenhammer 1986).

4.2.2 Bacteriocins Produced by Bifidobacterium

Bifidobacterium is an important component of human intestinal symbiotic bacteria, accounting for 3–7% of the total amount of microorganisms in adults and 91% in newborns. *Bifidobacterium* produces a lot of lactic acid and acetic acid in the metabolic process. The acidic environment formed by *Bifidobacterium* can inhibit the colonization and growth of pathogenic bacteria and has a positive effect on host health. Unlike *Lactobacillus*, only a limited number of studies have shown the production of bacteriostatic substances or bacteriocins in the strains of *Bifidobacterium*. *Bifidobacterium bifidum*, *B. longum*, *B. infantum*, and *B. thermophilus* have been reported to produce bacteriocins.

4.2.2.1 Bifidin

In 1984, Anand et al. reported that *B. bifidum* 1452 could produce bacteriocin bifidin, which was thermostable and had no loss of activity after heating for 30 minutes at 100 °C. Partially purified bifidin was obtained by ethanol-acetone precipitation and molecular sieve. The crude extract remained stable in refrigerator for 3 months at 5–8 °C. Sequence analysis showed that the antimicrobial peptide contained two kinds of amino acids, phenylalanine and glutamic acid, as well as trace amounts of aspartate, threonine, serine, glycine, isoleucine, and leucine (Anand et al. 1984). In vitro experiments confirmed that bifidin could inhibit the propagation of *Micrococcus flavus* and *Staphylococcus aureus*.

Bifidin I is an antibacterial substance isolated from the metabolites of *B. infantum* BCRC 14602 (Cheikhyoussef et al. 2010). It can inhibit lactic acid bacteria and some gram-positive bacteria such as *Staphylococcus*, *Streptococcus*, and *Bacillus*. It also inhibits gram-negative bacteria such as *Salmonella*, *Shigella*, and *Escherichia coli*, but it is not effective on yeast. The relative molecular weight was about 30,000. The maximum activity (1600 AU/mL) can be obtained after 16–20 h culture in MRS medium. The thermal stability is not affected in the range of pH 4–10.

4.2.2.2 Bifilong

Bifilong Bb-46 is produced by *B. longum* Bb-46. It can inhibit *E. coli*, *Salmonella typhimurium*, *Bacillus cereus*, and *Staphylococcus aureus*. It is sensitive to trypsin and pepsin, but not heat-resistant. The activity of 121 °C loses after heating for 15 min and is stable between pH 4 and 7 but decreases rapidly above pH 9.

4.2.2.3 Bifidocin B

Bifidocin B is the first isolated bifidobacterial bacteriocin, with a molecular weight of 33,000, belonging to class IIa bacteriocins (Yildirim and Johnson 1998). Bifidocin B, synthesized and secreted by *B. bifidum* NCFB 1454, was sensitive to protease but resistant to organic solvents, heat, and pH. Bifidocin B was not inactivated by heating for 15 min at 121 °C, and its activity was stable in the range of pH 2 to 10. Bifidocin B can inhibit some foodborne pathogens and spoilage bacteria, such as *Listeria*, *Enterococcus*, *Bacillus*, *Leuconostoccus*, *Phanerococcus*, and so on, but has no effect on other gram-positive bacteria and gram-negative bacteria. Bifidocin B is produced during exponential phase and reaches the maximum yield (3, 200 AU/mL) at the beginning of the stable period.

4.2.2.4 Thermophilicin B67

B. thermophilum RBL67 is isolated from infant feces. In a reaction system simulating the composition and activity of bifidobacteria in the proximal colon of infants, the addition of *B. thermophilus RBL67* resulted in a sharp decrease in the number of *Bifidobacterium* although the composition and activity of intestinal flora did not change much. This may be due to the killing of related microorganisms by thermophilicin B67 produced by *B. thermophilus* RBL67 (Zihler et al. 2011). When *B. thermophilus* RBL67 was mixed together with *Pediococcus acidilactici* UVA1, a bacteriocin-producing bacterium isolated from infant feces, the culture system was stable, and cell growth and bacteriocin production and activity were not affected (Mathys et al. 2009). In the reaction system simulating the microbial composition of pig intestinal tract, when mixing *B. thermophilus* RBL67 with prebiotics oligo-saccharides and galactooligosaccharides, the growth of *Salmonella enteric* subsp. *enterica* serotype N-15 could be inhibited, which presumably not only related to acetic acid but also to the bacteriocin produced by *B. thermophiles* (Tanner et al. 2014).

4.2.3 Bacteriocins Produced by Lactococcus, Enterococcus, and Other Species

Lactococcus is widely used in dairy industry as a starter. Bacteriocins produced by *Lactococcus lactis* include diplococcin, lactococcin, lacticin, and nisin. It is also found that some *Lactococcus* will produce a large amount of non-lantibiotics.

4.2.3.1 Nisin

Nisin is produced by *L. lactis*, which consists of 34 amino acid residues as intro-duced before. Nisin contains five rare amino acids, namely, ABA, DHA, DHB, ALA-S-ALA, and ALA-S-ABA. They form five inner rings through thioether bonds. Nisin has hydrophilic and hydrophobic properties, showing the characteris-tics of cationic peptides, isoelectric point in the alkaline range. The solubility of Nisin increases with the decrease of pH. It is almost insoluble in alkaline condition. The stability of Nisin is related to its solubility at different pH values. In hydrochlo-ric acid solution with pH 2.5 or lower, even boiling does not affect the activity. The activity remains stable after heating for 30 minutes at pH 2 and 121 °C. However, when pH exceeds 4, nisin inactivates rapidly in aqueous solution.

4.2.3.2 Lactococcin

Lactococcin A is produced by *Lactococcus lactis*. The producing bacteria can also produce an immune protein with an amphiphilic alpha helix and a molecular weight of 11,000. This immune protein interacts with Lactococcin A receptor protein to prevent bacteriocin from inserting into the cell membrane, so that lactococcin has no inhibitory effect on the producing bacteria (Hui et al. 1995).

4.2.3.3 Enterocin

Many enterococci produce antimicrobial peptides called enterocins, such as entero-cin A produced by *Lactococcus lactis* MG1614, enterocin AS-48 produced by *Enterococcus faecalis*, and enterocin B produced by *Enterococcus faecalis* W3.

Enterocin A belongs to class IIa bacteriocins. It contains 47 amino acids and 2 disulfide bonds. Enterocin B has a wide spectrum of antimicrobial activity and a stable thermostability (Aymerich et al. 1996). It also inhibits *Listeria monocytogenes*, *Enterococcus faecalis*, *Clostridium*, and *Staphylococcus aureus* (Casaus et al. 1997). Enterocin 1146 is also a bacteriocin produced by *Enterococcus faecalis*, which inhibits all *Listeria*, most *Clostridium butyricum*, and *Clostridium perfringens*. Enterocin 1146 can inhibit the growth of *Listeria* without affecting the fermentation of starter and thus has broad application prospects in fermentation industry (Parente and Ricciardi 1994).

4.2.3.4   Pediocin

Pediocin is produced by *Pediococcus*, which is often used to ferment vegetables, cheese, meat, and sausages. Most of the *Pediococcus* can be isolated from plants and plant products but also from pickles, fermented sausages, and so on. Similar to *Lactococcus*, the condition of nutritional demand of *Pediococcus* is severe. Pediocin is produced by *Pediococcus acidilactici*, *Pediococcus cerevisiae*, and *Pediococcus pentosaceus*. Pediocin A, pediocin PA-1, pediocin AcH, pediocin JD, pediocin Bac, and pediocin SJ-1 were found to be the bacteriocins produced by *Pediococcus lactis*. The amino acid sequences of most of the pediocins are highly homologous and share some common structural features, such as highly conserved N-terminus and poorly polarized C-terminus. The relative molecular weight of pediocin is about thousands of Dalton and are thermally stable.

Etchells et al. (1964) used both *Pediococcus cerevisiae* FBB-61 and *Lactobacillus plantarum* FBB-67 in pickled cucumber, and the results showed that the growth lag of *Lactobacillus plantarum* FBB-67 was as long as 10 days. In fact, *Lactobacillus plantarum* is more acid-resistant than *Pediococcus cerevisiae*. It is speculated that the product of *Pediococcus cerevisiae* may have certain inhibitory effect on *Lactobacillus plantarum*. Rueckert et al. (1979) found that the product of FBB-61 was a bacteriocin, which was stable at high temperature (100 °C for 60 min) and freezing point and sensitive to pronase and could not penetrate the semipermeable membrane. The product was then named as pediocin A. This bacteriocin has a wide range of bacteriostasis, which can be compared with nisin to a certain extent, including lactic acid bacteria, *Clostridium botulinum*, *Clostridium perfringens*, and *Staphylococcus aureus*. But it has no effect on gram-negative bacteria.

Pediocin PA-I is the most studied bacteriocin produced by lactic acid bacteria except nisin. It was encoded by a 9.3 kb plasmid. Pediocin SJ-1 is encoded by a 6.1 kb plasmid, which is very similar to pediocin PA-I in bacteriostatic spectrum, molecular weight, and genetic characteristics, in addition to being sensitive to alpha-amylase. AcH is produced by *Lactococcus lactis* H. The molecular weight of AcH is 2700. It is thermostable and sensitive to trypsin, papain, protease K, and chymotrypsin. It has a broad inhibition spectrum and can inhibit *Lactobacillus*, *Leuconostoc*, *Staphylococcus aureus*, *Listeria monocytogenes*, and so on. The inhibiting mechanism is to inhibit ATP synthesis, reduce energy supply, damage transport system, and eventually lead to cell death of sensitive bacteria.

4.2.3.5 Enterolysin

Enterolysin is secreted by *Enterococcus* during growth. Enterolysin A, produced by *Enterococcus faecalis* LMG 2333, is a class III bacteriocin with unstable heat and broad antibacterial spectrum (Nilsen et al. 2003). The precursor protein of enterolysin A contains 343 amino acids. After the removal of 27 amino acids, the final secretion of mature enterolysin A contains 316 amino acids. These bacteriocins can destroy the cell membrane of sensitive bacteria and have antibacterial effect on *Lactobacillus*, *Lactococcus*, *Pediococcus*, *Enterococcus*, *Bacillus*, *Listeria*, and *Staphylococcus*.

4.2.3.6 Leucocin

Leuconostoc is widely found in many food raw materials, dairy products, and wine fermentation. Leucocin is produced by *Leuconostoc* isolated from goat milk, cheddar cheese, retail mutton, and vacuum packaged meat. Mesenterocin 5, Leucocin A, Leucocin S, and carnocin have been reported, but they are only descriptions of general characteristics and lack of complete biochemical and genetic information. For example, mesenterocin 5 produced by *Leuconostoc mesenteroides* isolated from cheddar cheese is relatively heat-resistant (it can survive 30 minutes at 100 °C), and the relative molecular weight is about 4500. When *Leuconostoc mesenteroides* was inoculated on the medium for about 6 hours, mesenterocin 5 was produced, and the maximum yield was achieved in 10 hours. Mesenterocin 5 can inhibit the growth of *Listeria monocytogenes* but has no effect on the fermentation starter such as *Lactobacillus*. Leucocin A, a peptide bacteriocin with a relative molecular weight of 3930.3, can be produced during the whole incubation period (about 7 days) of *Leuconostoc gelidum* in the range of pH 4.0–6.5 at 25 °C.

4.3 Biosynthesis and Genetic Regulation of Typical Bacteriocin Produced by Lactic Acid Bacteria

Bacteriocin-producing lactic acid bacteria synthesize and secrete some bacteriocins in the logarithmic phase. Bacteriocin is usually transported to the medium by means of cell membrane permeation, but some types of bacteriocins remain in the cell under certain conditions. The production of bacteriocin is synchronous with the growth of bacteria, and the yield is closely related to the number of producing strain. Optimizing culture conditions can effectively increase the output of bacteriocin, such as adding sugar, vitamins, nitrogen, and other stimulants in the medium or adjusting pH and temperature to the optimal range. The production of bacteriocin can be increased by increasing the number of bacteriocin-producing lactic acid bacteria (Abbasiliasi et al. 2011; Espeche et al. 2014). PH condition of culture medium is one of the important factors affecting bacteriocin production, but different kinds of bacteriocins have different pH requirements (de Arauz et al. 2012). For *Lactococcus*, the Elliker medium without buffering ability is more conducive to the production of bacteriocin than the M17 medium with high buffering ability. During fermentation, high yield can be achieved by regulating pH. Different strains produce bacteriocins at different stages. For example, pediocin AcH is produced in large quantities at pH 5.0 or below 5.0 after entering a stable growth stage. On the contrary, nisin and leuconocin formed in large quantities during the logarithmic growth period, and the pH value was high.

Bacteriocin biosynthesis is regulated by genes that are generally located on chromosomes or plasmids of the producing bacteria (Nes et al. 1996). Operators of thioether antibiotics and bacteriocins II are mostly located on chromosomes, such as plantaricins EF and plantaricins NC8, produced by *Lactobacillus plantarum* 8P-A3. The regulatory genes of these two bacteriocins are located on chromosomes, and the size of the gene fragments is about 20 kb. Most of the gene clusters that regulate bacteriocin production are located on plasmids. Mandal et al. (2010, 2011) reported a new bacteriocin pediocin NV5 produced by *Lactococcus lactis* LAB 5, and plasmid elimination experiments showed that the pediocin NV5 gene cluster was located on a plasmid about 5 kb in size.

4.3.1 Biosynthesis and Regulation of Nisin

The gene controlling biosynthesis, immunity, and regulation of nisin is located on a 70 KB conjugated transposon named Tn5276. The regulatory synthesis gene cluster of nisin involves structural genes, mature genes, immune genes and regulatory genes which are controlled by gene cluster nisA/ZBTCIPRKFEG, and adjacent genes related to sucrose metabolism. The 11 genes of the gene cluster consist of 4 transcription units, namely, *nisABTCIPRK*, *nisI*, *nisRK*, and *nisFEG*. Among them, *nisI* and *nisRK* are constitutive transcription units, while *nisABTCIPRK* and *nisFEG* are regulated by *nisRK*. The functions of each gene in the nisin synthetic gene cluster have been identified (Riley and Chavan 2007a):

1. The structural gene *nisA* encodes a propeptide containing 57 amino acids.
2. The posttranslational modification gene *nisB*: *nisB* encodes the dehydratase NisB, which is located on the cell membrane and is responsible for dehydrating specific Ser and Thr in the propeptide to form Dha and Dhb. If the nisin precursor peptides were fused with a variety of non-wool thiobacillins and co-expressed with the nisBTC gene, the modified polypeptides could be produced, which indicated that the activity of NisB was not limited to the precursor of nisin, but could

dehydrate Ser and Thr in a variety of amino acid sequences. Ser and Thr in nisin precursor peptides are never dehydrated, suggesting that NisB has some flexibility in substrate selection, but the specific mechanism is still unclear.

3. The posttranslational modification gene *nisC*: *nisC* encodes cyclizase NisC, which is also located on the cell membrane. The nisin precursor dehydrated by NisB is further modified by NisC to form five intramolecular thioether rings. Deletion of *nisC* gene results in no formation of the wool sulfur ring, and the precursor cannot be secreted into the extracellular domain.

4. The transport gene *nisT*: NisT belongs to the ATP-binding cassette transporter family. A typical ABC transporter contains two transmembrane domains and two ATP-binding domains. While NisT contains only one transmembrane domain and one ATP-binding domain, so two NisT molecules form a complete ABC transporter. The homologous dimer form acts as a transporter. NisT is located on the cell membrane and is responsible for transporting the dehydrated and cyclized nisin precursor to the cell membrane. Deletion or inactivation of NisT results in the inability of nisin to secrete extracellularly and accumulate in the cytoplasm. In terms of substrate selection, NisT is not only capable of transporting fully modified nisin but also capable of transporting partially modified or unmodified nisin.

5. The precursor peptide excision gene *nisP*: NisP is an extracellular serine protease. Excision of the precursor sequence of nisin is the last step in the biosynthesis of nisin. Heterologous expression of NisP can also remove the precursor peptide from nisin precursor and produce active nisin molecules.

6. Immune-related genes *nisI* and *nisFEG*: Because nisin has strong antibacterial and bactericidal activities, *Lactococcus lactis* expressing nisin needs a specific mechanism to protect itself from nisin attacks. This function is achieved through NisI and NisFEG systems. When the two systems work alone, they can only provide weak nisin immunity for *Lactococcus lactis*, which shows that there is a strong synergy between the two systems. NisI exists in the form of membrane-bound lipoprotein or membrane-unbounded free form. Both forms of proteins can interact with nisin molecules and prevent them from contacting the lipid bilayer, acting as the "front guard" of the guard cell portal. NisF is a cytoplasmic ATP-binding protein; NisG and NisE are membrane proteins. These three proteins form ABC-type transporters. It is speculated that NisFEG proteins are responsible for clearing the nisin molecules that invade the membrane and returning them to the extracellular space.

7. Regulate genes *nisR* and *nisK*: They encode response regulators and histidine kinases, respectively, forming a two-component regulatory system for nisin biosynthesis. The sequence of NisR protein was similar to that of the transcriptional regulatory protein in the two-component regulatory system. The inactivation of NisR protein coding gene results in the inability of *Lactococcus lactis* to produce nisin. NisK is located on the cell membrane and can sense and bind mature nisin. When mature nisin exists outside the cell, NisK binds to nisin and initiates autophosphorylation of histidine, which activates specific signaling pathways, transfers phosphate groups to NisR, and activates nisin synthesis and the transcription

of immune-related genes. Inactivation of nisK gene did not induce nisA gene transcription.

Based on the existing research, Kuipers et al. put forward a model of nisin synthesis and regulation in 1993 (Kuipers et al. 1993). First, NisK is activated and self-phosphorylated after induction of extracellular signals (nisin); NisK, as a conducting protein, transfers phosphoryl groups to the response regulator NisR; phosphorylated NisR acts as a transcription activator to activate the transcription of *nisA/Z* and *nisF* promoters, leading to downstream gene transcription and synthesis of unmodified nisin precursors and progenitors; NisB and NisC are responsible for dehydration of pronisin to form thioether bonds and other posttranslational modifications; NisT is responsible for transporting the modified pronisin to the extracellular membrane; NisP is responsible for extracellular processing of the modified pronisin, removing the leading sequence and releasing the active mature nisin (Huo 2007).

4.3.2 Biosynthesis and Regulation of Pediocin

Pediocin belongs to class IIa bacteriocins. Its biosynthesis requires at least four genes: structural gene, immune gene, transporter coding gene, and transmembrane protein gene (Venema et al. 1995). The structural genes mainly encode the precursor peptides of pediocin, while the immune genes encode the immune proteins to protect producing strain from the attack of the secreted pediocin. The coding genes of the ABC transporters encode proteins responsible for the transmembrane transport of pediocin (Rodriguez et al. 2002). The secretion process is as follows: the binding of the precursor peptide's guiding sequence with the hydrolytic domain of ABC transporter triggers the release of energy from ATP hydrolysis, and the conformation of the transporter changes, so that the guiding sequence is separated from the precursor peptide. At the same time, mature pediocins are transported across the cytoplasmic membrane. An auxiliary protein is needed during the secretion process. As ABC transporters do not have the ability to hydrolyze proteins from the N-terminal, the separation of the guiding sequence from the precursor peptide is accomplished by a specific protease.

The regulation of pediocin depends on the synergism of inducible peptide, transmembrane histidine kinase (HK), and reaction regulator. HK is a receptor for the mature inducing peptide of pediocin. It releases phosphoryl groups through autophosphorylation inside the cytosol, and phosphoryl groups activate the response regulator into a transcription-activating factor that regulates gene transcription related to the biosynthesis of pediocin, including pediocin, immunity proteins, regulatory factors, etc. The synthesis of pediocin is controlled by transmembrane transport system. Biologically inactive precursors of pediocin are firstly synthesized in ribosomes. Then the precursors are cleaved at a specific site and the guided sequence is removed. Finally, bioactive tablets are formed and secreted out of cells. The amino acids of the ABC transmembrane transporter of pediocin have high homology. The C-terminal is a conserved ATP binding site, and the N-terminal is a target cell membrane binding region, containing a specific 150 amino acid extension. This structural region plays an important role in the removal of the lead peptide sequence and is a recognition signal for the removal of the precursor and the transmembrane transport of mature pediocin molecules out of the cytoplasm.

Take pediocin PA-1 as an example. This bacteriocin is encoded and secreted by plasmid. The synthetic gene cluster of pediocin PA-1 is composed of *pedA*, *pedB*, *pedC*, and *pedD*. The *pedA* gene regulates and encodes precursors containing 62 amino acids. *PedB* is a regulatory gene for immune-related proteins. The proteins encoded by *pedC* and *pedD* genes are involved in the translocation, processing, and secretion of bacteriocins. The transmembrane transport of class IIa bacteriocins, including Pediocin PA-1, is accomplished by the ABC transporter and an auxiliary protein. They are two transmembrane proteins that work together to form their own transport system (Havarstein et al. 1994). The number of amino acids in ABC transporters ranged from 715 to 724, with high homology at N and C ends. Venema et al. (1995) demonstrated that the N-terminal of PedD (the ABC transporter of Pediocin PA-1) plays a role in the removal of precursor peptides, and it does not participate in the secretion process (Rodriguez et al. 2002). The leader peptide is not only the signal of the propeptide but also responsible of the transmembrane transport of mature molecules. The protease region of the ABC transporter binds to the precursor peptide and then activates the hydrolysis of ATP and the conformational changes of the transporter protein, resulting in the removal of the precursor peptide and the transport of mature bacteriocin through the cytoplasmic membrane. The deletion mutation of any gene can lead to the failure to produce normal bacteriocin.

4.3.3   Biosynthesis and Regulation of Plantaricin

Lactobacillus plantarum C11 is the first strain to describe the synthesis and regulation of plantaricin at the gene level. In addition to *Lactobacillus plantarum* C11, similar bacteriocin biosynthesis gene clusters were found in the genomes of *Lactobacillus plantarum* V90, J51, J23, NC8, and WCFS1. These clusters are about 18–19 kb in length and about 25 genes, consisting of 5–6 operons. The conserved regions of the gene clusters are bacteriocin operons (*plnEF1*) and transport operons (*plnGHSTMVW*); the relatively conserved regions include one regulator and two to three bacteriocin operons. Each gene cluster also contains one or two unknown operons with relatively low conservatism.

The gene cluster of *Lactobacillus plantarum* C11 has five inducible operons. They are *plnABCD*, *plnEF1*, *plnJKLR*, *plnGHSTMVW*, and *plnMNOP*. *plnABCD* is a quorum-sensing transcription regulator, which can not only activate its own transcription but also control the transcription of other operons. *plnABCD* encodes a signal transduction system. *plnA* encodes a self-inducible peptide, *plnB* encodes a transmembrane protein-histidine protein kinase, and *plnCD* encodes two highly homologous reaction regulators. PlnC activates the target protein expression, and PlnD inhibits the target protein expression (Diep et al. 1994). *plnEF1* and *plnJKLR* synthesize immune proteins, so that the peptides have antibacterial activity. Transport operon *plnGHSTMVW* encodes the bacteriocin secretion pathway and works together to control bacteriocin secretion. The *plnH* encodes a transporter. *plnSTMVW* is highly homologous in this operon and encodes proteins belonging to the class II CAAX aminoproteinase family. This class of CAAX genes is very rare in other bacteriocin gene clusters. So far, their roles in bacteriocin biosynthesis are not obvious. *plnMNOP* encodes four hypothetical proteins. The N-terminal of the lead sequence of plnN transcription contains a double glycine structure. The *plnO*-encoded proteins are highly homologous to the glycosyltransferase family, while the *plnP*-encoded proteins are highly homologous to the type II aminotransferase family.

4.4 The Mechanism of Bacteriocin

Bacteriocin of lactic acid bacteria can achieve bacteriostatic or bactericidal effects through different mechanisms. The target of bacteriocin is the cell membrane of sensitive bacteria, which changes the cell membrane structure by utilizing different electrovalence in its protein structure (Reeves 2012). The effect of bacteriocin on target cells is divided into two steps. First, bacteriocin binds to specific and non-specific receptors on the surface of cell membrane of sensitive bacteria, at which stage bacteriocin still shows sensitivity to proteolytic enzymes. Irreversible specific changes occur on this step, such as formation holes on the membrane of susceptible cells, leading to the loss of ATP and K^+ ions or the obstacle of biochemical reaction. The balance of osmotic pressure inside and outside the cell is destroyed, and then the target cell is killed.

4.4.1 Antibacterial Mode of Nisin

Nisin is a typical type A lantibiotics. Type A lantibiotics inhibit bacteria mainly through two ways, inhibiting cell wall synthesis and forming pores in cell membrane. Lipid II is an important carrier in the process of cell membrane synthesis, responsible for transporting peptidoglycan components to the cell wall. Nisin can bind to lipid II and inhibit the formation of cell wall, which hinders the synthesis of cell membrane and phospholipid compounds and causes the release of intracellular substances, then leading to cell lysis (Riley and Chavan 2007b). This hypothesis has also been confirmed by experiments. As the adsorption of nisin on the surface of pathogenic bacteria is a prerequisite for its sterilization, it was found that adding activated carbon and other adsorbents in the system would significantly prolong the bacteriostatic time of nisin. Nisin has a strong adsorption effect, but it is obviously affected by pH. When the pH is 6.5, the adsorption rate can reach 100%, while at pH 4.5, the adsorption rate is only 43%. Nisin is positively charged and combines with sensitive bacterial cells to increase the permeability of cell membranes, which leads to the loss of nutrients and cell lysis. Therefore, there is a theory that the bacteriostatic mechanism of nisin is similar to that of cationic surfactant.

Nisin-induced pore-formation is a deeply studied mechanism. There are two mature theories: barrel plate model and wedge model. The first three steps of the two models are similar. Firstly, the N-terminal of nisin binds to the peptide glycan precursor lipid II and inserts into the membrane. At this time, the C-terminal of nisin crosses the cell membrane and switches the peptide to the transmembrane direction. When the transmembrane potential of nisin molecule is high enough inside and outside the cell, the position of nisin molecule can be changed perpendicular to the plane of the membrane. When the nisin molecules are overturned across the membrane, they are arranged in a circle like a barrel plate. This is barrel plate model. While the wedge model considers that when the transmembrane potential is high enough, the nisin molecule can change its position perpendicular to the plane of the membrane, resulting in the lipid surface bending, forming a wedge-like pore (Twomey et al. 2002). Pores allow hydrophilic molecules with relative molecular weight less than 500 to pass through, leading to potassium ion outflow from the cytoplasm, cell membrane depolarization and ATP leakage, extracellular water molecules inflow, whole cell wall degradation, cell autolysis, and death. This reaction mainly occurs in the compartment of differentiated daughter cells. There are two kinds of cell wall hydrolases involved in the reaction, N-ethylphthalide-L-alanine phthaliminase and N-ethylphthalide-glucosidase, which are strong cationic proteins that bind to negatively charged substances in the cell wall through electrostatic interaction. Cationic peptides and intramural inhibitors replace the latter by a process similar to cation exchange, and then enzymes activate and rapidly cleave cells.

Nisin mainly killed or suppressed gram-positive bacteria and spores, but had no obvious inhibitory effect on negative bacteria. This is because the cell wall of gram-positive bacteria is thick and its composition is relatively simple. While the cell wall of gram-negative bacteria is complex and compact, allowing only molecules with relative molecular mass below 600 to pass through. The relative molecular weight of nisin is about 3510, so it cannot pass through the cell wall and contact the cell membrane of gram-negative bacteria. When nisin is combined with chelating agent EDTA or surfactant, gram-negative bacteria such as *Salmonella* begin to become sensitive and can be inhibited or killed by nisin. The inhibitory effect of nisin was concentration dependent, which was related to the concentration of nisin and the cell concentration of sensitive bacteria. In addition, the physiological state of the sensitive bacteria also has some influence. The vegetative reproductive cells in the energy state are more likely to be killed than the vegetative reproductive cells in the static state. The sensitivity of spores to nisin is higher than that of bacterial cells, and the inhibition occurs before the expansion and growth of spores, which is very important in the preservation of hot-processed foods.

In conclusion, nisin mainly inhibits the microorganisms by adsorbing to the surface of cell membranes. At the same time, pH, ion concentration, lactic acid concentration, and nitrogen source types are also some factors affecting the adsorption of nisin. Nisin has two mechanisms; nisin-induced pore formation leads to the loss of proton motility (PMF), ion leakage, and ATP hydrolysis of sensitive bacteria and then to cell death. Nisin can also combine with lipid II, interfere with cell wall synthesis, and inhibit cell growth or survival. The dual mechanism of nisin enables it to work at nanomolar concentration.

4.4.2   Antibacterial Mode of Pediocin

The bacteriostasis mechanism of pediocin also mainly owes to the formation of pores on the cell membrane of sensitive bacteria. However, the adsorption and binding of pediocin to cell membranes are caused by electrostatic interaction. There is no need for the existence of membrane receptor proteins, so there is no selectivity. After being adsorbed on the surface of cell membrane, the hydrophobic region of the C-terminal of the pediocin peptide chain interacts hydrophobically with the tail of the phospholipid bilayer of cell membrane, which enables the C-terminal to insert into the cell membrane of sensitive bacteria, thus forming a pore (Bhunia et al. 1991).

4.4.3   Antibacterial Mode of Plantaricin

The antimicrobial mechanism of lantibiotics-type plantaricin is similar to that of nisin. Plantaricin of non-lantibiotics type mainly forms a hydrophilic channel on the cell membrane of susceptible bacteria. The formation of the channel is related to the "coupling molecule groups" on the surface of target cell membrane. However, this hydrophilic channel is self-directed by specific receptor proteins on the membrane and is not related to membrane potential, so it is also called non-energy-dependent. Compared with lantibiotics-type plantaricin, the function of this bacteriocin depends on destroying the stability of cell membrane function (such as energy conduction), rather than forming pores on the cell membrane. The inhibitory effect is related to the lipid composition and culture pH. In addition, the coupling molecular groups on the surface of the target cell membrane make it easier for the bacteriocin to interact with the target cell, thus improving the bacteriostatic effectiveness.

4.5   The Application of Lactic Acid Bacteria Bacteriocin

The food industry is a major production system in China. During food preservation, human or environmental factors can lead to corruption and deterioration, which not only cause economic losses but also affect health and even cause irreversible physical damage. Although some antioxidants and preservatives are added in food production to inhibit the growth of microorganisms and retard spoilage, most of these additives are synthetic compounds, and improper or excessive consumption will lead to physical damage. Therefore, it is urgent to develop a natural, nontoxic, harmless, and effective bacteriostatic agent. Bacteriocin produced by lactic acid bacteria is a kind of polypeptide substance. One advantage of this substance is that it can be degraded and digested by protease in digestive tract without residues. On the other hand, bacteriocin can inhibit most spoilage bacteria and foodborne pathogens (Hoover and Steenson 2014). The most widely used bacteriocin is nisin. Nisin has been widely used in food production, which effectively inhibits the growth of harmful bacteria, improves the quality of food processing, and prolongs the storage time of food.

4.5.1   Application of Nisin in Food Industry

Nisin is a highly effective and nontoxic polypeptide antimicrobial substance extracted from the *Lactococcus lactis*. Nisin is a natural preservative and antimicrobial agent recognized and used by most countries (Tolonen et al. 2004). Hirsch et al. first applied nisin to food preservation research. They found that nisin inhibited the growth of *Clostridium* in cheese production but had some adverse effects on the growth of fermentation medium and the ripening process of cheese. In 1952, McClintock et al. tried to add a certain amount of nisin to the mixture in cheese production and found that the quality of cheese had been greatly improved. On this basis, researchers gradually tried to apply nisin into other food production. In 1953, nisin was first commercially produced in the United Kingdom with the trade name of Nisaplin. In 1969, the Food Additives Committee (FAC) and the World Health Organization (WHO) evaluated the safety of nisin and confirmed that nisin could be used as a food additive in food. At present, more than 50 countries and regions around the world have approved nisin as a safe biological preservative, mainly used in processing cheese, milk products, and canned products (Jia 2009).

4.5.1.1   Application in Meat Products

In traditional meat production, nitrate and nitrite are commonly used as chromogenic agents to produce salted red and salted flavor and to inhibit the growth of *Clostridium botulinum*. However, nitrite can be converted to nitrosamines under specific conditions, and nitrosamines are carcinogens, so nitrate and nitrite in meat products have carcinogenic risks. Studies have shown that nisin can effectively control the growth of *Clostridium botulinum*, and nisin itself is acidic, which can reduce the pH value of surrounding media, thus reducing the residual nitrite content and the formation of nitrosamines (Rayman et al. 1983).

Rayman et al. found that nisin could also be used as an effective substitute to reduce the amount of nitrate and nitrite used in ham (Rayman et al. 1981). After adding 3000 IU/mL nisin to ham, the nitrate content decreased from 0.02% to 0.003%, while the quality of ham remained unchanged. In addition, in the traditional meat processing process, excessive heat treatment can significantly change the texture and appearance of meat products. However, the storage life of meat products with nisin can be prolonged by only 45% of the original heat treatment, and the final product has little difference in color, aroma, and taste compared with the traditional processed products. Nisin can also increase the sensitivity of some bacteria to heat and have some auxiliary bactericidal action. After soaking meat products in a certain concentration of nisin aqueous solution, reheating and sterilizing can ensure that the sterilization temperature is reduced, the sterilization time is shortened, and the original flavor of meat products is maintained on the premise of reaching the shelf life. When the addition of nisin was 0.3 g/kg in sausage processing, the overwhelming majority of gram-positive bacteria were inhibited, and the product color, aroma, and taste were not affected. Pork silk processing with 0.48 g/kg nisin could replace the preservative potassium sorbate and improve product quality.

Nisin is often used in combination with other preservatives to expand the scope and enhance the effect of bacteriostasis in practical production. The method is to mix preservatives into a solution, then mix them directly with meat products, or inject into meat products. It can also be coated on the surface of meat products. The operation is simple and convenient.

4.5.1.2 Application in Milk Products

Nisin has been successfully used in hard cheese, pasteurized milk, canned concentrated milk, yoghurt, butter, milk dessert, ice cream, and other products. Dairy products are rich in nutrients and are extremely susceptible to spoilage. Pasteurization and refrigeration can prolong the shelf life, but the spores of sporogenous bacteria are not killed and can germinate under suitable conditions. The shelf life of dairy products can be prolonged by adding 30–50 mg/kg nisin, which can be doubled at 35 °C. The addition of 80–100 mg/kg nisin in canned refined milk can reduce the sterilization time by 10 minutes. The addition of 20 mg/kg nisin in UHT milk can completely inhibit the growth of spore-producing bacteria in sterilized milk (Maisnierpatin et al. 1992). Adding 40 IU/mL nisin to yoghurt can delay the post-acidification process for 3 days and keep the number of live bacteria above 10^7 CFU/mL. The sensory quality of yoghurt is good.

Nisin is the first antiseptic used in cheese. The mixed application of nisin-resistant bacteria and nisin-producing bacteria in cheese starter can increase the quality of cheese by more than 90% compared with 41% by conventional methods. In processing solid and semisolid cheese, nisin helps to reduce the swelling caused by butyric acid. When soft cheese is stored under uncontrollable temperature, adding nisin can inhibit the growth of anaerobic bacteria and prolong the storage time of products (Davies et al. 1997).

4.5.1.3 Application in Pickles

Bottled pickles are a kind of food with long storage period and convenient consumption. Generally, pasteurization is used in bottled pickles. *Clostridium* and a few heat-resistant gram-positive bacteria will remain in food. Traditional methods of inhibiting bacteria mainly depend on high osmotic pressure (high salt and sugar), hypoxic environment, and chemical preservatives. At present, salt content in all kinds of pickles is high, but high salt foods are easy to induce hypertension and other diseases. The addition of 100 mg/kg nisin in some pickles not only inhibits the re-fermentation of lactic acid bacteria and the growth of *Staphylococcus* and *Bacillus* but also reduces the use of salt and the risk of high salt (Tolonen et al. 2004). In addition, adding nisin to pickles is better than adding conventional sodium benzoate and potassium sorbate. Some countries forbid sodium benzoate to be used in food. Therefore, it is of great economic significance to use nisin for preservation and bacteriostasis of pickles.

4.5.1.4 Application in Cans

Nisin is soluble and stable and has high bacteriostatic activity under acidic conditions, so it can be used for preservation of canned foods with high acidity (pH < 4.6). For example, after sterilization of canned tomatoes, there are still a small number of acid-resistant gram-positive bacteria such as *Clostridium pasteurianum* and *Bacillus leaching*, which can cause product spoilage. But adding 100–200 IU/g of nisin can effectively inhibit the growth and reproduction of these bacteria. In canned potatoes and mushrooms with low acidity (pH > 4.6), some thermophilic bacteria still survive after high heat treatment, which leads to canned spoilage. And high heat treatment has a certain adverse effect on the sensory quality of food. Adding appropriate amount of nisin can not only effectively inhibit the growth of spores but also reduce the heat treatment time of cans, maintain their freshness, and prolong their shelf life. For example, the common spoilage bacteria in canned mushrooms are *Bacillus stearothermophilus* and *Clostridium nigrificans*. The former increases the acidity of the contents, while the latter produces gas and makes the lid of the cans inedible. If sterilization is only carried out by heating high temperature or prolonging time, the color of the content will be darkened, and the elasticity of the tissue will become worse. The sensory properties of canned food can be greatly improved by the combination of heating and adding nisin, even though the storage period at room temperature remains unchanged for 2 years.

Although Nisin has good bacteriostasis performance, its essence is a kind of biological polypeptide substance, which will be destroyed in varying degrees when sterilized at high temperature and will degrade continuously with the prolongation of storage period. If the concentration of nisin is lower than 0.002%, its effect will be significantly weakened. Therefore, the stability and residues in cans should be noticed, so as to ensure that nisin is always within the effective concentration range during the shelf life of cans.

4.5.1.5 Application in Alcohol Production and Beverages

Lactobacillus and *Pediococcus cerevisiae* are common contaminating bacteria in brewage. The abnormal growth and reproduction of Lactobacillus will lead to beer turbidity, acidification and stickiness. While *Pediococcus cerevisiae* makes the agglutination of *Saccharomyces cerevisiae* worse. Almost all the spoilage bacteria in beer can be inhibited when the concentration of nisin is above 100 IU/mL. 100 IU/mL of Nisin can inhibit the spoilage of lactic acid bacteria, but the yeast is almost unaffected. Therefore, in the production of alcoholic beverages such as beer, fruit wine, and strong ethanol, in addition to the addition of yeast, adding appropriate amount of nisin can be used to inhibit the growth of gram-positive bacteria. Like the application of nisin in other food production, the addition of nisin not only reduces the time and temperature of pasteurization and the damage to product quality but also prolongs the shelf life of alcoholic products such as beer, which can be doubled. This effect is more obvious for non-pasteurized alcoholic products.

Malic acid-lactic acid fermentation caused by lactic acid bacteria often occurs in wine making, that is, the transformation of malic acid in wine into a biochemical reaction of lactic acid and CO_2. Generally, the measure of inhibiting malic acid-lactic acid fermentation is to add enough SO_2 (up to 100 mg/L) in wine. But SO_2 will destroy the typical flavor of wine, and it is strictly restricted to use due to the inhibition of normal fermentation bacteria growth. It has been reported that the addition of nisin in the brewing process can replace part of the role of SO_2 and not only will not affect the composition and flavor of the finished wine but also can inhibit unnecessary malic acid-lactic acid fermentation and eliminate the adverse effects caused by lactic acid bacteria.

Adding appropriate amount of nisin to fruit juice products before pasteurization can not only reduce the heat processing strength and increase the residue of nisin but also prevent the growth of miscellaneous bacteria such as *Bacillus*, thus preventing the corruption of fruit juice products.

4.5.2 Application of Bacteriocin in Medical Treatment

More than 2500 kinds of antibiotics have been found now. Most of them have adverse effects, only dozens of antibiotics commonly used in clinic. Antibiotics mainly interfere with the metabolic process of bacteria to inhibit their growth and reproduction or directly kill them. The extensive use of antibiotics and burst of bacterial resistance bring new hidden dangers to human health. Therefore, alternatives to antibiotics will become a new hotspot in drug research. Bacteriocin fills this gap because of its good bacteriostatic effect and safety characteristics.

Bacteriocins has a narrow antimicrobial spectrum and has a certain specificity and targeting. Methicillin-resistant *Staphylococcus aureus* (MRSA) is an important pathogen of nosocomial infection. Studies have found that some bacteriocins can inhibit MRSA (Aunpad and Na-Bangchang 2007). Bacteriocin is also effective in some skin diseases such as acne. Oral cavity is a necessary place for pathogenic bacteria in food. The special environment of oral cavity can easily lead to dental caries. Therefore, the oral mouth wash containing bacteriocin provides a new possibility to prevent this kind of oral disease. Bacteriocin also has great potential in the prevention and treatment of cow mastitis. At present, bacteriocin or bacteriocin-producing strains are usually used as an adjuvant therapy at animal level. So far, no pure bacteriocin can be directly used as a drug in clinical practice. For *Helicobacter pylori* and *Neisseria*, the efficacy of mutacin B-Ny266 and nisin A is comparable to that of vancomycin and oxacillin. Mersacidin, a B-type wool thiobacillin, is very effective in the treatment of staphylococcal infections and shows the ability to eliminate *Staphylococcus aureus* in rat models.

4.5.3 Application of Bacteriocin in Feed Industry

The application of bacteriocin in feed industry is an extension of food industry application. Bacteriocins can be added as an additive to the fodder. They have good thermal stability and therefore can tolerate high-temperature sterilization in feed processing. At the same time, bacteriocin can inhibit the exogenous pathogenic bacteria in animal intestinal tract, but has no killing effect on the inherent bacteria in animal intestinal tract, reduce the harm of pathogenic bacteria to animals, promote the healthy growth of animals, and improve the utilization rate of feed. In pig feed research, it is found that adding of bacteriocin in feed can reduce the diarrhea rate of piglets and increase the survival rate of piglets, which has significant economic benefits.

4.5.4 Application Prospect of Bacteriocin

Nisin is a highly effective, safe, and nontoxic natural preservative, which meets the development requirements of food preservatives in the future. It can be widely used in food, especially in food which needs heat treatment. It not only has good antiseptic and bacteriostatic effect but also can weaken the intensity of heat treatment, reduce processing costs, and improve the flavor and quality of products. In recent years, the combination of nisin and non-heat treatment technology - high hydrostatic pressure or pulsed electric field, has become a research hotspot in food and other industries. Nonthermal compound bacteriostasis technology not only shows good bacteriostasis effect but also prolongs the shelf life of food, so that the nutrition, taste, and other varieties of food are maintained.

Nisin in food can be affected by some external factors or food media, which will lead to the decline or even disappearance of bacteriostasis. For example, if food is not processed by heat or the heat treatment is not enough, the protease of microorganisms, plants, or animal organisms in food can degrade nisin during shelf life, which is below the minimum inhibitory concentration and will lose its effect. Nisin has better bacteriostasis effect in liquid and homogeneous food, but it is less effective in solid and heterogeneous food. Therefore, it is necessary to increase the dosage of nisin in solid and heterogeneous food. If Nisin is combined with other antibacterial substances to form a compound preservative, it can play a broadspectrum antibacterial effect. From a single type to a compound preservative, the application prospect of nisin will be broader.

References

Abbasiliasi S et al (2011) Effect of medium composition and culture condition on the production of bacteriocin-like inhibitory substances (blis) by lactobacillus paracasei la07, a strain isolated from BUDU. Biotechnol Biotechnol Equip 25(4):2652–2657

Anand SK, Srinivasan RA, Rao LK (1984) Antibacterial activity associated with Bifidobacterium bifidum. Cult Dairy Prod J 19:6–8

Anderssen EL et al (1998) Antagonistic activity of lactobacillus plantarum C11: two new two-peptide bacteriocins, plantaricins EF and JK, and the induction factor plantaricin a. Appl Environ Microb 64(6):2269–2272

Arakawa K et al (2009a) Effects of gassericins a and T, bacteriocins produced by lactobacillus gasseri, with glycine on custard cream preservation. J Dairy Sci 92(6):2365–2372

Arakawa K et al (2009b) Negative effect of divalent metal cations on production of gassericin T, a bacteriocin produced by lactobacillus gasseri, in milk-based media. Int Dairy J 19(10):612–616

Aunpad R, Na-Bangchang K (2007) Pumilicin 4, a novel bacteriocin with anti-MRSA and anti-VRE activity produced by newly isolated bacteria *Bacillus pumilus* strain WAPB4. Curr Microbiol 55(4):308–313

Aymerich T et al (1996) Biochemical and genetic characterization of enterocin a from *Enterococcus faecium*, a new antilisterial bacteriocin in the pediocin family of bacteriocins. Appl Environ Microbiol 62(5):1676–1682

Bhunia AK et al (1991) Mode of action of pediocin AcH from Pediococcus acidilactici H on sensitive bacterial strains. J Appl Bacteriol 70(1):25–33

Brotz H et al (1997) The lantibiotic mersacidin inhibits peptidoglycan biosynthesis at the level of transglycosylation. Eur J Biochem 246(1):193–199

Casaus P et al (1997) Enterocin B, a new bacteriocin from *Enterococcus faecium* T136 which can act synergistically with enterocin a. Microbiology-Uk 143:2287–2294

Chatterjee S et al (1992) Mersacidin, a new antibiotic from Bacillus fermentation, isolation, purification and chemical characterization. J Antibiot 45(6):832–838

Cheikhyoussef A et al (2010) Bifidin I–A new bacteriocin produced by Bifidobacterium infantis BCRC 14602: purification and partial amino acid sequence. Food Control 21(5):746–753

Cintas LM et al (2000) Biochemical and genetic evidence that *Enterococcus faecium* L50 produces enterocins L50A and L50B, the sec-dependent enterocin P, and a novel bacteriocin secreted without an N-terminal extension termed enterocin Q. J Bacteriol 182(23):6806–6814

Collado MC, Hernandez M, Sanz Y (2005a) Production of bacteriocin-like inhibitory compounds by human fecal Bifidobacterium strains. J Food Prot 68(5):1034–1040

Collado MC et al (2005b) Antimicrobial peptides are among the antagonistic metabolites produced by Bifidobacterium against helicobacter pylori. Int J Antimicrob Agents 25(5):385–391

Cotter PD, Hill C, Ross RP (2005) Bacteriocins: developing innate immunity for food. Nat Rev Microbiol 3(10):777–788

Davies EA, Bevis HE, Delves–Broughton J (1997) The use of the bacteriocin, nisin, as a preservative in ricotta–type cheeses to control the food–borne pathogen *Listeria monocytogenes*. Lett Appl Microbiol 24(5):343–346

de Arauz LJ et al (2012) Culture medium of diluted skimmed milk for the production of nisin in batch cultivations. Ann Microbiol 62(1):419–426

de Kwaadsteniet M, ten Doeschate K, Dicks LMT (2008) Characterization of the structural gene encoding Nisin F, a new lantibiotic produced by a Lactococcus lactis subsp lactis isolate from freshwater catfish (Clarias gariepinus). Appl Environ Microbiol 74(2):547–549

Diep DB et al (1994) The gene encoding plantaricin a, a bacteriocin from lactobacillus plantarum C11, is located on the same transcription unit as an agr-like regulatory system. Appl Environ Microbiol 60(1):160–166

Dobson AE, Sanozky-Dawes RB, Klaenhammer TR (2007) Identification of an operon and inducing peptide involved in the production of lactacin B by lactobacillus acidophilus. J Appl Microbiol 103(5):1766–1778

Draper LA et al (2013) The two peptide lantibiotic lacticin 3147 acts synergistically with polymyxin to inhibit gram negative bacteria. BMC Microbiol 13(1):1–8

Eijsink VGH et al (1998) Comparative studies of class IIa bacteriocins of lactic acid bacteria. Appl Environ Microbiol 64(9):3275–3281

Espeche MC et al (2014) Physicochemical factors differentially affect the biomass and bacteriocin production by bovine Enterococcus mundtii CRL1656. J Dairy Sci 97(2):789–797

Etchells JL et al (1964) Pure culture fermentation of brined cucumbers. Appl Microbiol 12(6):523–535

Fimland G et al (2005) Pediocin-like antimicrobial peptides (class IIa bacteriocins) and their immunity proteins: biosynthesis, structure, and mode of action. J Pept Sci 11(11):688–696

Flynn S et al (2002) Characterization of the genetic locus responsible for the production of ABP-118, a novel bacteriocin produced by the probiotic bacterium lactobacillus salivarius subsp salivarius UCC118. Microbiology-Sgm 148:973–984

Garneau S, Martin NI, Vederas JC (2002) Two-peptide bacteriocins produced by lactic acid bacteria. Biochimie 84(5-6):577–592

Gross E, Morell JL (1971) Structure of nisin. J Am Chem Soc 93(18):4634–4635

Hastings JW, Sailer M, Johnson K, Roy KL, Vederas JC, Stiles ME (1991) Characterization of Leucocin A-UAL 187 and cloning of the Bacteriocin Cene from Leuconostoc gelidum. J Bacteriol 173(23):7491–7501

Havarstein LS, Holo H, Nes IF (1994) The leader peptide of colicin V shares consensus sequences with leader peptides that are common among peptide bacteriocins produced by gram-positive bacteria. Microbiology-Uk 140:2383–2389

Heinrich P et al (1987) The molecular organization of the lysostaphin gene and its sequences repeated in tandem. Mol Gen Genet MGG 209(3):563–569

Hindre T et al (2004) Regulation of lantibiotic lacticin 481 production at the transcriptional level by acid pH. FEMS Microbiol Lett 231(2):291–298

Hoover DG, Steenson LR (2014) Bacteriocins of lactic acid bacteria. Academic Press, New York

Hui FM, Zhou LX, Morrison DA (1995) Competence for genetic transformation in *Streptococcus pneumoniae*: organization of a regulatory locus with homology to two lactococcin a secretion genes. Gene 153(1):25–31

Huo, G. C. Research and application of lactic acid bacteria. 2007

Jia SR (2009) Biological preservatives. China Light Industry Press, Beijing

Joerger MC, Klaenhammer TR (1986) Characterization and purification of helveticin J and evidence for a chromosomally determined bacteriocin produced by lactobacillus helveticus 481. J Bacteriol 167(2):439–446

Kawai Y et al (2009) DNA sequencing and homologous expression of a small peptide conferring immunity to Gassericin a, a circular Bacteriocin produced by lactobacillus gasseri LA39. Appl Environ Microbiol 75(5):1324–1330

King BF, Biel ML, Wilkinson BJ (1980) Facile penetration of the *Staphylococcus aureus* capsule by lysostaphin. Infect Immun 29(3):892–896

Klaenhammer TR (1993) Genetics of bacteriocins produced by lactic acid bacteria. FEMS Microbiol Rev 12(1-3):39–85

Konisky J (1982) Colicins and other bacteriocins with established modes of action. Annu Rev Microbiol 36(1):125–144

Kuipers OP et al (1993) Characterization of the nisin gene cluster nisABTCIPR of Lactococcus lactis. Eur J Biochem 216(1):281–291

Magnusson J, Schnürer J (2001) Lactobacillus coryniformis subsp. coryniformis strain Si3 produces a broad-Spectrum Proteinaceous antifungal compound. Appl Environ Microbiol 67(1):1–5

Maisnierpatin S et al (1992) Inhibition of listeria-monocytogenes in camembert cheese made with a nisin-producing starter. Lait 72(3):249–263

Mandal V, Sen SK, Mandal NC (2010) Assessment of antibacterial activities of pediocin produced by Pediococcus acidilactici lab 5. J Food Saf 30(3):635–651

Mandal V, Sen SK, Mandal NC (2011) Isolation and characterization of pediocin NV 5 producing Pediococcus acidilactici LAB 5 from vacuum-packed fermented meat product. Indian J Microbiol 51(1):22–29

Marciset O et al (1997) Thermophilin 13, a nontypical antilisterial poration complex bacteriocin, that functions without a receptor. J Biol Chem 272(22):14277–14284

Mathiesen G et al (2005) Characterization of a new bacteriocin operon in sakacin P-producing lactobacillus sakei, showing strong translational coupling between the bacteriocin and immunity genes. Appl Environ Microbiol 71(7):3565–3574

Mathys S, Meile L, Lacroix C (2009) Co-cultivation of a bacteriocin-producing mixed culture of Bifidobacterium thermophilum RBL67 and Pediococcus acidilactici UVA1 isolated from baby faeces. J Appl Microbiol 107(1):36–46

Miescher S et al (2000) Propionicin SM1, a bacteriocin from Propionibacterium jensenii DF1: isolation and characterization of the protein and its gene. Syst Appl Microbiol 23(2):174–184

Morgan SM et al (2005) Sequential actions of the two component peptides of the lantibiotic lacticin 3147 explain its antimicrobial activity at nanomolar concentrations. Antimicrob Agents Chemother 49(7):2606–2611

Mortvedt CI et al (1991) Purification and amino-acid-sequence of lactocin-S, a bacteriocin produced by lactobacillus-sake-L45. Appl Environ Microbiol 57(6):1829–1834

Mulders JWM et al (1991) Identification and characterization of the Lantibiotic Nisin-Z, a natural Nisin variant. Eur J Biochem 201(3):581–584

Müller E, Radler F (1993) Caseicin, a bacteriocin from Lactobacillus casei. Folia Microbiol 38(6):441–446

Muriana PM, Klaenhammer TR (1991) Purification and partial characterization of lactacin F, a bacteriocin produced by lactobacillus acidophilus 11088. Appl Environ Microbiol 57(1):114–121

Nakamura K et al (2013) Food preservative potential of gassericin A-containing concentrate prepared from cheese whey culture supernatant of lactobacillus gasseri LA39. Anim Sci J 84(2):144–149

Nes IF et al (1996) Biosynthesis of bacteriocins in lactic acid bacteria. Antonie Van Leeuwenhoek 70(2-4):113–128

Neumann VC et al (1993) Extracellular proteolytic activation of bacteriolytic peptidoglycan hydrolases of Staphylococcus simulans biovar staphylolyticus. FEMS Microbiol Lett 110(2):205–212

Nilsen T, Nes IF, Holo H (2003) Enterolysin a, a cell wall-degrading bacteriocin from *Enterococcus faecalis* LMG 2333. Appl Environ Microbiol 69(5):2975–2984

Paik SH, Chakicherla A, Hansen JN (1998) Identification and characterization of the structural and transporter genes for, and the chemical and biological properties of, sublancin 168, a novel lantibiotic produced by Bacillus subtilis 168. J Biol Chem 273(36):23134–23142

Parente E, Ricciardi A (1994) Influence of pH on the production of enterocin 1146 during batch fermentation. Lett Appl Microbiol 19(1):12–15

Rayman MK, Aris B, Hurst A (1981) Nisin: a possible alternative or adjunct to nitrite in the preservation of meats. Appl Environ Microbiol 41(2):375–380

Rayman K, Malik N, Hurst A (1983) Failure of nisin to inhibit outgrowth of Clostridium botulinum in a model cured meat system. Appl Environ Microbiol 46(6):1450–1452

Reeves P (2012) The bacteriocins, vol 11. Springer, New York

Riley MA, Chavan MA (2007a) Bacteriocins. Springer, Berlin/Heidelberg

Riley MA, Chavan MA (2007b) Bacteriocins: ecology and evolution. Springer, Berlin/Heidelberg

Rodriguez JM, Martinez MI, Kok J (2002) Pediocin PA-1, a wide-spectrum bacteriocin from lactic acid bacteria. Crit Rev Food Sci Nutr 42(2):91–121

Rueckert PW et al (1979) Mammalian and microbial cell-free conversion of anthracycline antibiotics and analogs. J Antibiot 32(2):141–147

Schneider TR et al (2000) Ab initio structure determination of the lantibiotic mersacidin. Acta Crystallograph Sect D-Biol Crystallograph 56:705–713

Tanner SA et al (2014) Synergistic effects of Bifidobacterium thermophilum RBL67 and selected prebiotics on inhibition of Salmonella colonization in the swine proximal colon PolyFermS model. Gut Pathog 6(1):44

Tolonen M et al (2004) Formation of nisin, plant-derived biomolecules and antimicrobial activity in starter culture fermentations of sauerkraut. Food Microbiol 21(2):167–179

Twomey D et al (2002) Lantibiotics produced by lactic acid bacteria: structure, function and applications. Anton Leeuw Int J Gen Mol Microbiol 82(1-4):165–185

Tyne DV, Martin MJ, Gilmore MS (2013) Structure, function, and biology of the *Enterococcus faecalis* Cytolysin. Toxins 5(5):895–911

Upreti GC, Hinsdill RD (1975) Production and mode of action of lactocin 27: bacteriocin from a homofermentative lactobacillus. Antimicrob Agents Chemother 7(2):139–145

van den Hooven HW et al (1996) Surface location and orientation of the lantibiotic nisin bound to membrane-mimicking micelles of dodecylphosphocholine and of sodium dodecylsulphate. Eur J Biochem 235(1–2):394–403

Venema K et al (1995) Functional analysis of the pediocin operon of Pediococcus acidilactici PAC1. 0: PedB is the immunity protein and PedD is the precursor processing enzyme. Mol Microbiol 17(3):515–522

Whitford MF et al (2001) Identification of bacteriocin-like inhibitors from rumen Streptococcus spp. and isolation and characterization of bovicin 255. App Environ Microbiol 67(2):569–574

Wirawan RE et al (2006) Molecular and genetic characterization of a novel nisin variant produced by *Streptococcus uberis*. Appl Environ Microbiol 72(2):1148–1156

Yildirim Z, Johnson MG (1998) Characterization and antimicrobial spectrum of bifidocin B, a bacteriocin produced by Bifidobacterium bifidum NCFB 1454. J Food Prot 61(1):47–51

Yonezawa H, Kuramitsu HK (2005) Genetic analysis of a unique bacteriocin, Smb, produced by Streptococcus mutans GS5. Antimicrob Agents Chemother 49(2):541–548

Zendo T et al (2003) Identification of the lantibiotic Nisin Q, a new natural nisin variant produced by Lactococcus lactis 61-14 isolated from a river in Japan. Biosci Biotechnol Biochem 67(7):1616–1619

Zihler A et al (2011) Protective effect of probiotics on Salmonella infectivity assessed with combined in vitro gut fermentation-cellular models. BMC Microbiol 11:264

Chapter 5
Lactic Acid Bacteria Starter

Wei Chen and Feng Hang

5.1 Introduction

5.1.1 Starters

5.1.1.1 Conception

A starter culture can be defined as a microbiological preparation containing numerous cells of at least one microorganism, which is added to a raw material to produce fermented food and can accelerate and control the fermentation process (Leroy and Vuyst 2004).

5.1.1.2 Function

Application of lactic acid bacteria (LAB) starter in foods involved in the production of organic acids (mainly lactic acid), aromatic compounds, proteolytic enzymes, and bioactive materials.

5.1.1.2.1 Lactic Acid Fermentation

Conversion of lactose to lactic acid is one of the most common and important traits for LAB starters. The produced acid leads to pH value reduction, undesired bacteria inhibition, milk curding, and aroma formation. In yoghurt production, post-acidification is a phenomenon that the acidity increase during storage, which due to β-galactosidase is still active at 0–5 °C. Post-acidification is a common but

W. Chen (✉) · F. Hang
Jiangnan University, Wuxi, China
e-mail: chenwei66@jiangnan.edu.cn

© Springer Nature Singapore Pte Ltd. and Science Press 2019
W. Chen (ed.), *Lactic Acid Bacteria*,
https://doi.org/10.1007/978-981-13-7283-4_5

undesirable characteristic of yoghurt. The characteristic results in an unacceptable taste by consumers and so shortens yoghurt's shelf life. Post-acidification is mainly associated with species, termination time, storage temperature, and storage time. Therefore, it is very important criteria for evaluating the fermentability of a strain. LAB with low and medium post-acidification is chosen in practice.

5.1.1.2.2   Flavor Substance

Giving good flavor to the fermented product is another important function of the starter. The flavor substances are mainly aldehydes, ketones, alcohols, and acids formed by the bacterial metabolism involved in protein hydrolysis, fat hydrolysis, and glycolysis. For example, the flavor substances unique to yoghurt is composed of acetaldehyde, diacetyl, 3-hydroxybutanone, and volatile acid produced by *Lactobacillus delbrueckii* subsp. *bulgaricus*. The characteristic flavor of fermented meat products are mainly ketones, volatile alcohols, and organic acids produced by LAB fermentation. The special flavor of fermented vegetables is mainly contributed by organic acids (e.g., lactic acid) and amino acids. All in all, since fermented dairy products have various flavors, a good starter should be selected according to the actual demand of the product.

5.1.1.2.3   Proteolysis

In cheese production, the proteolytic ability of the starter is usually investigated. For example, in cheddar cheese, the selected strains with strong proteolytic ability will be added to milk as adjunct starter to accelerate protein hydrolysis and shorten the ripening time. Proteolysis is also involved in the production of yoghurt. *L. delbrueckii* subsp. *bulgaricus* has a certain proteolytic ability and hydrolyzes proteins into peptides and free amino acids. The proteolytic ability of LAB is generally weak, and the bitter peptides accumulated by excessive hydrolysis of proteins cause bitterness in the product. Therefore, in the case the shelf life of the product is short, proteolysis will not be considered; in the case the shelf life is long, a starter culture with a moderate proteolytic ability is usually selected.

5.1.1.2.4   Bioactive Compounds

Recently, the bioactive compounds produced by LAB have been a research focus, including exopolysaccharides (EPSs), γ-aminobutyric acid (GABA), bacteriocin, etc.

EPSs

EPSs synthesized by LAB are extracellular carbohydrate biopolymers, which can be divided into two groups: homopolysaccharides (e.g., dextran, leven, and polyfructosans) and heteropolysaccharides according to the monosaccharide composition.

EPSs may be secreted into the medium as ropy EPSs or may be attached to cell surface of the microorganism in the form of capsular EPSs. EPSs have a variety of health-promoting effects such as cholesterol lowering, antitumor, bacterial adhesion, immunomodulating, etc. EPSs also function as natural thickeners, stabilizers, emulsifiers, and texturizers in food industrial, which affect the texture of the fermented products. LAB are well known as polysaccharide producers, such as *Streptococcus thermophilus*, *L. delbrueckii* subsp. *bulgaricus*, and *Lactococcus lactis*, which are all conventional dairy starter bacteria. Their ability of producing EPS has attracted more researches interest.

GABA

GABA is synthesized by glutamate decarboxylase (GAD), which catalyzes α-decarboxylation of L-glutamate or its salts to GABA. This enzyme has been found in LAB. Though *Lactobacillus* is the best GABA producer, *Lactococcus*, *Streptococcus*, and *Bifidobacterium* can also synthesize GABA. It acts as a neurotransmitter in the central nervous system of most vertebrates and plays additional roles on the overall human physiology such as lowering the blood pressure in mild hypertensive patients, acting as smooth muscles relaxation system, and regulating immune system. In addition, GABA also play roles on anti-heat stress, reproductive performance, endocrine hormone, carcass, and meat quality in animal husbandry. As a new-type functional component, GABA has been applied in food industry and health care. Hence, high GABA-producing LAB could be used as starters for producing functional fermented foods.

Antimicrobial Substances

LAB can produce a variety of metabolites, including organic acids, hydrogen peroxide, bacteriocins, etc., which prevent the proliferation of food spoilage bacteria and pathogens and ensure food preservation. Bacteriocins are proteinaceous (proteins or peptides) substances with bacteriocidal activity, primarily active against mostly Gram-positive bacteria and especially closely related organisms, such as nisin, diplococcin, and class IIa bacteriocins. In recent years, nisin (lantibiotic, class I), one of the most extensively studied bacteriocins, has been commercially used in food to control pathogens, such as *Listeria monocytogenes*.

5.1.1.3 Advantages

The advantages of commercial starter cultures are as follows: (1) they have only one or more than one kind of microorganisms; (2) the number ratio of different strains is fixed in commercial mixed starters; (3) they have distinctive characteristics of antibacteriophage and thermostability; (4) the cells are harvested at appropriate period in growth curve, and if the fermentation products are in log-phase, the lag-phase should be short; (5) fermentation parameters (time and temperature) have been determined.

5.1.1.4 Main Application

5.1.1.4.1 Starters in Food Industry

LAB is important in food industry and widely used in dairy industry because of (1) conversion of lactose into lactic acid, (2) production of gas, and (3) resistance to heat even high temperatures in the production of Swiss cheese. LAB plays a great role in enhancing the properties of fermented foods, and especially it enhances the nutritional, technological, organoleptic values of fermented foods and beverages. Nutritive values of fermented product are superior to the raw foods because of different acids and vitamins secreted by starters. Microorganisms in starters convert lactose into lactic acid, which acts as a preservative, and make fermented milk products have desirable texture and characteristic body through fermentation process. Furthermore, starters can secret preservatives such as bacteriocins, which can preserve product and increase their shelf life and safety. For instance, the sensory qualities of cheese were improved by *Lactococcus lactis* subsp. *cremoris*, and *Propionibacterium freudenreichii* was used for the formation of eyes (pores) and color in Swiss cheese.

5.1.1.4.2 Dietary Supplement

Probiotics is a rapid increase section in dietary supplement. The global probiotic market was valued approximately 62.6 billion US dollars in 2014 and is estimated to reach 96.0 billion US dollars by 2020. Due to the low pH in the stomach, digestive enzymes, and bile in the intestinal tract, the viable cells of probiotics will decrease. Health benefits of probiotics are determined by the number of viable cells that reach the gut. Hence, microencapsulation systems are developed for effective delivering and controlled releasing of viable probiotic cells to the gut.

5.1.1.4.3 Functional Food

During yoghurt fermentation, LAB can hydrolyze protein to generate bioactive peptides and synthesize functional EPSs, which facilitate the production of functional yoghurt or drinks, such as Actimel and Yakult.

5.1.2 LAB Starters

The carefully selected stain or strains as starter cultures can be isolated from natural and healthy products. Till now, LAB starters have been the most widely used starter cultures. Most of LAB has the ability to produce lactic acid, various amino acids, and other nutrients, which endow the food special flavor, improved quality, and intestinal microbial flora regulation. They have been widely used in large-scale food production, such as yoghurt, cheese, fermented meat, and vegetables.

5.1.2.1 The Commonly Used Strains

LAB are a heterogeneous group of microorganisms that convert carbohydrates into lactic acid, which are generally characterized as anaerobic or facultatively anaerobic, Gram positive, catalase negative, nonspore-forming, and non-motile. The first pure culture of LAB (*Bacterium lactis*, now known as *Lactococcus lactis*) was isolated by Joseph Lister from yoghurt in 1873. Depending on their hexose fermentation pathway, LAB can be divided into three physiological groups: homofermentative LAB, heterofermentative LAB, and facultative heterofermentative LAB. Homofermentative LAB can convert sugars exclusively into lactic acid through the Embden-Meyerhof-Parnas (EMP) pathway, whereas heterofermentative LAB (e.g., *Lactobacillus brevis*, *Lactobacillus fermentum*) convert sugars to lactic acid and other by-products such as acetic acid, ethanol, and/or carbon dioxide through the phosphoketolase (PK) pathway. The third group, facultative heterofermentative LAB (e.g., *Lactobacillus plantarum*), metabolizes hexose sugars through EMP but uses pentose sugars through the PK pathway (Zhang et al. 2016). The homofermentative LAB can rapidly decrease pH and increase lactic acid relative to other fermentation products; therefore the vast majority of industrial LAB are homofermentative bacteria. For instance, *S. thermophilus*, *Lactobacillus bulgaricus*, *Lactobacillus casei*, *Lactobacillus rhamnosus*, *Lactobacillus helveticus*, *Lactobacillus acidophilus*, etc. are selected as starter cultures for yoghurt and cheese production.

5.1.2.1.1 LAB Species Used in Different Foods

LAB starters are widely applied in various food fermentations, such as dairy products, drinks, vegetables, beans, meat, fish, etc. LAB is implicated to play a varied role in food fermentation, and the predominated species are diverse in various food types. The usually used LAB strains in different fermented food are listed in Table 5.1.

5.1.2.1.2 LAB with Special Function

Probiotics are living microorganisms, which are beneficial to human health. Using probiotics is becoming more and more popular as the increasing reports show the diverse benefits of probiotics to human health. Worldwide, foods containing probiotics are major contributors to the rapid-growing functional food market. The traditional probiotic products are fermented dairy products (yoghurt drinks, yoghurt, cheese, etc.), probiotics powder, and infant formula.

Table 5.1 LAB strains and some probiotics in different fermented foods

Materials	Food	LAB
Dairy	Yoghurt	*Lb. delbrueckii* subsp. *bulgaricus*, *S. thermophilus*
	Probiotic yoghurt	*Lb. casei*, *Lb. acidophilus*, *Lb. rhamnosus*, *Lb. johnsonii*, *B. lactis*, *B. bifidum*, *B. breve*
	Kefir	*Lb. kefir*, *Lb. kefiranofacies*, *Lb. brevis*
	Butter and buttermilk	*Lc. lactis* subsp. *lactis*, *Lc. lactis* subsp. *lactis* var. *diacetylactis*
		Lc. lactis subsp. *cremoris*, *Leuc. mesenteroides* subsp. *cremoris*
	Nonporous hard cheese	*Lc. lactis* subsp. *lactis*, *Lc. lactis* subsp. *cremoris*
Vegetables	Sauerkraut	*Leuc. mesenteroides*, *Lb. plantarum*, *Pediococcus acidilactici*
	Pickles	*Leuc. mesenteroides*, *P. cerevisiae*, *Lb. brevis*, *Lb. plantarum*, *Lb. pentosus*
	Fermented olives	*P. acidilactici*, *P. pentosaceus*, *Lb. plantarum*
	Fermented vegetables	*Lb. fermentum*
	Juice	*Lb. casei*, *Lb. plantarum*, *Lb. xylosus*, *Lb. sakei*
	Olive	*Lb. brevis*, *P. pentosaceus*, *Lb. plantarum*, *Lb. pentosum*
Grains	Fermented doughs	*Lb. sanfranciscensis*, *Lb. farciminis*, *Lb. fermentum*, *Lb. brevis*, *Lb. plantarum*, *Lb. amylovorus*, *Lb. reuteri*, *Lb. pontis*, *Lb. panis*, *Lb. alimentarius*, *Weissella cibaria*
Meat	Fermented sausage (EU)	*Lb. sakei*, *Lb. curvatus*
	Fermented sausage (US)	*P. acidilactici*, *P. pentosaceus*
Fish		*Lb. alimentarius*, *Carnobacterium piscicola*

Lb Lactobacillus, B Bifidobacterium, Lc Lactococcus, Leuc Leuconostoc, P Pediococcus

Probiotics are generally consumed as a part of food. Studies have suggested that probiotic food must contain at least 10^6 cfu/g in order to give healthy benefit. This means eating 100 g product everyday, we can get 10^8–10^9 cfu live cells. In the guidelines for the evaluation of probiotics in food, the World Health Organization (WHO) stressed that in order to ensure the beneficial effect of probiotics, the number of live cells must be more than 10^6 cfu/g at the end of the shelf life. Currently, probiotic strains commercially available are shown in Table 5.2.

5.1.2.2 Categories of LAB Starters

5.1.2.2.1 According to the Number of Strains

Single-strain Starter

This starter contains only one strain. For instance, in the fermented dairy industry, **S.** *thermophilus* and **L.** *bulgaricus* are the most common yoghurt's starter cultures. Generally, they are fermented separately and then mixed together in appropriate

Table 5.2 Commercially available probiotic strains

Sources	Strains
Chr. Hansen	*Lactobacillus acidophilus* LA1/LA5
	Lactobacillus delbrueckii subsp. *bulgaricus* Lb12
	Lactobacillus paracasei CRL431
	Bifidobacterium animalis subsp. *lactis* Bb12
Danisco	*Lactobacillus acidophilus* NCFMs
	Lactobacillus acidophilus La
	Lactobacillus paracasei Lpc
	Bifidobacterium lactis HOWARUTM/Bl
DSM food specialties	*Lactobacillus acidophilus* LAFTIs L10
	Bifidobacterium lactis LAFTIs B94
	Lactobacillus paracasei LAFTIs L26
Nestle	*Lactobacillus johnsonii* La1
Snow brand milk	*Lactobacillus acidophilus* SBT-20621
	Bifidobacterium longum SBT-29281
Institut Rosell	*Lactobacillus rhamnosus* R0011
	Lactobacillus acidophilus R0052
Yakult	*Lactobacillus casei* Shirota
	Bifidobacterium breve strain Yakult
Fonterra	*Bifidobacterium lactis* HN019 (DR10)
	Lactobacillus rhamnosus HN001 (DR20)
Probi AB	*Lactobacillus plantarum* 299V
	Lactobacillus rhamnosus 271
Danone	*Lactobacillus casei* Immunitas
	Bifidobacterium animalis DN173010
Essum AB	*Lactobacillus rhamnosus* LB21
	Lactococcus lactis L1A
BioGaia	*Lactobacillus reuteri* SD2112
Morinaga Milk Industry Co. Ltd.	*Bifidobacterium longum* BB536
Lacteol Laboratory	*Lactobacillus acidophilus* LB
MediPharm	*Lactobacillus paracasei* F19

proportions to produce yoghurt (Aghababaie et al. 2015). Additionally, *L. lactis* subsp. *cremoris* or *L. delbrueckii* subsp. *bulgaricus* is utilized as culture starter in the buttermilk production. Therefore, the strain selection of culture starter depends on the raw materials used and the sensory and texture of products desired.

Multi-strain Starter

This starter usually contains more than one strain. The strains may belong to the same species or species having similar characteristics. For instance, a commercial multi-strain starter can be consisted of *L. lactis* subsp. *cremoris/lactis* and *Leuconostoc mesenteroides* subsp. *cremoris*.

Mixed Starter

This kind of culture starter contains many unknown strains. The proportion of different strains varies with the product. For example, the kefir grains are a natural mixed starter, which is a symbiotic association of yeasts and LAB. The bacteria and yeasts present in the kefir grains include *L. delbrueckii* subsp. *bulgaricus*, *L. helveticus*, *L. acidophilus*, *Lactobacillus kefiranofaciens*, *L. lactis* subsp. *cremoris*, *Kluyveromyces marxianus*, *Kluyveromyces lactis*, *Saccharomyces cerevisiae*, *Torulaspora delbrueckii*, etc. The mixed culture starters utilized in the production of Swiss cheese are composed of *L. bulgaricus*, *S. thermophilus*, and *Propionibacterium*. *L. bulgaricus* and *S. thermophilus* are involved in the flavor formation, while the flavor and pore texture of mature cheese are due to *Propionibacterium*.

5.1.2.2.2 According to the Fermentation Temperature

Mesophilic Starter

The optimal growth temperature of LAB in mesophilic starter is between 20 and 30 °C, such as *Lactococcus* spp., *Leuconostoc* spp., and *Streptococcus* spp. These starters are very effective in combination or separate use in the fermentation of milk. Mesophilic starters are widely used to produce different kinds of cheeses (such as Brie, Cheddar, Gouda, etc.). *L. lactis* in mesophilic starters are further classified into four types as follows:

Type O: The characteristic of this type of starter is lactic acid produced by homo-fermentation. *L. lactis* subsp. *lactis* and *L. lactis* subsp. *cremoris* are commonly used type O starter.

Type D: This type of starter includes *L. lactis* subsp. *lactis* biovar *diacetylactis* besides LAB in type O starter. The obvious characteristic of this type of starter is diacetyl production, which is a source of characteristic flavor for buttermilk. Meanwhile, it also produces carbon dioxide that affects various subtle flavors.

Type L: This type of starter contains *L. mesenteroides* besides LAB in type O starter. The special substances during fermentation are diacetoacetic acid, acetaldehyde, and a small amount of carbon dioxide.

Type LD: This type of starter can produce a subtle blend of flavor and aroma, which is composed of *L. lactis* subsp. *lactis* biovar *diacetylactis* and *L. mesenteroides*.

Thermophilic Starter

The optimal growth temperature of LAB in thermophilic starters is between 40 and 45 °C, which are micro-aerobic microorganisms in nature. Thermophilic starters often include two strains: *S. thermophilus* and *L. delbrueckii* subsp. *bulgaricus*. They are mainly used for the production of yoghurt. The contribution of these two strains to yoghurt offers lactic acid and acetaldehyde aroma, respectively.

5.1.2.2.3   According to the Preservation Method

In the actual production or in lab, it is extremely important to preserve pure culture to keep morphological, physiological, biochemical characteristics and genetic traits stable. There are three preservation methods to maintain survival and original characteristics of pure culture for a long time.

Liquid Starter

Seed culture is usually stored in liquid medium (whole milk, skim milk, whey, etc.) in small scale. As a result, a large amount of starter is required in the production process and needed a step-by-step expanding culture. The advantages of liquid starter are convenience of preparation, low price, and suitability for production in small scale. However, the shortcomings are as follows: the continuous subculture has a large workload and high cost; the strain is easy to mutate and degenerate; the storage time is short; the quality between batches is unstable.

Powder Starter

Powder starter are obtained by drying with or without concentration through various methods, including vacuum drying, spray drying, freeze-drying, and concentrated freeze-drying. The advantages of this starter are the small workload, convenient storage, and transportation. However, cells are susceptible to damage during the manufacturing process. When the starter is dried in a frozen state, it is a lyophilized starter. The cryoprotectants are added during the preparation process to effectively reduce the impairment and lethality of the cells during the lyophilization. Currently, concentrated or non-concentrated lyophilized starter is widely used in commercial production or laboratory.

Frozen Starter

Frozen starter is the concentrated cells frozen at various conditions, including deep freezing at −80 to −30 °C and ultralow temperature freezing at −196 °C in liquid nitrogen. Commonly used cryoprotectants are 10% glycerol, 5% methanol, 5% or 10% dimethyl sulfoxide, 5% glucose, etc. The obvious advantage for this starter is the good effect and its long-time preservation, whereas the equipment requirements are high.

5.1.3   Preparation Technologies of LAB Starters

Industrial production processes of LAB starters consist of fermentation, cooling, concentration, cryoprotection, freezing, lyophilization (or other drying methods), preservation, and packaging. The aims of concentration and lyophilization are to reduce the culture volume and decrease the costs of storage and transportation. The subsequently obtained DVS (direct vat set) or DVI (direct vat inoculation) can immediately resume fermentation activities after the direct inoculation to food matrices. Recently, the viable counts in the concentrated freeze-dried LAB starters can reach 10^{11}–10^{12} CFU/g.

5.2 Preparation of Liquid LAB Starters

5.2.1 Strain Selection

Starters used in dairy industry have been converted from traditional microorganisms to the strains with specific functions (i.e., probiotics). In order to obtain specific function of LAB starter, the strains of the desired traits and phenotypes are often isolated by high-throughput screening methods. Based on microbiology and genomics, the systematic biology methods are used to study the relationship between the phenotypes and genotypes and to isolate the ideal probiotic strain. The potential strains, applied to starter, also need to consider the satisfaction, functional properties, sensory properties, nutrients, and probiotic effects imparted to the product (Carminati et al. 2015).

Choosing the right strain is the primary task in developing a starter while also considering production performance, stability, shelf life, and viability during production and processing. In addition, temperature, pH, dissolved oxygen, metabolites, culture medium, and fermentation conditions can affect the stability and viability of the starter with varying degrees. The general requirements for selecting starter strains are as follows: (1) safety and nontoxicity, (2) with certain special functions (such as acid production, aroma production, hydrolysis ability, etc.), and (3) a certain ability to respond to stress. The quality of starter bacteria should be assessed before use. The appropriate strains should be selected according to the purpose of production. The selection should be based on product characteristics, such as acid production, aroma production, viscous substances, post-acidification, or protein hydrolysis. A good starter should have better fermentation capacity, higher viable count, and stronger vigor.

5.2.2 Microorganism Propagation

5.2.2.1 Culture Medium

The principle of starter culture is to obtain vigorously growing cells that can be used for product fermentation. The process involves adding various nutrients and controlling some parameters such as pH, temperature, aeration, etc. Choosing suitable proliferation medium is essential for the successful preparation of starter culture, since the nutrient absorption and culture conditions determine the yield of cells and its metabolites. In the design of culture medium, it should be noted that the addition of phosphate can both enhance the buffering capacity and anti-bacteriophage activity, but some LAB such as *L. bulgaricus* may be inhibited by phosphate; the addition of growth factors, such as yeast extract, could promote the early growth of *Lactococcus*, and Mg^{2+} could promote the proliferation of LAB; the addition of $CaCO_3$ facilitates in maintaining the survival of liquid starter culture, e.g., the activity of *L. acidophilus* remained unchanged for a long time when 1.5% $CaCO_3$ was added to the culture medium.

The culture of LAB in the laboratory generally uses MRS or M17 media. Due to complicated formula and excess nutrients, it is not suitable for large-scale industrial production. At present, the industrial production of LAB starter culture mostly uses semisynthetic media, which is generally based on milk and whey (Table 5.3), supplementing carbon sources, nitrogen sources, vitamins, minerals, antioxidants, phage inhibitors, neutralizers, and the like. The preparation methods of the medium mainly includes the following: (1) adding the growth promoter to skim milk, and controlling the pH at the same time; (2) adding a growth promoter to the hydrolyzed milk while controlling the pH; (3) adding a growth promoter to the whey matrix while controlling the pH; (4) adding a growth promoter to lactose while controlling the pH; and (5) ameliorating MRS, glucose yeast paste medium, etc. Three strains of *L. plantarum* were cultured by using modified MRS medium (glucose 20 g/L, yeast extract 20 g/L, sodium acetate 10 g/L, sodium citrate 10 g/L, dipotassium hydrogen phosphate 5 g/L, pH 6.2). It was found that the proliferation rate of the strain increased by 50–241%, the biomass reached the maximum, and the bacteriocin production increased. In addition, skim milk is used in combination with proteolytic enzymes while adding tomato juice, and $CaCO_3$ could be used as the basic medium for *S. thermophilus* and *L. bulgaricus* (Zacharof and Lovitt 2013). Krzywonos and Eberhard (2011) carried out large-scale cultivation of *L. plantarum* MiLAB393 by adding 10% beet syrup and 0.8% yeast extract to wheat stillage (removed solid particles) to a cell density of 1.6×10^{10} cfu/mL. Aguirre-Ezkauriatza et al. (2010) cultured *L. casei* with medium fed-batch which is made of 10% protein and 90% lactose obtained from ultrafiltration of goat milk whey, and the viable count after freezing and concentration reached $5.0 \times 10^9 - 2 \times 10^{10}$ cfu/g.

Lactose is the main carbon source for LAB fermentation. Low concentration carbon source such as maltose, sucrose, or glucose can be used to promote the

Table 5.3 Composition of lactic acid bacteria starter culture medium

Carbon source	Nitrogen source	Vitamins and minerals	Antiphage substances	Antioxidants	Neutralizing agents
Lactose, maltose, sucrose, glucose	Milk protein, whey protein, casein protein, hydrolysate protein	Yeast extract, corn extract	Phosphate, citrate	Ascorbic acid, ferrous sulfate	Carbonate, phosphate, hydroxide, oxidate

growth of LAB. Milk protein, whey protein, or casein hydrolysate can be used as nitrogen sources. Yeast extract and corn extract not only can be used as nitrogen sources but provide vitamins and minerals. Phosphates can be used as acid neutralizers and phage inhibitors to prevent excessive acid production.

The metabolic pathway of LAB will be affected by the composition of the medium leading to the different survival rate of the bacteria in the subsequent drying process. The ratio of unsaturated fatty acid (UFA) to saturated fatty acid (SFA) and the content of cyclic fatty acid ΔC19:0 (3%) in the cell membrane of *L. delbrueckii* subsp. *bulgaricus* CFL1 cultured in MRS were significantly higher than cultured in whey protein. The ratio of UFA to SFA was 0.5 and the content of cyclic fatty acid ΔC19:0 is 1%. The glass transition temperature (Tg) and freeze-thaw temperature could be significantly reduced by increased contents of UFA and cyclic fatty acid ΔC19:0 in cell membrane lipids. This improved the cell membrane fluidity, cell freeze resistance, and cell stability during dehydration process (Gautier et al. 2013). The freeze-dried survival rate of *L. delbrueckii* subsp. *bulgaricus* ND02 cultured in MRS containing 4% yeast extract was only 12%, while that cultured without yeast extract, the survival rate was 55.3% (Shao et al. 2014). Compatible solutes (glycerol, non-reductive disaccharides, amino acids, etc.) can be accumulated as metabolic end products in cells by regulating metabolic pathways or providing a certain stress environment. These substances can protect cells from osmotic stress during dehydration by balancing osmotic pressure difference between inner and outer membranes of cells. In addition, extra protection effects on heat and other stresses can also be generated.

The survival rate of the bacteria can be significantly affected by carbohydrates during the drying process of the starter. For example, when *L. delbrueckii* subsp. *bulgaricus* is cultured in a medium containing lactose, sucrose, and trehalose, the cells can adapt to osmotic stress during freeze-thaw process. Similarly, the survival rate of *Lactobacillus sakei* was increased during spray drying when sucrose was added to the medium. The protective effect of mannose on *L. bulgaricus* was better than fructose, sucrose, and glucose, but glucose was much worse than fructose and sorbitol. The protective effects of lactose, sucrose, and trehalose on the drying process of *L. delbrueckii* subsp. *bulgaricus* were also studied. The protective effects of sucrose and trehalose were comparable. Carbon source (sugar type) in the medium can significantly affect the cell morphology and physiological characteristics of the bacteria and then show the different affections on resistance to external stress. Sugars in the medium which affect the metabolism of intracellular mannitol, trehalose, and glutamate can significantly enhance the survival rate of the cells in the process of drying and dehydration.

The concentration of culture medium will affect the survival rate of LAB during drying. *L. plantarum* growing in the diluted MRS medium has a higher viability than that growing in nutrient-enriched MRS medium after drying. The viability of the bacteria was reduced by the presence of NaCl in the medium during drying process.

5.2.2.2 Cell Expansion Process

The liquid starter production process is a gradual expansion of the bacterial culture process. The general workflow is usually divided into several steps like stock culture activation, the mother starter, the intermediate working starter, and the bulk starter. The starter can be classified into the following sections based on the culture stage:

1. **Stock culture**: pure culture of strains obtained from the laboratory
2. **Mother starter**: a starter obtained from stock culture activation
3. **Intermediate working starter**: the intermediate working starter is cultured by increasing the amount of the medium before large industrial production of bulk starter
4. **Bulk starter:** a starter used to obtain the desired product

The production of the starter comprises the following steps (Fig. 5.1): (1) medium preparation, (2) medium sterilization, (3) cooling the medium to desired temperature, (4) inoculation, (5) fermentation, (6) cooling finished product, (7) concentration, (8) frozen, (9) drying, and (10) packaging and storage. The liquid starter usually includes the steps from (1) to (6), and the concentrated starter includes the steps from (1) to (10).

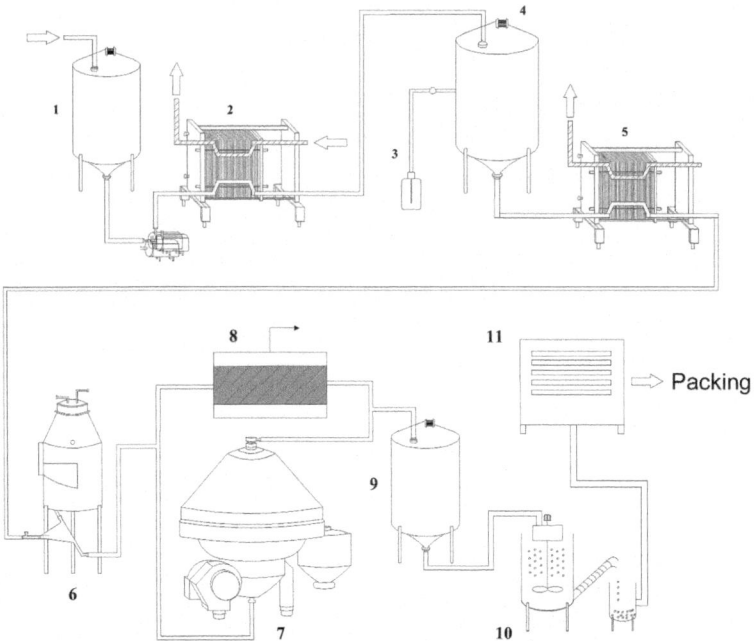

Fig. 5.1 Process flowchart a of typical starter culture. (1) Medium feeder; (2) sterilization; (3) starter seed liquid; (4) fermentation tank; (5) cooling; (6) buffer tank; (7) centrifugation; (8) ultra-filtration; (9) concentrator; (10) freeze granulation; (11) freeze-drying

There are several key control points in the cell expansion process of the starter.

1. The first stage of preparing the starter is heat treatment of the medium (90–95 °C for 20–45 min). In order to destroy the phage, eliminate the bacteriostatic substances, and kill the original microorganisms; medium is heated.
2. Cool the sterilized medium to the proper temperature. The inoculation temperature depends on the type of starter. The inoculation temperature of the mesophilic starter is 20–30 °C, whereas the inoculation temperature of the thermophilic starter is 35–45 °C.
3. After cooling to the desired temperature, a fixed amount of starter is added to the medium for expending culture. The inoculum amount, culture temperature, and incubation time must remain constant throughout the cell expansion process.
4. Cooling of the culture medium. When the starter reaches a predetermined acidity, a cooling operation is performed to control the growth rate and metabolism of the cells and to ensure high viability.

5.2.2.3 Culture Conditions

The culture conditions, including the medium composition, medium pH, and cell growth phase, have a significant effect on the composition of the cell membrane fatty acid and the viability of the bacteria in the subsequent drying process.

5.2.2.3.1   Culture Temperature

Culture temperature is one of the most vital factors affecting the proliferation ability of LAB during fermentation. The suitable growth temperatures for most LAB are 37–43 °C. Although some LAB (such as *S. thermophilus* and *L. delbrueckii* subsp. *bulgaricus*) can grow at 45 °C, the optimum growth temperatures are 40–42 °C. The temperature during the process determines the survival of LAB. Therefore, the duration for LAB to contact with high temperature should be avoided or shortened, and the temperature should not exceed 45–50 °C.

5.2.2.3.2   pH

The acid stress treatment during fermentation generally increases the survival rate of the cells during the drying process. Studies have shown that *L. acidophilus* has different tolerance to freezing cultured in the medium with different pH values, and the survival rate of cells treated with pH stress is improved during freeze-drying and cold storage. *L. acidophilus* cells that are obtained under fermentation conditions without pH control (fermentation end point pH 4.5) have stronger resistance to freezing, lyophilization, ethanol, and H_2O_2 than that cultured with pH-static culture (pH 6.0). Under the conditions of free acidogenic fermentation, the survival rate of

L. delbrueckii subsp. *bulgaricus*, after spray drying, is higher than when the growth of the cells is maximum (pH 6.5). When *L. delbrueckii* subsp. ND02 was cultivated at pH 5.1 (MRS medium without yeast extract), the lyophilized survival rate of the harvested cells was 68.3%, while the lyophilization survival rate of the cells was 51.2% when cultured at pH 5.7 (Shao et al. 2014). The characteristics of strains also have different responses to pH stress and fermentation mode. Acid stress can improve the survival rate of *L. paracasei* F19 during low-temperature vacuum drying. *Bifidobacterium lactis* Bb12 had higher cell stability under optimal growth (constant pH 6.8) and acid stress (constant pH 5.0) but poor cell stability under free acidogenic fermentation, while *L. delbrueckii* TMW1.1377 showed no significant difference in cell stability under free acidogenic fermentation and optimal fermentation conditions (constant pH 6.3), but in the case of acid stress (constant pH 4.5), the bacterial cell survival rate during drying increased (Foerst et al. 2012).

Reports on the effect of pH on the survival rate of bacteria in the subsequent drying process are not consented. Some studies have reported that the survival rate can be improved, while others believe that the opposite is true. The reason for this phenomenon is that other factors (such as strains, stresses caused by salt and sugar during drying, drying conditions and protective carriers, etc.) are not considered. When resuspended in phosphate buffer solution for low-temperature drying (20–40 °C), the survival rate of cells is observed to decrease, while in the more severe dry conditions (spray drying, inlet air temperature, 200 °C; outlet temperature, 70 °C), the protective effect of sucrose on *L. bulgaricus* is very likely to be concealed by the significant protection provided by the skim milk.

5.2.2.3.3 Oxygen

For oxygen-sensitive LAB, oxygen contact during fermentation is a major factor in the inactivation of the cells. Many methods have been used to reduce the oxygen content of the fermentation process, and the most common method is fermentation under vacuum conditions.

5.2.2.4 Co-culture

Most naturally or industrially produced fermented foods are produced through simple or complex microbial populations. The microbial population not only interacts with the fermentation substrate through a series of molecular and physiological mechanisms but also interacts with each other. Co-culture of a variety of microorganisms contributes to metabolizing each other, which not only allows for more complex biological processes (multifunctionality) but also tolerates the variability of the fermentation environment (robustness). The robustness of the strain not only ensures that the starter maintains high vigor and guarantees product quality but also effectively resists phage infection (Smid and Lacroix 2013).

The most typical co-culture of starters is yoghurt starter composed of *S. thermophilus* and *L. bulgaricus*. During the cultivation process, *S. thermophilus* has lower nutrient requirements, and the temperature is suitable for proliferation, which leads to the logarithmic growth of *S. thermophilus* and the growth inhibition of *L. bulgaricus*; in the midterm, *S. thermophilus* enters the stable phase, yet *L. bulgaricus* grows logarithmically; in the late stage, *S. thermophilus* enters secondary growth, and *L. bulgaricus* entered the stable phase. The growth rate, acid production ability, flavor substances, extracellular polysaccharides, and proteolysis ability of the two strains were improved after co-cultures. The reason is that formic acid, CO_2, and purine produced by *S. thermophilus* can stimulate the growth of *L. bulgaricus* (de Souza et al. 2012); amino acids and small peptides produced by *L. bulgaricus* can further promote the proliferation of *S. thermophilus*. It was found that the post-acidification ability of *S. thermophilus* and *L. casei* increased after co-culture, due to the intracellular substances released from *S. thermophilus* cell lysates that promoted the growth of *L. casei* (Ma et al. 2015). Synergistic symbiosis between LAB and propionic acid bacteria in Swiss cheese is also common. A substance such as lactic acid produced by fermentation of LAB is used as a carbon source for growth and metabolism of propionic acid bacteria; meanwhile, propionic acid bacteria can increase pH value by consuming lactic acid, decrease acid stress, and thus promote proliferation of LAB.

5.2.2.5   Stress Treatment

In the production and application of starter, LAB are often treated with a variety of stresses (such as acid stress, salt stress, cold stress, heat stress, etc.) via the various abrupt changes of growth environment, and the cells will adjust metabolic pathway and gene expression to adapt to the new environment (i.e., shock response). Studying the stress response of LAB under various stresses has great practical significance for reducing the damage and improving the survival rate of the starter (Beales 2004). The heat resistance of LAB can be enhanced by mild heat treatment. Studies have shown that heat shock treatment can increase the heat resistance of *Lactococcus* and *Lactobacillus* by 300 times. The freeze tolerance of cells of *L. acidophilus* RD758 and *L. bulgaricus* CFL1 can be enhanced by cooling the cells at the fermentation end point to 28 °C for 8 h or adjusting the pH to 5.25 for 30 min.

5.2.2.5.1   Cold Stress

Growing under low temperature, the metabolism and physiological morphology of LAB will change, such as the responses of cell membrane and related enzymes activity to low temperature, the change of fatty acid composition and cell membrane permeability, intracellular Ca^{2+} and protein loss caused by temperature, etc., which in turn affects the normal physiological metabolism of LAB. It has been confirmed that a large amount of cold-induced proteins are induced in LAB when the temperature is suddenly lowered. These proteins can affect the fluidity of the cell membrane and the superhelix, transcription, and translation of DNA. The authors speculate that these proteins may act as a transcriptional enhancer to express other cold shock genes or RNA chaperones and unfold to produce RNA molecules. The survival of *L. delbrueckii* subsp. *bulgaricus* after freeze-drying was increased in different degrees (8.11–16.82%) after cold stress treatment (10 °C, 2 h), and the expression of two cold shock-inducing genes (*csp*A and *csp*B) increased (Shao et al. 2014).

5.2.2.5.2 Heat Stress

LAB usually undergoes heat stress under high-temperature conditions. Heat stress is a stress response in which an organism resists a high-temperature environment above a normal physiological temperature environment. In the process of heat stress, LAB enhance the cell adaption to the environment mainly by inducing the transient expression of heat shock proteins and changing the physiological structure of cells. At present, considering the mechanism of heat damage to cells, heat shock (72 °C/15 s) is usually used to induce LAB transiently produce heat shock proteins such as *Dna*K, *Gro*ES, *Gro*EL, *Hsp*84, *Hsp*85, *Hsp*100, Clp, *Htr*A, and *Fts*H, which can improve the heat stress resistance and thus improve the survival rate of LAB (Serrazanetti et al. 2009). The freeze-dried survival rates of *L. delbrueckii* subsp. *bulgaricus* can be improved in different degrees (5.62–13.48%) after heat stress treatment, and the expression of six heat shock-inducing genes (*gro*ES, *hsp*, *hsp*20, *hsp*40, *hsp*60, and *hsp*70) increased (Shao et al. 2014). Heat stress can increase the survival rates of *L. lactis* and *Lactobacillus paracasei* NFBC338 by 700-folds after heat treatment and 18-folds after spray drying. The obvious difference between the two processes may be caused by dehydration damage during spray drying (Fu and Chen 2011).

5.2.2.5.3 Acid Stress

A large amount of lactic acid is accumulated during the high-density culture of LAB produces, which greatly suppresses the cells' further proliferation. Lactic acid is passively diffused into the cytoplasm and rapidly dissociated into protons and corresponding polar groups, which cannot pass through the cell membrane. The accumulated intracellular proton causes the cytoplasm pH decrease, reduces the substance across the membrane dependent on the proton driving force, and affects the energy supply for multiple transmembrane transport. Meanwhile, the internal acid conditions also greatly reduce the activity of acid-sensitive enzymes and can cause permanent damage on proteins and DNA, which in turn has a harmful effect on cells. At present, it has been reported that acid stress has many effects on LAB, such as proton pump ATPase, deamination, amidation, and decarboxylation to produce biogenic amine to control pH. One of the most effective acid stress responses is glutamate decarboxylase, which consumes intracellular protons through carboxylation, resulting in an increase in intracellular pH (Tsakalidou and Papadimitriou 2011).

5.2.2.5.4 Salt Stress

In the high-density culture, the medium pH is usually controlled by adding lye to eliminate the feedback inhibition of acid on the cells and to obtain the maximum biomass of LAB. The addition of lye produces a large amount of salt, which causes salt stress on cells. Under salt stress, the cells mainly adapt to hyperosmotic environment through the synthesis of stress protein and the changes of lipid composition in cellular membrane.

5.2.2.5.5 Osmotic Stress

In hypertonic solution, the dehydration leads to changes in cell volume, endogenous substances, and transmembrane pressure. NaCl is usually used to induce osmotic stress in LAB. In order to resist the adverse effects of high osmotic stress, cells can synthesize or absorb compatible substances including amino acids and derivatives thereof, polyols and sugars. As the compatible solute accumulates in the cell, it acts to protect the cells (Chun et al. 2012). At present, the use of suitable compatible substances to improve the stability of bacterial cells is a common method in the industrial production of LAB; different cells have different compatible substances (e.g., betaine, carnitine, and aspartic acid). However, the mechanism is unclear. Recent studies have shown that betaine is not a suitable compatible substance for LAB, mainly due to the formation of crystals of betaine during lyophilization, which has fatal damage to *Lactobacillus coryniformis* Si3 cells, resulting in a sharp decline of survival rate (Bergenholtz et al. 2012).

5.2.3   *High-Density Culture*

High-density cell culture refers to a culture method in which the cell concentration reaches ten times than that of the normal culture. The principle of high-density cell culture is devised to increase cell density in culture medium compared to common expanding culture. Thereby, the cell yield is improved, the volume of the biochemical reactor is decreased, the culture period is shortened, the downstream separation and concentration are simplified, and the amount of generated wastewater is reduced.

In laboratory and industrial applications, high-density culture of LAB cells is generally achieved by improving culture conditions and/or changing control methods. Factors affecting high cell density culture include culture environment, restricted substrate concentration, accumulation of growth inhibitory substances, carbon-to-nitrogen ratio, dissolved oxygen, and culture methods.

Culture environment includes medium composition and culture conditions, wherein the carbon-nitrogen ratio in the medium composition is particularly important for the growth of LAB. A pH-static culture (e.g., chemical neutralization or buffer salt) is an effective method for improving cell yield in high-density cell culture of LAB.

5.2.3.1 Buffer Salt Method

In LAB starter preparation, the accumulated lactic acid leads to an increase in the acidity of the culture medium and inhibits the further growth of the cells. A buffer salt, without an effect or with a promoting effect on LAB cells, is added to the culture medium, and then culture medium pH values is buffered and controlled to be static, thereby promoting the proliferation of the LAB cells. Commonly used buffer salt systems are citrate, acetate, K_2HPO_4, and KH_2PO_4.

5.2.3.2 Chemical Neutralization

The buffer salt can work within its buffering capacity; once exceeding its capacity, the buffer system will be out of function; and the pH of the culture medium will continue to decrease for the accumulated lactic acid, eventually leading to the growth inhibition of LAB. The pH of the medium was maintained at about 6.0 by a continuous neutralization method to obtain a high concentration of bacterial cells. A schematic diagram of the continuous fermentation for LAB in laboratory is shown in Fig. 5.2. The alkali solution is preciously delivered to the fermenter to ensure the

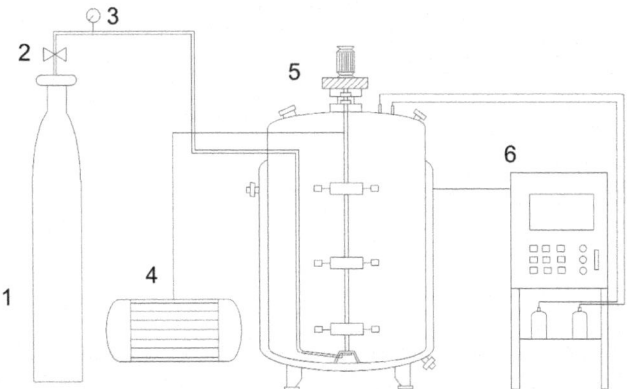

Fig. 5.2 Schematic diagram of lactic acid bacteria feed culture fermenter. (1) Nitrogen bottle; (2) gas flow regulating valve; (3) gas flowmeter; (4) constant temperature water tank device; (5) fermentation tank; (6) control panel

culture medium pH suitable for LAB cell growth by monitoring the change of pH. This method and device were used for the proliferation of LAB. Compared with the ordinary culture method, the device maintains pH at the optimum growth for three LAB strains, and finally the biomass and production efficiency are both improved (Zacharof and Lovitt 2013).

By changing the culture method, the effects of substrate deficiency and accumulation of harmful products on the high-density culture of LAB can be effectively solved. The culture method is associated with the type of bioreactor selected. The types of bioreactors that can be used for high-density cell culture include stirred tank reactors, reactors with external or built-in cell retesntion devices such as dialysis membrane reactors, airlift reactors, and shaken ceramic flask. In industrial production, high-density cell culture is generally carried out using a stirred tank and a feeding process. The abovementioned method is simple in operation, high in production potential, and convenient for the regulation of various parameters (Riesenberg and Guthke 1999).

5.2.3.3 Fed-Batch Culture

Fed-batch culture, also known as continuous-flow culture, is the most mature and widely used culture method. This culture method dynamically adds fresh culture nutrient solution according to the cell growth condition and the medium characteristics and increases the viable cell counts. Fed-batch control types include feedback control and non-feedback control. The basis of feedback control is the physiological model. It achieves real-time feed regulation by controlling physiological parameters such as cell and substrate concentration, pH, and dissolved oxygen during the culture process. Since the microbial growth feedback takes a certain time, the feedback control usually has hysteresis; moreover, fluctuations in sugar concentration or accumulation of harmful substances may occur during the cultivation. Non-feedback control is based on the feeding fluid kinetic models, which are classified as constant feeding, variable feeding, intermittent feeding, and exponential feeding fed-batch culture (Chang et al. 2014). The biomass and product concentration reached 88.8 g/L and 3.7 g/L in a LAB high-density culture by two-time batch feeding, which was 107% and 257% of those than the normal fluid flow. The biomass and product concentration reached 212.9 g/L and 10.6 g/L by eight times batch feeding, which was 101% and 429% of those than the normal fluid flow (Chang et al. 2011).

5.2.3.4 Continuous Culture

Continuous culture of high-density cells is achieved by the sterilized medium continuously fed into the fermentation medium at a certain speed. Meanwhile, the cells in the medium are recovered by sedimentation, centrifugation, membrane filtration, etc., and then resuspended into the medium. Therefore, a large amount of fermentation products and metabolites in high-density cell culture inhibited cell growth and are eliminated.

Commonly used continuous high-density cell culture separation devices are as follows (Chang et al. 2014).

5.2.3.4.1   Hollow Fiber Membrane

This separation device was originally applied to an enzyme immobilization. In cell culture, small molecules are able to pass through the cell membrane, while macro-molecules are trapped, separating the product from the cells. When the cells prolif-erate excessively, the hollow fiber membrane does not separate the product well, and the low oxygen supply is also a disadvantage, and the proliferation of the aero-bic LAB cannot be satisfied. When the fermenter with hollow fiber membrane is used for the submerged fermentation of *L. lactis* subsp. *lactis*, the ultrafiltration system is continuously operated at fermentation time of 5–12 h, and the whey, lac-tose, and yeast extracts are supplemented; at the fermentation of 12–30 h, the medium was removed/recirculated every 30 min for 10 min; after 30 h fermentation, it was found that the viable counts obtained by this system was 2×10^9 cfu/mL, which was 2.4-fold higher than that of the fermentation apparatus without the hol-low fiber membrane system (Ramchandran et al. 2012).

5.2.3.4.2   Dual Hollow Fiber Bioreactors

Aerobic microorganism cannot grow well in common hollow fiber systems because oxygen supply is not sufficient. Dual hollow fiber bioreactors (DHFBRs) are used to promote growth of aerobic microorganism. It includes a silicone tubing system for supplying oxygen (gaseous substrates) and a polypropylene tubing system for supplying liquid nutrients. However, it is impossible to work continuously because microorganism (fungal) has so high strength that the tubes may be distorted. Therefore, industrial applications are not possible unless porous ceramic or stain-less structures can be used.

5.2.3.4.3   Depth Filter Perfusion Systems

In depth filter perfusion systems (DFPSs), nutrients can be delivered directly to cells, and there are high-density cells suspended in the macropores of the depth fil-ters because the membrane pore size is two to three times larger than the diameter of cells. The cell density could be as high as $1–3 \times 10^8$ cells/cm^3 of the DFPS volume (Lee et al. 2008). Compared to hollow fiber or dual hollow fiber immobilized cell systems, installing several units of commercially available DFPS units could make scale-up easier.

5.2.3.5 Cell Retention Culture

Cell retention culture was carried out to achieve high cell concentration and high productivity. There are three methods: sedimentation, centrifugation, and membrane filtration. Membrane filtration modules used for cell recycle and concentration are located outside the bioreactor. The membrane bioreactor allows continuous cultivation of the cells at high growth rate and finally to high cell density. In this system, membrane filtration continuously removes inhibitory metabolites, and meanwhile fresh culture media is fed (Fig. 5.3). The batch culture of four LAB was achieved in a stirred tank reactor containing culture medium. There was a pH electrode in the bioreactor, so the pH value of medium was automatically monitored. Meanwhile, 5.0 mol/L NaOH solution was used to adjust and control the pH of medium. Weight change of culture was measured by a digital balance every 3 h to ensure the consumption rate of alkaline solution. Nitrogen gas containing no oxygen was continuously provided after microfiltration during the culture. Temperature was precisely controlled by a heat exchanger coupled with a temperature-controlled water bath. Culture and feed medium were previously formulated in the reservoir tank and then sterilized by a heat exchanger. The aseptic medium was pumped directly into the membrane bioreactor (Jung and Lovitt 2010). Cell retention culture of *Leuconostoc citreum* was carried out in this bioreactor equipped with an internal ceramic filtration system. With dilution rate of 0.07 h^{-1}, the OD of culture medium containing *L. citreum* increased to 75, which was 6 times higher than that in batch culture. The productivities of mannitol and lactic acid were 2.7–3 times higher than those in batch culture (Sung et al. 2012).

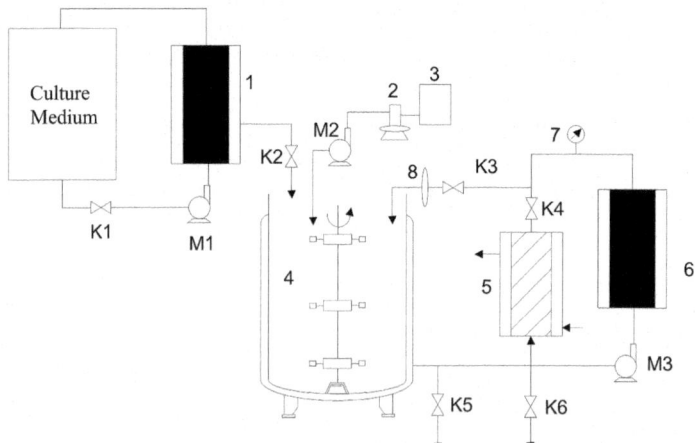

Fig. 5.3 Schematic diagram of cell culture process in membrane bioreactor. (1) Ceramic membrane filtration equipment (substrate, pore size: 0.2 μm); (2) weighing sensor; (3) alkaline tank; (4) fermentation tank; (5) heat exchanger; (6) ceramic membrane filtration equipment (product, pore size: 0.2 μm); (7) flowmeter; (8) pH meter; K1–K6, valves; M.1–M3, centrifugal pumps

5.2.3.6   Dialysis Culture

Dialysis culture is a culture method in which a semipermeable membrane is used to effectively remove harmful low molecular weight metabolites in a culture solution and provide sufficient nutrients to the reactor. The advantages of the culture method are as follows: (1) dialysis culture can increase the accumulation of biomass and increase the cell density; and (2) compared with microfiltration and ultrafiltration, the dialysis membrane does not block during dialysis and can work for a long time. However, the dialysis culture equipment has large investment and high technical requirements, and is not suitable for actual industrial production.

5.3   Preparation of Powder LAB Starters

5.3.1   Cell Collection and Concentration

5.3.1.1   Collection Time

The tolerance of LAB against external environmental stress changes with the different growth stages and culture medium. This property makes it difficult to standardize the fermentation conditions, but the culture conditions can be optimized to increase the cell tolerance to stress. Cells collected during stationary phase are generally considered to have the higher stress tolerance than those in lag phase and logarithmic growth phase, which is mainly due to the natural response of cells to nutrient deficiencies and metabolic toxic accumulation during fermentation (Fu and Chen 2011). The collection time of the LAB cells during stationary phase, from the initial stage to the late stage, has no effect on the concentration, fermentation, and acid production capacity of the concentrated starter before lyophilization but can alleviate the decrease in the acid-producing ability of the starter after lyophilization. Increasing the fermentation temperature only reduces the decrease in acid production capacity of the lyophilized LAB cells collected in early stationary phase (Velly et al. 2014). The results showed that the significant increase in the cyclopropanation of membrane of unsaturated fatty acids in stationary phase was correlated with membrane rigidity, cell viability during lyophilization, and storage (Velly et al. 2015). Therefore, cells in stationary phase are widely collected and used in subsequent drying processes.

5.3.1.2   Cell Collection by Centrifugation

Based on the difference in density between the cells and medium, the LAB cell collection is usually carried out by centrifugation, in which the cell density is concentrated 10–40-fold. Detail information on the effects of the centrifugation step on cell damage and metabolic activity is limited, but it is generally considered that this process has damage to the cells. However, results of some studies have shown that centrifugation have no damage to bacterial cells. For instance, the speed and duration time of centrifugation had no effect on the acid production rate of **L.** *bulgaricus* CFL1. On the contrary, the survival rates of the cells during lyophilization and preservation increased. In addition, there were no difference of cells activity for **L.** *mesenteroides* between batch and continuous centrifugations.

5.3.1.3 Cell Collection by Membrane Filtration

In case of high-viscosity cultured media, especially strains producing extracellular polysaccharides lead to high viscosity, a higher centrifugation speed is required to achieve desired recovery. Membrane separation technology (e.g., mircofiltration, ultrafiltration) is a potential alternative to centrifugation methods, but the investment of the equipment is high, which limits its large-scale application in industry.

5.3.2 Protectants for Cells

In order to improve the viability of LAB during freeze-drying and spray drying, the cells harvested from the culture medium still need further treatment before drying. Generally, the cells collected by centrifugation are resuspended to the solution containing protectants in the laboratory, while homogenization and atomization are required in the industrial production process. The type of protectants was depended on the drying process. Since the temperature of fluidized bed and vacuum drying is low (5–35 °C), the protectants were generally in solid form, and the cells were mixed with them directly or dried without protectants to reduce the cost and to improve the drying efficiency. Directly mixing the solid protectants with the cells, the *Aw* will decrease. Different *Aw* will affect the stress tolerance of the cells and the survival rate during the subsequent drying process, so the effects of the solid protectants are worse than that of the liquid protectants. On the contrary, spray-drying feed is in the form of liquid; the higher moisture content helps drying material to maintain the temperature around the wet bulb temperature, thereby avoiding the excessive heating of cells.

The protectants in liquid form include skimmed milk, whey protein, glycerin, betaine, ribose, lactose, dextrin, and polyethylene glycol, as well as antioxidant (eliminating reactive oxygen). These protectants have two protective mechanisms for the cells: (1) stabilizing the cell structure in the process of drying and rehydration, e.g., trehalose, and (2) forming the physical barriers to alleviate stress during drying (heat, osmotic stress) and rehydration (osmotic stress). Furthermore, there are reports that protectants can improve the stability of LAB during storage (Foerst et al. 2012). The viable counts of **L.** *paracasei* F19 protected by sorbitol without reduction are stored at 20 °C for 3 months (water activity, *Aw* 0.07–0.33) and the attenuation significantly reduced stored at 37 °C (*Aw* 0.07–0.33). The protective mechanism of sorbitol on cell stability was attributed to the elevation *Tg* (e.g.,

37 °C, $Aw = 0.07$, Tg increased from -32 °C to 12 °C). The protective mechanism of trehalose was different, which did not exert cells stability during storage.

Protectant formulation is an important step to stabilize cells during drying process. Using proper formulated protectant can reduce the stress during freeze-drying and storage. At present, LAB cells are usually freeze-dried in the presence of sugar and polymers. High viscosity amorphous matrix can be formed by sugar and polymer, which can reduce the mobility of molecules and thus limit the degradation reaction controlled by diffusion. Sugar can replace water to form hydrogen bonds with biological molecules, which will be damaged in the drying process, and thus exert protective effect on cells. Proper cryoprotectant is added to the concentrated cells before freeze-drying and spray-drying process, to reduce the transmembrane osmotic pressure and to improve the environmental adaptation ability. Cryoprotectant also can be added in the culture medium to enhance the cell adaption ability to environmental conditions (Basholli-Salihu et al. 2014). The survival rate of cells with protectant can reach approximately 70% (Savini et al. 2010).

5.3.2.1   Sugar Alcohols and Amino Acids

Carbohydrates have a protective effect on LAB cells during freeze-drying process, by increasing the Tg and avoiding the formation of intracellular ice crystal which will do damages to cells. Trehalose is an effective protectant in the freeze-drying process of LAB cells because of its remarkable increase in Tg, strong ion-dipole, and hydrogen bonding with biological macromolecules. In addition, sorbitol, mannitol, sucrose, lactose, and mannose also have the similar effects. Trehalose is one of the best protectants for cells (microorganisms, plants, and lower invertebrates) against osmotic and heat stress. The notable protective effects of trehalose are summarized as the following three aspects: (1) replacing water molecules to maintain the original conformation of lipid bilayer in cell membrane; (2) higher Tg and higher tendency to maintain glass state than other nonreducing disaccharides (e.g., sucrose), in which chemical reactions such as free radical oxidation that caused further cell damage are inhibited compared with crystal state; and (3) effectively resist protein denaturation and stabilize cell molecules structure (e.g., nucleic acid, mitochondria, and ribosome). During freeze-drying and fluidized bed drying, trehalose and sucrose were superior to glucose and maltodextrin in maintaining cell viability (Strasser et al. 2009).

Cell death rate was positively correlated with water absorption and membrane fluidity during freeze-drying. Replacing maltodextrin with glucose (D-or L-) could decrease the membrane fluidity and water absorption and thus improve the survival rate of cell during freeze-drying. One percent sorbitol can increase the survival rate of *L. helveticus* in vacuum-drying process. Amino acids have similar protective effects to carbohydrates, e.g., phenylalanine, arginine, and glycine inhibit the proteins denaturation during freeze-drying. When 20% (w/v) trehalose was used as a protective agent, the survival rate of *L. rhamnosus* GG during spray drying was 68.8%, and further supplemented with sodium glutamate, the survival rate reaches 80.8% (Sunny-Roberts and Knorr 2009).

5.3.2.2 Proteins and Polysaccharides

Proteins have higher Tg than carbohydrates, so skim milk, whey protein, serum, bovine serum albumin, sodium caseinate, and peptone are good dehydration protectants. Skim milk is an effective protective agent for LAB in drying process, especially in spray drying. The results showed that 10–20% (w/v) reconstituted skim milk was the best bacterial cell protectant, and the protective effect was better than that of Arabic gum, gelatin, starch, maltodextrin, polyglucose, and other prebiotics (Fu and Chen 2011). The protective mechanism of skim milk is not fully understood. It was previously thought that the protective effect of skim milk was related to lactose, which was similar to that of nonreducing disaccharides such as trehalose and sucrose in maintaining the integrity of cell membranes. However, it was not confirmed by later experiments.

Recently, it is generally believed that the proteins in skim milk can form a protective layer on the surface of the cell and reduce cell damage during spray drying; calcium ions improve the rehydration ability of bacterial cells after spray drying and then enhance the survival rate of bacteria. Some studies have also shown that the reconstituted skim milk can achieve higher bacterial survival rate, but the protective effect on bacterial functional characteristics (acid and bile salt tolerance and cholesterol assimilation ability) is worse than that of maltodextrin (Reddy et al. 2009a). Concentrated whey protein (WPC80) and sorbitol are effective cryoprotectants for *L. bulgaricus* LB6. Arabia gum, gelatin, and pectin are also effective protectants for reducing LAB cell injury during spray drying and can improve the stability of preservation. For instance, the survival rate of *L. casei* NFBC338 in spray drying (outlet temperature 100–105 °C) protected by Arabic gum was elevated more than 10-fold.

5.3.3 Cell Encapsulation Technology

The production performance and physiological function of LAB depend largely on their survival rate; however various environmental factors cause a decrease in cell viability during process and storage. When a sufficient amount of LAB was administrated, these cells need to survive after through the host's upper digestive tract and then reach the intestinal tract to the colon. The extremely acidic environment and multiple digestive enzymes in the stomach dramatically reduce the number of viable cells. The microcapsule technology originated from the pharmaceutical industry achieves the effect of controlling release of the drug by encapsulating. Encapsulation provides a physical barrier against the harsh environment factors to reduce the inevitable losses in processing, storage, and digestion.

Microcapsules are one of the most effective methods for protecting LAB cells encapsulated in a suitable film-forming material. This method can reduce the influence of phage and adverse environmental factors on the cells; improve the cell survival rate during lyophilization, freezing, and storage; and help the starter in a uniform and convenient powder form.

According to the processing method, LAB microcapsules can be divided into extrusion technology (Fig. 5.4), emulsification technology (Fig. 5.4), and aerosol technology (Fig. 5.5).

Extrusion technology is a widely used method for preparing microcapsules. The basic principle is to drip a colloidal solution (e.g., alginic acid) containing LAB cells into a cross-linking agent (e.g., $CaCl_2$) to form gel particles. Droplets are formed in the following manner: electrostatic atomization, spinning, rotating disk

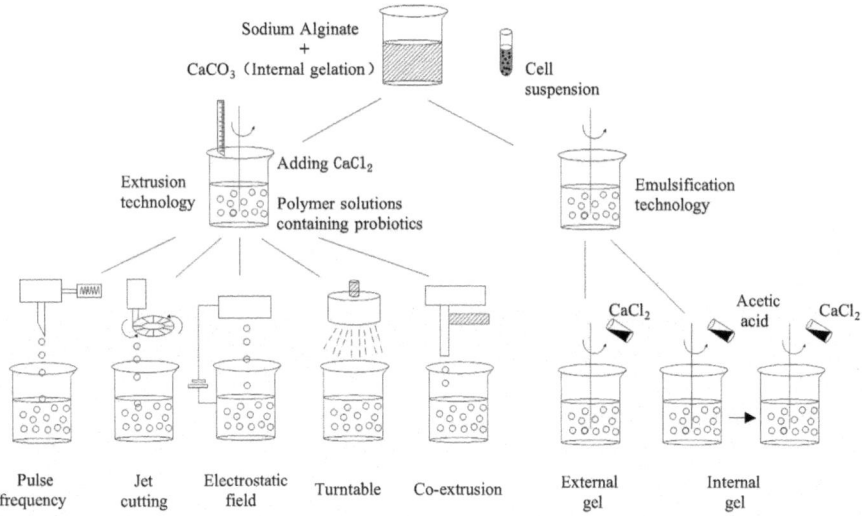

Fig. 5.4 Schematic diagram of extrusion and emulsification technology of *Lactobacillus* microcapsules (Martin et al. 2015)

Fig. 5.5 Schematic of the double aerosol technique for encapsulation of probiotics in alginate microbeads

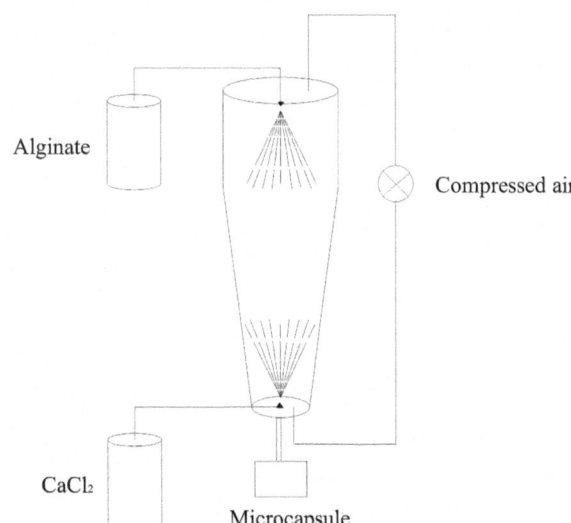

atomization, jet cutter, and etc. Extrusion technology can prepare microcapsules with large (5 mm) and small particle size (200 μm or less), which affects food sensory. This method is uncontinuous and difficult for large-scale production.

The basic principle of the emulsification technology is to disperse the colloidal and cell solution in vegetable oil and then add an ionic solution to induce gel formation. However, the process of removing the oil after gel formation requires the introduction of an additional separation step, resulting in a time-consuming and uncontinuously process.

The aerosol technique is used to prepare microcapsules by using two separate aerosols. In this method, a millimeter-size alginate solution containing microorganism was dropped into the cross-linking $CaCl_2$ bath, in order to produce alginate microbeads with a diameter of 10–40 μm, which is further dried by spray drying or freeze-drying.

The advantage of the aerosol technique is that it can be suitable for encapsulating probiotic bacterial cultures with high productivity since it does not need to heat in processing. Both macrobeads produced by extrusion method (approximately 2 mm diameter) and microbeads produced by aerosols technique can protect *L. rhamnosus* GG cells in high acid and bile environment well (Sohail et al. 2011).

Another classification is according to the wall material, which is as follows:

It is a great challenge to choose encapsulated materials for encapsulation technology. The materials should be food grade and have the function of embedding and protecting probiotics. Food-grade natural polymers are perfect materials for encapsulation, nontoxic and harmless, with good biocompatibility and the ability to form gels.

5.3.3.1 Alginate and Its Mixtures

Alginate is a linear heteropolysaccharide, which has two basic structural units including D-mannuronic (M) and L-guluronic acids (G). Alginate is extracted from different kinds of algae, which leads to the widely various difference in monosaccharide composition and sequence. In the same way, the different monosaccharide composition and sequence of M-units and G-units strongly lead to different functional properties of alginate as supporting material. In this sense, when two G-units are aligned side by side, a hole which could bind selectively divalent cations will be formed.

The mixture of bacterial cell suspension and sodium alginate solution is dripped into a solution containing a multivalent cations (usually Ca^{2+} in the form of $CaCl_2$). These droplets will form gel spheres instantaneously, entrapping the cells in a three-dimensional structure. The reason is that a polymer cross-linking occurs after the exchange of sodium ions from the guluronic acids with divalent cations, resulting in a chain-chain association that constitutes the so-called egg-box model.

The viscosity of the sodium alginate solution and the distance between the syringe and the calcium bath were the determined factors for the size and shape of the spheric bead. The viscosity of the gel can be increased by rising the concentration while the particle size of beads are decreased. In addition, another important factor regulating droplet size is the extruder orifice diameter. And the particle size of beads was also influenced by the composition of the alginate, which was certificated by the fact that low guluronic alginates can result in the formation of small beads.

Though calcium alginate has been widely used for the encapsulation of probiotic bacteria, alginate microparticles were found having some drawbacks. For instance, the susceptibility to acidic environments caused the loss of the mechanical stability in the solutions containing monovalent ions or chelating agents. And the fast diffusion of moisture and other fluids across the beads can be augmented by the porous structure of the particles. Because of this fact, the barrier properties were reduced against detrimental and unfavorable factors. By mixing other polymer compounds with alginate, coating substances with different characteristics, and adding modification on the structure of alginate, the defects mentioned above can be overcome methodically.

5.3.3.1.1   Starch

Starch is a polysaccharide synthesized by plants, which is composed of α-D-glucose units linked by glycosidic bonds. In the small intestine, resistant starch (RS) cannot be digested by pancreatic enzymes (e.g., amylases). Because of this property, RS can be used as an ideal wall material to encapsulate probiotic cells into starch granules. In a mixture of alginate and resistant starch as wall materials, the prepared microcapsules containing high cell viability can promote the delivery of probiotic cells into the intestine in viable and a metabolically active state.

5.3.3.1.2   Chitosan

Chitosan is obtained by deacetylation of chitin extracted from crustacean shells. It is a linear polysaccharide with positive charge and can be dissolved in water when pH below 6. Chitosan can form a gel by ionotropic gelation like alginate. The inhibitory effects of chitosan were also exhibited on the different species of LAB, so that it is not suitable for wall materials and preferred as a coating material.

By coating semipermeable chitosan layers on the microcapsules, the physical and chemical stabilities of alginate particles can also be improved. The deteriorative effects of calcium chelating and antigelling agents can be reduced by the structure of alginate particles. As the structure of the beads is getting more denser and much stronger, the breaking of microcapsules and releasing of cells can be avoided.

Low molecular weight chitosan is preferred rather than high molecular weight chitosan, since it diffuses faster into the alginate matrix, resulting in the formation of spheres with higher density and strength. Probiotic cells (*S. cerevisiae* Y235) were entrapped in alginate-chitosan microcapsules by emulsification-internal gelation technique. The entrapped low-density cells (5×10^6 cfu/mL) were then cultured to reach as high cell density as that of directly entrapped high-density cells (10^9 cfu/mL). After being freeze-dried without cryoprotectant, the survival rate of entrapped low-density cells was significantly higher (Song et al. 2014). The secreted macromolecules (e.g., proteins and exopolysaccharides) decreased substance permeability and increased density of capsule wall due to cell growth and metabolism and thus affected cell aggregation pattern by the changes of cell membrane composition, which might contribute to the resistance improvement of cultured cells under harsh simulated gastric fluid condition.

5.3.3.1.3 Polyamino Acids

Another coating material is polyamino acids. Poly-L-lysine (PLL) can be mixed with alginate matrix, which forms strong complexes and achieves the advantages mentioned previously for chitosan. It is also investigated in the generation mechanism of multilayer shells of PLL on the capsules constructed with alginate. After repeating the procedure for several times, the first layer of PLL produces positive charge with the negative charge given by the second alginate layer on the surface of the particle. At the same time, polycationic polymers, polyethylenimine, and glutaraldehyde can be used as alternatives.

5.3.3.2 Xanthan Gum and Its Mixtures

Besides alginate, xanthan (a heteropolysaccharide) is the most commonly used gum. It is consisted of poly-pentasaccharide groups containing two glucose, two mannose, and one glucuronic unit. And the polymer backbone of xanthan is formed by 1→4 linked β-(D)-glucose units (Cook et al. 2012).

Emulsion method was used to make microcapsules. In this progress, the discrete water phase containing xanthan gum was cross-linked with calcium chloride while suspended in oil. Compared to alginate, the microcapsules can protect cells from bile and acid (at pH 2). In this sense, the microcapsules were the same with alginate in cross-linking with calcium ions. To produce microcapsules which are stable in acid environment, xanthan can also be used to combine with gellan. In the process of microencapsulation, the survivals of *B. infantis* and *B. lactis* in simulated gastric juice (SGJ) were vastly improved. The study showed that the number of viable cells decreased by only 0.67 log cells/mL compared to that of non-encapsulated *B. infantis* decreased from 10 log cells/mL to 0 over half an hour. Xanthan was also used as a surfactant in an emulsion of the *Pediococcus acidilactici* encapsulation. The procedure of encapsulation improved the viability of cells in SGJ, and the releasing can be achieved in simulated intestinal juice, which can promote a possibility for effective delivery in the intestine.

5.3.3.3 Carrageenan and Its Mixtures

κ-Carrageenan is a neutral polysaccharide in additive of food commonly, and it can be extracted from marine macroalgae. It is necessary to keep the temperatures range from 60 to 90 ° C for carrageenan to dissolve effectively at high concentrations such as 2–5%. After adding probiotics at 40–45 ° C to the solution of polymer, gelation occurs with the cooling to room temperature. In order to prevent swelling and induce gelation, K^+ ions (in the form of KCl) are utilized to stabilize the gel. It has been reported that KCl has inhibitory effect on sorts of LAB. Because of this reason, Rb^+, Cs^+, and NH^{4+} ions can be used as alternatives to KCl. Compared to K^+ ions, these ions produce stronger gel beads.

The proportion of carrageenan and locust gum are recommended to be 1:2 in order to give a strong gel for microencapsulation in some reports. Because of its low susceptibility to the organic acids, the mixture can ensure the high efficiency of lactic acid fermentation. As a result, the mixture has been widely used in microencapsulating for probiotic cells. However, gel formation of κ-carrageenan and locust bean is dependent on calcium ions, which have adverse effects in the viability of *Bifidobacterium* spp.

5.3.3.4 Sodium Carboxymethyl Cellulose

Sodium carboxymethyl cellulose (NaCMC) and its derivative are cellulose which can be dissolved in water. NaCMC contains linked glucopyranose residues of different levels of carboxymethyl substitution. Because of its abilities in resisting gastric acid and properties in dissolving to intestinal, NaCMC has been widely utilized in delivery of drugs and probiotics. Microcapsules were prepared using *Lactobacillus reuteri* suspension in NaCMC and rice bran. The cell-polymer suspension was cross-linked with aluminum chloride. This method improved the viability of *L. reuteri* after heat exposure (Chitprasert et al. 2012).

5.3.3.5 Gum Acacia

Gum acacia (or gum Arabic) containing three main components can protect cells under acidic conditions. A polysaccharide moiety is the largest component of gum acacia. It comprised a 1→4 linked galactose backbone, arabinose, and rhamnose branches and was terminated by a glucuronic acid monomer. A covalently attached arabinogalactan-protein complex was the second largest component. The proportion of the complex was about 10% of the total polymer. And a glycoprotein of 1% total weight is the smallest fraction. Spray-dried *L. casei* with gum acacia was reported to be a good protection to pig gastric fluids in vitro, and the survival of *Bifidobacterium* was increased in the drying process.

5.3.3.6 Pectin

Pectin is a heteropolysaccharide used as a food gelling agent and can be extracted from fruits. It was also used in the fields of medicines and dietary fiber. The microcapsules were prepared with pectin and coated with whey. It was determined that these microcapsules in the stomach and the small intestine can remain intact in the process. However, it is still unclear for the protective effect of the microcapsules of *Lactobacillus* after exposure to gastric conditions.

5.3.3.7 Cellulose Acetate Phthalate

Cellulose acetate phthalate (CAP) is insoluble in acid (pH $\leq$5) and soluble when the pH $\geq$6 in the existence of phthalate groups. This can be its advantage to provide an effective method to deliver large numbers of viable cells to the colon by preparing probiotic microencapsulation with CAP. It has been reported that preparing an emulsion with starch and oil with the addition of CAP can improve the viability of cells in simulated gastric fluid and in the process of spray drying.

5.3.3.8 Proteins and Its Mixtures

The structures of proteins are very large and complex, but proteins can be gelled by enzymatic cross-linking, heat controlled sol-gel transition, and chemical cross-linking. Proteins are important wall materials for the encapsulation of probiotics. In this way, proteins like casein, bovine serum albumin, and soy protein are widely used as commercial available proteins.

5.3.3.8.1 Gelatin

Gelatin is a protein derived by partial hydrolysis of collagen. It has a special structure and versatile functional properties, in this way, the viscosity of its solution is high, and the solution can form gel when cooling. It can react with anionic polysaccharides synergistically such as gellan gum. As the two polymers mentioned above both carry net negative charges and repel one another, they can be miscible at pH higher than six. However, if the pH is under gelatin's isoelectric point, the net charge on the gelatin changes to positive, resulting in the interaction with the gellan gum which is negatively charged. The gel structure formed by the interaction of gelation and toluene diisocyanate is tolerant against cracking and breaking as being applied to the encapsulation of *L. lactis* subsp. *cremoris*.

Genipin, a naturally occurring fruit extract, is used mainly for covalently binding amino groups in proteins. It can be cross-linked with *Bifidobacterium adolescentis* in gelatin to produce the encapsulation by the covalent binding of proteins. The time of dissolution increased with the concentration of genipin, to achieve the higher level of cross-linking. The rate of cell death was largely reduced in simulated gastric juice with the cross-linking of gelatin and genipin. Microcapsules coating malto-dextrin also provide better protection.

5.3.3.8.2   Lactoprotein

Since non-milk-based materials as embedding materials for LAB cells may be resisted or not allowed in some cases, lactoprotein is widely welcomed as wall material for LAB cells microcapsules.

Milk

Coating of milk microspheres with carrageenan-locust bean gum has good protective effect on bacterial cells, but the structure of the microcapsules is irregular, and the mechanical properties are poor (Shi et al. 2013a). After the cells were incubated in artificial gastric juice of pH 2.5 and bile salt of 1% for 2 h, the survival rates were 100% and 3.12%, respectively, for encapsulation of *L. bulgaricus* prepared by extrusion of milk, alginate, and cell mixture. The results showed that alginate-human-like collagen microspheres provided significant protection for probiotic (Shi et al. 2013b).

Casein

Casein as a water-resisting substrate (pH <6) can be used as a protectant for LAB cells when passing through the gastric fluid in some cases. The glutamine and lysine in casein were cross-linked by adding transglutaminase into the mixed solution of casein and cells to form a microencapsulation structure of LAB. Compared with non-encapsulated bacteria, the acid resistance of *Lactobacillus paracasei* and *Bifidobacterium lactis* encapsulated by this method was 20% (pH 2.5) higher. Chymosin can selectively hydrolyze κ-casein and cause the casein micelles in milk unstable and aggregate to form a gel structure after the non-covalent binding and cross-linking between the polypeptide chains. The protein in skim milk can also form microcapsule structure under the action of chymosin, and the survival rate of *B. lactis* under acidic condition (pH 2.5) can be increased from 0.1% (not encapsulated) to more than 10%. The microencapsulation of casein and pectin could significantly improve the survival rates of *L. acidophilus* and *B. lactis* at pH 1.

Whey Proteins

Whey liquid is a by-product of cheese production, in which there are lactose, minerals, vitamins, proteins, etc. Whey proteins isolated from whey are mixtures of globular protein, which are high-quality dietary protein source. Because of their interaction activity, whey proteins are usually utilized to interact with active molecules.

And whey proteins are used as encapsulating agent for probiotics due to their superior gelling and emulsification properties. Therefore, they can be used as wall materials for microencapsulation. Another advantage of whey protein as an encapsulation wall material of microcapsules is that it can be digested in vivo to produce bioactive peptides. The microcapsule of LAB cells was formed by spray drying after the whey protein, and milk fat was emulsified. Although the survival rate of the cells was related to the characteristics of the strain, the embedded cell survival rate in the simulated digestion experiment was improved. The *L. rhamnosus* GG entrapped by whey protein isolate (WPI) has a good ability to tolerate porcine gastric juice. It is confirmed by fluorescence activated cell-sorting system that the cells treated by gastric juice not only survive but also maintain vitality. The microcapsules with diameter of less than 8 μm can be prepared from the mixture of *B. animalis* subsp. *lactis* Bb12 and whey protein by electrospray or electrospinning technology. Microcapsules made of whey protein and pectin (whey protein as the wall material, pectin as the surface coating) can improve the acid tolerance of *L. rhamnosus*, but the introduction of whey protein seemed to reduce the protective effect of microcapsules, which were mainly attributed to the weak interaction and the high expansion capacity between the coating and the wall material.

5.3.3.8.3 Legume Proteins

Protein, isolated from bean, and alginate can form a dense gel structure to prepare microcapsules which are sensitive to acid. Also *B. adolescentis* shows excellent protection under synthetic stomach juice (Khan et al. 2013). Among legume proteins, pea proteins are increasing attracting interest due to their high nutritional value, digestibility, bioavailability, and long-term health benefits. Pea protein isolate (PPI) is usually utilized in food industry to enrich the protein content of products. It is a food-grade material which can encapsulate acid-sensitive probiotic. By encapsulating *L. casei* ATCC393 in PPI-alginate hydrogel capsules, the encapsulation rate of $(85.69 \pm 4.82)\%$ was obtained via extrusion technology.

Compared with the freeze-dried *L. casei* free cells, the capsules did not show any protective effect during freeze-drying. The freeze-dried encapsulated *L. casei* stored for 84 days ($-15\ ^\circ$C) displayed a survival rate of $(59.9 \pm 17.4)\%$, but the freeze-dried capsules were expected to have a weaker buffering effect than the fresh capsules in simulated gastric fluid (Xu et al. 2016). Due to its excellent functional attributes, nutritional importance, and low sensitization, chickpea protein can be used as a suitable wall material for probiotic microcapsules. The chickpea is dominated by two salt-soluble globulins-type storage proteins: legumin and vicilin. The chickpea protein-based alginate microcapsule can be prepared by emulsion-based encapsulation technique. The simulated gastric juice experiments show that the microcapsule has good protective effect for *B. adolescentis* with the diameter less than 100 μm. And there were no perceived adverse effects on the sensory attributes of this ingredient when incorporated in foods (Wang et al. 2014). The chickpea protein-alginate microcapsules prepared by extrusion technology can make *B. ado-*

lescentis tolerant to synthetic gastric juice and simulated intestinal fluid, but the larger particle size is not suitable for food industry (Klemmer et al. 2011).

5.3.4   Vacuum Freeze-Drying

Vacuum freeze-drying is the most satisfactory and convenient method for long-term preservation of cultures and preparation of LAB starter. By reducing the Aw of starter to less than 0.2, it can preserve bacterial cells at temperatures above zero for a long time, reduce the cost of transportation and preservation, and maximize the viability of cells. However, this method has the disadvantages of long time-consuming and high-energy consumption.

5.3.4.1   Production Process

Vacuum freeze-drying is mainly composed of three processes: prefreezing, primary drying, and secondary drying process. During the primary drying process, a quantity of solvent (generally water) is reduced with the substance solution prefrozen. After sublimation firstly and desorption in secondary drying process, the Aw cannot support biological or chemical reaction any longer (Reddy et al. 2009b).

The prefreezing process is the first and shortest step in the freeze-drying process, which controls the size, distribution, and morphology of ice crystals. Low pressure (vacuum) and heating are required for sublimating ice into vapor. At the same time, the input of heat should be controlled strictly in order to ensure the collapse temperature for product. Pressure gradient of vapor should be realized through trapping condensate by cold trap to ensure the removing of water from the sample. Usually a content of unfrozen water (20–30%) still remained after the sublimation of ice, and it is not conducive for freeze-dried products in the process of storage. Thus, the sample product was heated to the temperature above zero (20–30 °C) for the desorption of unfrozen water in order to reduce moisture content (<5% dry weight) and Aw (<0.2).

Factors such as temperature, final moisture content, and storage conditions (temperature, relative humidity, and air) can affect the stability of freeze-dried LAB cells. To ensure a high survival rate and vitality recovery rate of freeze-dried LAB, the factors mentioned above should be taken into consideration in the process of freeze-drying. The following procedures and relevant parameters are generally used for preparing LAB lyophilized powder under pilot test:

1. The bacterial cell suspension added with cryoprotectants was packed aseptically into the labeled aseptic flask (the filling depth of the suspension was about 0.5 cm). The glass flask was placed on the sterilized stainless-steel tray. The sterile stopper was half-filled into the glass flask, which facilitated the vapor evacuation during the subsequent sublimation and desorption process. And the thermocouple sensor is placed on the bottom of the glass flask.

2. The tray with the glass flask was placed on the shelf that has been precooled at 4 °C, and the shelf temperature was gradually lowered at a rate of 0.5 °C/min until the temperature reaches −50 °C. Maintain this temperature until the sample temperature is stable (about 2 h) to ensure that the sample is completely solidified.
3. After the sample is frozen, the condenser temperature was lowered to −65 °C, the vacuum pump switch was opened, and the chamber pressure was reduced to 20 Pa. Then the shelf temperature was gradually increased to −20 °C at a rate of 0.25 °C/min to start the sublimation phase of the ice.
4. The process parameters (e.g., temperature and pressure) were monitored in real time during the freeze-drying process to detect functional faults in time and determine the end point of sublimation. When the temperature of the product is significantly increased and the vapor pressure drops, it marks the end of the sublimation phase.
5. The secondary drying process was started after 3–6 h of sublimation. The shelf temperature was gradually raised to 25 °C at a rate of 0.25 °C/min. And the chamber pressure was reduced to the minimum.
6. After 10 h of desorption, dry air was injected to replace the vacuum state, and then the stopper is quickly plugged (if the stopper system is available, it can be carried out in the freeze-dryer), and the screwtop is covered.

5.3.4.2   Cell Injury Mechanism

Even freeze-drying products at very low temperatures cannot completely guarantee the survival rate of bacterial cells; the survival rate of different strain varies in the lyophilization and depends on the cell sensitivity to environmental stress. Cell damage during lyophilization is mainly caused by the formation of ice crystals, osmotic pressure-induced cell membrane damage, denaturation, and dehydration resulting in changes in intracellular hydrophilic macromolecular properties. Freezing leads to the formation of ice crystals and the concentration of cytoplasm solutes. Low-rate freezing is usually used with the addition of protectants (see Sect. 5.3.2 for details). Dehydration may lead to irreversible changes in cell membrane lipids and the structure of sensitive proteins with a serious reduction in cell viability. In addition, oxidation is also an important way of cell damage during drying and storage.

5.3.5   Spray Drying

5.3.5.1   Production Process

Spray drying is a process in which the material is emulsified or dispersed, homogenized, and then atomized into a drying box containing hot air with water rapidly evaporated. Spray drying can also be further classified according to the different feeding ways and air flow directions (Fig. 5.6).

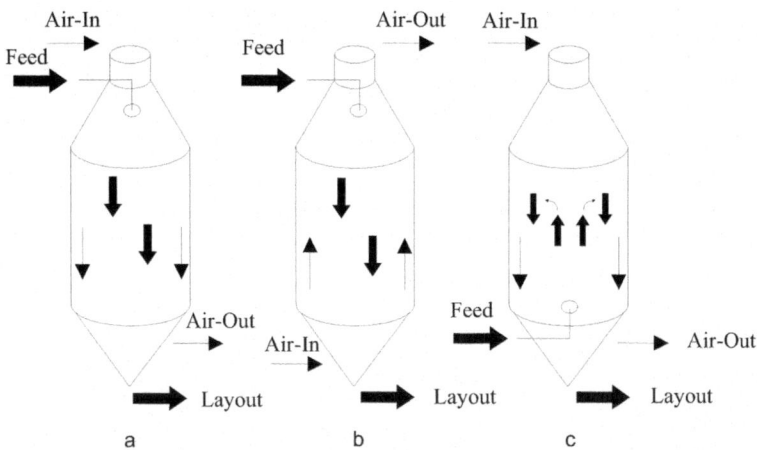

Fig. 5.6 Three different jet modes in spray drying. (**a**) advection; (**b**) countercurrent; (**c**) mixing

In the atomization process, the inlet temperature is usually as high as 220 °C. Due to the high water content and short contact time (about 0.1 min), it is generally considered that the temperature has little effect on the survival rate of the cells, which is mainly affected by the atomization pressure (Fu and Chen 2011). In addition, the higher atomization pressure results in a smaller droplet size and a longer residence time in the drying tower, thereby exacerbating thermal damage and dehydration damage of cells. In order to obtain higher survival rate in LAB starter prepared by spray drying, a lower spray pressure is needed to reduce the residence time of the cells in the drying tower.

Spray drying has high drying rate and continuous production capacity, and its energy consumption is about 1/10–1/6 of vacuum freeze-drying. It has been widely used in dairy, chemical, and pharmaceutical fields. The survival rate of LAB cells was too low (Ananta et al. 2005) due to direct contact with high-temperature hot air (120–180 °C), but there was also report that the survival rate is over 80% (Reddy et al. 2009b). The survival rates were related to the characteristics of cells and spray-drying parameters. In order to improve the survival rates during spray drying, the following strategies are usually employed:

1. Collect the cells in stationary phase. Bacteria in stationary phase usually have higher survival rates than that in the logarithmic growth phase.
2. Enhance the bacterial cell tolerance by heat shock or osmotic shock.
3. Improve the survival rate of the bacterial cells by protectants. Nonreducing sugars and other suitable solutes are effective protectants against dehydration damage. Reconstituted skim milk is also an excellent protectant in spray-drying process. For instance, trehalose and protein can increase the survival rate of *L. plantarum* TISTR2075 in spray-drying process to 57.70% and 25.31%, respectively (Lapsiri et al. 2012).

4. In spray-drying process, the higher outlet temperature will result in an increase in the lethality of the bacterial cells. At the same inlet air temperature, increasing the feed rate and lowering the outlet temperature can increase the survival rate of LAB. For instance, *Lactobacillus faecalis* CIDCA8348, *L. plantarum* CIDCA83114, and *S. lipolytica* CIDCA812 isolated from the kefir granules dried by spray drying, the cell viability decreased as the outlet temperature increased (70 °C, 75 °C, 80 °C, 85 °C); the highest survival rate achieved by *L. plantarum* is 7.94–10%, and the lowest heat resistance was *S. cerevisiae* (Golowczyc et al. 2010). When the outlet temperature was 65 °C, the survival rates of *L. lactis* D11 and *Lactobacillus pseudo-plantarum* UL137 were $2.95 \pm 0.07\%$ and $14.7 \pm 0.3\%$, respectively (To and Etzel 1997).

Although the survival rates of cells are associated with the characteristics of cells, researchers have been aimed to improve the survival rate of cells during spray drying. Some strains can obtain high survival rate in the laboratory, but for most strains, the problem of low viability in spray drying has been unsolved yet. Development of new spray-drying technology with lower drying temperature and more moderate processing are effective ways to improve the survival rate of cells, such as two-step spray drying. It was considered that this method (inlet temperature 80 °C, outlet temperature 48 °C, material temperature 45 °C) is an alternative to vacuum freeze-drying for producing probiotic powder, as the cost is 25% of vacuum freeze-drying (Chávez and Ledeboer 2007).

5.3.5.2 Cell Injury Mechanism

In spray drying, the lethal factors of LAB cells include thermal injury, osmotic stress, and capillary force induced by dehydration, oxidative stress induced by oxygen contact, and the decrease of intracellular pH and the increase of salt concentration caused by water loss. It is generally considered that dehydration and thermal damage are the two main factors responsible for cell death. In the case of the droplet temperature higher than 65 °C, the synergistic effect of dehydration and thermal damage resulted in the cell inactivation. When the droplet temperature reduced to 55 °C, the decrease of dehydration rate improved the survival rate of the cells (Ghandi et al. 2012).

5.3.5.2.1 Dehydration Damage

Aw is the primary condition for maintaining cell structure, stability, and intracellular biochemical reactions. The cell damage caused by dehydration in spray drying is the main injury of cytoplasm and membrane.

Damage on cytoplasm: The water loss leads to an increase in solution concentration and viscosity. When the viscosity increases to 10^{14} Pa·s, the solution begins to undergo a glass transition. The glassy state is a non-fixed, non-long-range ordered metastable state which has the characteristics of solid state, but the arrangement of molecules is more random than that of solid state. The glass state avoids destructive reactions, changes in molecular structure, and loss of chemical composition. The Tg is used to describe the vitrification state and refers to the temperature at which the amorphous system changes from a glassy state to a rubbery state or a rubbery state to a glassy state. The Tg has a certain correlation with the water, intracellular sugar, and protein content in the cytoplasm. Generally, the Tg increases with the decrease of water content and the increase of small molecules (e.g., polysaccharide or protein) in the cytoplasm.

Damage on cell membrane: The changes of cell membrane fluidity during the process of dehydration are the key factors of cell death. It has been conformed that LAB cell membrane is injured and membrane permeability is changed in spray drying, resulting in cells being very sensitive to the external environment (especially high concentrations of NaCl) and the loss of intracellular substances. The lipid bilayer is the basic structure of the cell membrane, and its stability mainly depends on van der Waals force and repulsive force. When water is lost (moisture content is less than 0.5 g/g dry weight), the van der Waals force between the carbon chains of the lipid molecules is rearranged and causes the cell membrane to change from the liquid crystal to the gel state. For LAB starter, the moisture content needed to maintain normal physiological activities of cells is about 0.5 g/g dry weight, whereas for long shelf life, moisture content needed is to be below 0.1 g/g dry weight (Aguilera and Karel 1997). When the water content is lower than the monolayer water content, the bacterial cells begin to be unstable or inactivated. For example, when the water content of **B**. *lactis* is less than 6%, the cell viability is less than 0.01%. Membrane phase transition temperature (Tm) is defined as the midpoint of the phase transition temperature of the cell membrane, and Tm decreases with the increase of water content. In the absence of protectant, the Tm of **L**. *bulgaricus* increases from 35 to 40 °C after spray drying. In addition, the cell membrane may undergo a transition from a lipid bilayer to inverted hexagonal phases during drying. The function of cell membrane and membrane proteins depends on its fluidity. The removal of water leads to a decrease in membrane fluidity and cell external moisture, resulting in the damage or functional changes of the membrane enzymes and proteins.

5.3.5.2.2 Thermal Damage

Cell death caused by thermal damage is mainly caused by damage or inactivation of cell membrane and intracellular thermosensitive substances (DNA, RNA, proteins or enzymes and ribosomes, etc.). The lethal mechanism of cell membrane during thermal injury is different from that of dehydration, as cell membrane permeability changes in dehydration. However, lipid oxidation is also one of the injury mechanisms of cell membrane in spray drying. The upregulated expression of genes in **S**. *cerevisiae* by heat-drying was mainly related to the redox balance system. The increase of glutathione content and lipid oxidative damage also indicated that oxidative stress was significant. During respiration, part of oxygen is converted into reactive oxygen species (ROS), which can be eliminated by normal cells through their own regulatory system. The increase of cytoplasmic concentration, the changes of pH and membrane fluidity result in the failure of dehydrated cells to remove ROS through normal chemical reactions. ROS can induce peroxidation of cell membranes and decomposition of phospholipids (França et al. 2007). For example, the content of unsaturated fatty acids in the cell membrane of *L. bulgaricus* after spray drying decreased from 13.29% to 6.84% (Teixeira et al. 1996).

Identification of key denatured substances is one way to discover the mechanism of the heat lethality of LAB during convective heat transfer and drying processes such as spray drying. It has been confirmed that ribosome denaturation responsible for gene transcription and protein synthesis is one of the important reasons for LAB inactivation in spray drying. Whether there are other important substances related to cell thermal death needs further confirmation. The factors affecting the survival rates of LAB cells include the inlet air temperature, the outlet air temperature, the retention time, and the dehydration rate. Among them, the outlet air temperature is the most influential factor.

5.3.5.3   Rehydration After Drying

The rehydration of the LAB starter after drying can revive the cells. The process includes four steps, wetting, submersion, dispersion, and dissolving, in which the wetting of starter particles is the control step. The proportion of cells restored to viable state was significantly affected by rehydration conditions (temperature, volume of rehydration medium, and rehydration time), physical and chemical properties of rehydrated substances, osmotic pressure, pH, and nutrients of rehydration solution.

Rehydration temperature is a key factor that significantly affects cell revival rate of lyophilized and spray-dried LAB starter cultures. Many studies have demonstrated that there is no optimal single-point rehydration temperature, and in any case, the temperature should not exceed 30 °C. The optimal rehydration temperature for thermophilic strains is 30–37 °C, and that of mesophilic strains is 22–30 °C.

The ratio of the dried starter added to the rehydration medium and its composition will significantly affect the revival rate of the strain. When the starter powder is added to the rehydration medium at 1:3, the revival rate of the cells is four to ten times higher than that of 1:50. In the case of lyophilized protectants (e.g., skim milk, peptone, tryptone meat extract, etc.) that were in the rehydration solution, the cells resurrection rate was significantly higher than that of the rehydration solution composed of phosphate and sodium glutamate, as the former solution could control the rehydration rate and avoid cell damage caused by osmotic stress. The cell revival rate in rehydration also depends on the characteristics of the strains, so it is necessary to standardize the rehydration process for different strains and products.

Table 5.4 Comparison of advantages and disadvantages of new drying technology

Drying technology	Advantages	Disadvantages
Low-temperature vacuum drying	Low equipment investment and production energy consumption, about half of the vacuum freeze-drying	High moisture content of bacteria powder, not conducive to preservation, uncontinuous production
Low-temperature vacuum spray drying	Lower drying temperature	High moisture content of bacterial powder, low survival rate
spray-chilling drying	The lowest cost of microcapsule preparation	Low microencapsulation rate, easy leakage of core material
Spray freeze-drying	Higher cell survival rate	Higher-energy consumption, longer production time
Fluidized bed drying	Lower-energy consumption, Higher storage stability	Batch and uncontinuous production

5.3.6   Other Drying Technologies

Due to the shortcomings of spray drying and vacuum freeze-drying in the preparation of starter, new drying technologies have been developed in recent years, including low-temperature vacuum drying, low-temperature vacuum spray drying, spray-chilling drying, spray freeze-drying, fluidized bed drying, etc. The advantages and disadvantages of the different technologies are shown in Table 5.4.

5.3.6.1   Low-Temperature Vacuum Drying

Vacuum drying can lower boiling point of water through vacuum negative pressure condition; conventional vacuum drying (temperature are 30–80 °C) is not suitable to produce LAB starter cultures. The temperature of sample can be near 0 °C through further improving vacuum, and this method is called low-temperature vacuum drying (Foerst et al. 2012). The investment and energy consumption of equipment for low-temperature vacuum drying is half of vacuum freezing drying, which have been used to dry thermosensitive food component and microbe. This method can avoid heat damage to cells, and the damage mechanism is dehydration. The high residual moisture content is the disadvantage of this method, which is harmful for the preservation of bacterial cells (Santivarangkna et al. 2007). When *L. paracasei* F19 was vacuum-dried by this method (1500 Pa, 15 °C) without protectants, the survival rate of cells was 29±4%, and *Aw* was 0.25. In the presence of sorbitol as protectant, the survival rate of cells was 54±6%, and *Aw* was 0.43; as in the case of trehalose as protectant, the survival rate of cells was 70±7%, and *Aw* was 0.47 (Foerst et al. 2012).

The survival rates of *L. paracasei* F19, *B. lactis* Bb12, and *L. delbrueckii* TMW1.1377 could achieve to 50.6%, 1.1%, and 0.03%, respectively, after the process of low-temperature vacuum drying, and the final moisture content of freeze-dry powder was about 6–7%.

The explanation for the varied survival rates in strains may be the difference in cell membrane component and death mechanism. The survival rates of *L. paracasei* F19 cells had no significant difference under low-temperature vacuum drying and freeze-drying. However, the survival rate of *B. lactis* Bb12 cells under low-temperature vacuum drying was lower than that under freeze-drying, and the survival rate of *L. delbrueckii* TMW1.1377 cells under low-temperature vacuum drying was ten times higher than that under freeze-drying. Generally, the production of low-temperature vacuum drying is discontinuous, and the production period is 10–20 h, which is a limitation for its further application in LAB starter culture preparation (Gong et al. 2014).

5.3.6.2 Low-Temperature Spray Drying

5.3.6.2.1 Low-Temperature Vacuum Spray Drying

To reduce the heat stress and oxidative stress during spray drying, low-temperature vacuum drying makes some modification on the basis of spray drying, such as ultrasonic spray nozzle, heating methods, and vacuum in drying tower, to achieve vacuum drying at low temperature. This drying method has some difference with conventional spray drying; the schematic diagram of equipment is shown in Fig. 5.7 (Kitamura et al. 2009). Heater is consisted of far-infrared heater and warm water heater. Far-infrared heater can efficiently heat the air and feeding materials, and warm water heater prevents the evaporated moisture from condensing on the surface of the inner glass wall of the drying tower. When this system was used to produce LAB starter cultures, the survival rate of cells was 25.7%, but moisture content was nearly 10.9% (Kitamura et al. 2009). The use of ultrasonic auxiliary feeding can enhance the cell survival rate to 70.6 ± 6.2%, and the *Aw* was 0.21 (Semyonov et al. 2011). Once the two problems (i.e., cell survival rate and moisture content) are solved, low-temperature vacuum spray drying will be widely applied in the production of LAB starter cultures.

Fig. 5.7 Schematic diagram of low-temperature vacuum spray dryer. (1) Nozzle; (2) far-infrared heater (35–120 °C); (3) drying tower; (4) water heater (35–70 °C); (5) water bath; (6) vacuum pump

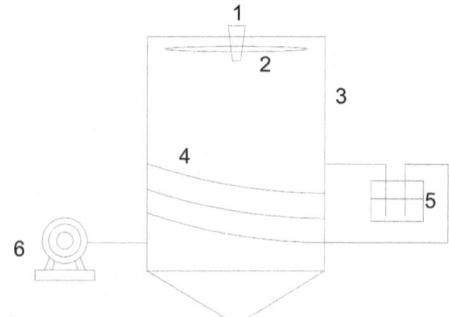

5.3.6.2.2 Spray-Chilling Drying

Spray-chilling drying has similar process to spray drying, but the mechanisms are different. Molten matrix containing LAB cells (usually using fat as supporter) is atomized to form drops which are quickly solidified to microcapsules when they are in contact to the cold air. Spray-chilling drying is considered to be the cheapest encapsulation technology that has the possibility of industrial scale manufacture. There are some disadvantages in this method, including low encapsulation capacity and expulsion of core material during storage. These shortages were caused by the arrangement characteristic of crystalline structure and polymorphic combination in the process of solidification and crystallization for many lipid materials. It is efficient in protecting the cells to pass through gastric and intestinal fluids by using wall material which combined interesterified fat with palm and palm kernel to produce microparticles containing *B. lactis* and *L. acidophilus* through spray-chilling technology. And the product can be stored at low temperatures (de Lara Pedroso et al. 2012).

5.3.6.2.3 Spray Freeze-Drying

Spray freeze-drying combine the process of freeze-drying and spray drying. The principle of this method is that the concentrated LAB cell suspension is sprayed into cold vapored liquid nitrogen (Amin et al. 2013) or into liquid nitrogen (Dolly et al. 2011), and the formed frozen and individual droplets will be further lyophilized or dried. Freezing by liquid nitrogen can maximally decrease cell damage, and the cell injury during this process mainly occurs in the drying process. Spray freeze-drying technique can overcome difficulties in long time-consuming and size control during vacuum freeze-drying and high temperature during spray drying.

In *Lactobacillus paracasei* F19 cells protected by trehalose and encapsulated by spray freeze-drying, the survival rate was over 60%, and water content can be decreased through fluidized bed drying (Semyonov et al. 2010). The technical advantages of this method are that (1) droplets size can be controlled to get larger specific surface area; (2) viability of bacterial cells is improved; and (3) microcapsules can be coated in later drying process (such as fluidized bed) and thus further improve the resistance ability of cells against external environment. However, disadvantages of this method are as follows: (1) it has higher-energy consumption, which is 30–50 times than spray drying; and (2) it needs long production period and continuous production cannot be achieved.

L. plantarum MTCC 5422 is prepared by spray freeze-drying and encapsulated with different wall materials such as (1) whey protein isolate (WPI) and sodium alginate (SA), (2) WPI and fructooligosaccharide (FOS), (3) denatured WPI (DWPI) and SA, and (4) DWPI and FOS. The prepared microcapsules were spherical in shape and exhibited good flowability and low hygroscopicity. Microencapsulation by spray freeze-drying did not affect the cell viability as indicated by embedding rate (87.92–94.86%). Among the different wall materials, DWPI combined with FOS had higher encapsulation efficiency and better stability during storage. This technique can overcome long drying time and lack of control on the microcapsules size and shape. Hence, it is believed that spray freeze-drying can be a suitable technique for probiotic microencapsulation (Rajam and Anandharamakrishnan 2015).

5.3.6.3 Fluidized Bed Drying

Fluidized bed drying based on fluid technique has been used to dry solid granules. During the process, granules were in the state of being divided, flowed, moved up and down, mixed, and impacted. Compared with spray drying, fluidized bed drying adopts more moderate temperature and can reduce the cell inactivation in maximum degree, so this method has been applied in the production of bakers' yeast and wine yeast. However, this method has hitherto rarely been used in the fields for producing LAB starter cultures. Other reports showed that fluidized bed dryer was usually used to coat LAB cells. Suspension of probiotic cells and absorbers or matrix molecules were blended to get the mixture in vibrating fluidized bed for the production of capsules forming by adherence. Moderately hot air flow was supplied to offer a requirement environment to dry off the aqueous medium from the bottom. After this process, the desired level of Aw can be achieved in the powdered matrix. Comparatively, this process is economic, but being a batch process is the main limitation.

When cellulose granules containing LAB was coated through fluidized bed technique (outlet temperature was 45 °C, fluidized temperature was 30 °C), the survival rate of LAB cells was about $(11.0 \pm 6.4)\%$ and moisture content was 3.6–3.8% (Strasser et al. 2009). Recently, it is reported that fluidized bed drying can improve stability of probiotics. Compared with freeze-drying, the survival rate of *L. casei* CRL 431 dried by fluidized bed drying was 2.5 log cfu/g higher after 52 weeks of storage at 25 °C. Osmotic stress treatment and fortification with vitamin E can further improve the survival of *L. casei* CRL 431 during storage. However, inulin did not enhance the survival of bacterial cells. Similar results were also observed in *L. acidophilus* ATCC4356, *L. rhamnosus* ATCC53103, *B. lactis* Bb12, and *B. lactis* HN019; the storage stabilities were higher than those of lyophilized cells (Nag and Das 2013). There are three methods that can be used in a whole milk powder matrix for drying *Lactobacillus paracasei* subsp. *paracasei* CRL 431, i.e., spray drying, freeze-drying, and fluidized bed drying. Among them, greater protection can be achieved by fluidized bed drying in the same conditions for cell viability. The reason of this phenomenon is the larger agglomerates of fluidized bed drying technology with lower porosity which can attribute to the low moisture absorption. A lower bacterial death rate can be achieved with the lower absorption of water, resulting in more rigid structure maintained and molecular transport limited for particularly oxygen. It makes producing probiotics powder possible to have adequate storage life in a milk powder matrix by fluidized bed drying. Commercial products can be obtained with further investigation and technology optimization. In this way, this technology could be brought to utilization for a broad range of probiotics in the future (Poddar et al. 2014).

After a two-step fluidized bed granulation and top-spray coating, *L. reuteri* DSM 20016 was granulated and dried by a top-spray fluidized bed at 50 °C with sweet whey; the microcapsules were coated with an aqueous shellac solution by top-spray fluidized bed granulation. The survival rate of cells was $43 \pm 4.5\%$ after granulation and was $22.69 \pm 2.41\%$ and $14.86 \pm 1.29\%$ after 15 min and 30 min shellac coating, respectively. After 28 days storage at 4 °C, the survival rates of cells between the coated and uncoated cells were comparable. However, the coated cells had higher tolerance in stimulated gastric acid and intestinal juice than the uncoated cells. After incubation, the final survival rates of coated cells were $76.74 \pm 24.36\%$ and $17.74 \pm 10.51\%$ for the uncoated cells (Schell and Beemann 2014). Therefore, combination of fluidized bed granulation, top-spray coating technique, and suitable microencapsulation material can embed LAB cells successfully, and this method can be used to produce LAB microcapsules.

5.3.7 Storage Stability of Dried LAB Starter Cultures

Generally, freeze-dried starter cultures usually contain at least a dose of 10^{11} cfu/g viable cells. However, viable cells in the starter cultures will decrease along with storage time. The cell survival rate in powder starter culture is mainly affected by water activity ($Aw > 0.25$), storage temperature, and oxygen content. The Aw (or moisture content) affecting storage stability is also influenced by relative humidity. When the relative humidity is lower than 11%, the LAB cells can keep glassy state (no liquid crystalline state) in long time, which was proved to facilitate the cell viability during storage. However, in a high relative humidity condition, water molecule was a plasticizer for carbohydrate acting as a protectant for cells, which enhances its membrane fluidity. The enhancement of membrane fluidity directly results in the crystallization of amorphous carbohydrate (no liquid crystalline state) and the loss of cell viability immediately (Miao et al. 2008). The reduction of Aw will increase the stability of cell viability during the storage. For instance, *Lactobacillus paracasei* subsp. *paracasei* LBC81 cells encapsulated by alginate system incorporating potato starch using fluidized bed drying were stored at 25 °C for 7 weeks, the survival rate of the cells decreased rapidly when Aw over 0.536, and the survival rate can reach around 70% (viable counts about 8.9×10^8 CFU/g) when the capsules were stored at Aw between 0.436 and 0.536 (Jiménez et al. 2015). The cells of *L. acidophilus* and *L. lactis* subsp. *cremoris* were encapsulated by emulsion method, using sodium alginate as wall material and glucose and mannitol as protectants. The microcapsules were dried by freeze-drying or spray-drying process ($T_{outlet} = 50$ °C) and stored in an aluminum foil pouch using NaOH ($Aw = 0.07$), LiCl ($Aw = 0.11$) and silica gel as desiccant for 10 weeks at 25 °C. At low Aw (0.07 and 0.11), the survival rate of freeze-dried cells (89–94%) was slightly higher than that of spray-dried (86–90%), and both of them were higher than that kept in foil pouch using silica gel as desiccant (81–87%). Furthermore, low Aw can facilitate the preservation of cell acid and bile tolerance, surface hydrophobicity, and β-galactosidase activity (Dianawati et al. 2013).

For LAB viability preservation, storage temperature is extremely important. The increase of temperature will result in various degradation reactions (e.g., Maillard reaction and oxidation reaction), sugar crystallization, and water release and thus cause the *Aw* to increase and cells to die (Velly et al. 2014). Generally, probiotics in low *Aw* products are stable when stored at 4 °C. However, the viable cells in the solid foodstuffs (e.g., milk powder, cereals, confectionery, and chocolate) containing probiotics will decrease as temperature increases and shelf life extends, which are commonly sold at ambient temperature. For instance, viable cells of **L.** *paracasei* subsp. *paracasei* CRL 431 decreased by 1.26, 1.45, and 1.8 log cfu/g, respectively, during 24 weeks of storage at 25, 30, and 37 °C (Nag and Das 2013).

5.4 Preparation Technology of Frozen LAB Starter

5.4.1 Frozen LAB Starter

The frozen LAB starter is prepared with concentrated cell suspension after high-density culture and concentration (e.g., centrifugation, membrane filtration, etc.), then frozen under the temperature below −70 °C with cryoprotectants, and cryopreserved. Cryogenic granulation is a manufacturing technique for preparing spherical or hemispherical particles by freezing liquid experimental materials with liquid nitrogen at extremely low temperatures. The diameter of particles prepared by cryogenic granulation is about 0.5–5 mm (Deb and Ahmed 2013). Cryogenic granulation technique is often used in the production of thermosensitive materials, such as bacterial cultures, probiotics, etc. Since low-temperature granulation can instantaneously and effectively freeze the material, the technique can prevent the phase separation of solution, suspension, and colloid during the freezing process to ensure the uniformity and stability of the material.

5.4.2 Selection of Cryoprotectants

In the process of freezing, intracellular ice crystals will form and cause mechanical damage to cell membrane, leading to cell rupture and apoptosis. Therefore, in order to ensure a higher survival rate and a longer preservation, the selection and proportion of cryoprotectants become the key factor to prepare the frozen starter.

According to the different penetration sites in cell, cryoprotectants can be divided into permeable, semipermeable, and impermeable cryoprotectants.

Permeable protectants can penetrate into both cell walls and cell membranes. Common protectants of this type are dimethyl sulfoxide (DMSO) and glycerol, e.g., glycerol can penetrate into the cell by forming hydrogen bonds between hydroxyl groups and intracellular macromolecules. In the case of dehydration, macromolecules such as proteins, carbohydrates, and fats in cells can maintain their original structure and prevent the formation of large amounts of ice crystals in the cells by these cryoprotectants.

Semipermeable protectants can only penetrate into cell walls but not into cell membranes, such as monosaccharides, disaccharides, and amino acids. Carbohydrate protectants have significant protective effects on cell during dehydration and freeze-drying. After osmosis, they combine with water molecules, increase the viscosity of solution, and slow down the formation of ice crystals. In addition, by increasing the concentration of cytoplasm and balancing transmembrane pressure of the cell, the cell damage of dehydration and shrinkage is reduced.

Impermeable protectants cannot penetrate into cell walls and cell membranes, such as skimmed milk powder, starch, yeast powder, and β-cyclodextrin. This type of cryoprotectants partially dehydrates cells by adsorbing to the cell surface to form a mucus layer. These protectants inhibit the growth of ice crystals by increasing the viscosity of the cytoplasm and the ice crystals around the cells in irregular shape.

Different cyroprotectants have different protective effects on different strains, and single protectant cannot satisfy the requirements of freeze-drying and bacterial resistance to adverse external conditions. Therefore, the combination of protectants is the most commonly used in bacterial cell cryopreservation. If the proportions and concentrations of protectants are coordinated, it will accelerate drying and maintain a high cell survival rate during drying and preservation. Generally, for prokaryotes, the commonly used cryopreservations are dimethyl sulfate, DMSO, and glycerin. For the cryopreservation of LAB strains, the most commonly used cryopreservation is the mixture of skim milk, trehalose and dextran or sucrose, and bovine serum albumin.

5.4.3 Liquid Nitrogen Freezing Technology

Most cryobiologists believe that cryogenic vitrification is the most promising method for the preparation and long-term preservation of biomaterials. The key to preventing the occurrence of ice crystals is to use the ultrafast freezing to make intracellular water rapidly pass through the dangerous temperature zone (−40 to −10 °C) of ice crystal formation and growth, which means that intracellular and extracellular water has solidified to a vitrification state before intracellular water form a large number of ice crystals, thus avoiding the harm of intracellular ice crystal. Therefore, according to the characteristic of liquid nitrogen, rapid freezing can achieve and form a uniform distribution of fine ice crystals or glassy state inside and outside the cell; the degree of cell structure damage will be greatly reduced. Because of the low temperature, the freezing rate is increased with the reducing of residual liquid. Meanwhile, the growth of the ice crystals is controlled, and the frozen cells can basically maintain the original properties.

As the rising of the frozen granular starter is prepared with liquid nitrogen, there are only a few equipment suppliers for this technology, such as Germany Engineering Alliance (GEA); American Novus International, Inc.; and so on. Although there are differences in the structure, materials, and production process, the equipment for preparing frozen starter with liquid nitrogen is generally composed of six important parts: feeding and mixing system, liquid nitrogen cooling system, nitrogen exhaust system, pelleting module, cleaning system, and hot air sterilization unit. In addition, the equipment can also be modified according to actual demand to meet the actual production needs.

Due to the recycling utilization of liquid nitrogen, there are special requirements for the components of pipeline, adapter, valve, and pump materials of each system. The pump needs to be equipped with a flexible regulator, which can not only conveniently deliver the prepared particles to the separator but also adjust the liquid nitrogen flow rate. This has an important effect on the properties and yield of the frozen starter particles. The frozen LAB starter is prepared with the steps as follows: the liquid nitrogen and the concentrated cell suspension are mixed and form granular particles through the pelleting system; the liquid nitrogen is separated from the starter particles in the separator; the gaseous nitrogen produced in the production process is discharged by the exhaust system, and the remaining liquid nitrogen is recycled; finally, the starter granules with diameter of 4–6 mm were sieved. And the diameter of the frozen starter particles can be adapted to the actual production needs.

References

Aghababaie M, Khanahmadi M, Beheshti M (2015) Developing a kinetic model for co-culture of yogurt starter bacteria growth in pH controlled batch fermentation. J Food Eng 166:72–79

Aguilera J, Karel M (1997) Preservation of biological materials under desiccation. Crit Rev Food Sci Nutr 37(3):287–309

Aguirre-Ezkauriatza E, Aguilar-Yáñez J, Ramírez-Medrano A et al (2010) Production of probiotic biomass (*Lactobacillus casei*) in goat milk whey: comparison of batch, continuous and fed-batch cultures. Bioresour Technol 101(8):2837–2844

Amin T, Thakur M, Jain S (2013) Microencapsulation-the future of probiotic cultures. J Microbiol Biotechnol Food Sci 3(1):35

Ananta E, Volkert M, Knorr D (2005) Cellular injuries and storage stability of spray-dried *Lactobacillus rhamnosus* GG. Int Dairy J 15(4):399–409

Basholli-Salihu M, Mueller M, Salar-Behzadi S et al (2014) Effect of lyoprotectants on β-glucosidase activity and viability of *Bifidobacterium infantis* after freeze-drying and storage in milk and low pH juices. LWT-Food Sc Technol 57(1):276–282

Beales N (2004) Adaptation of microorganisms to cold temperatures, weak acid preservatives, low pH, and osmotic stress: a review. Compr Rev Food Sci Food Saf 3(1):1–20

Bergenholtz ÅS, Wessman P, Wuttke A et al (2012) A case study on stress preconditioning of a *Lactobacillus* strain prior to freeze-drying. Cryobiology 64(3):152–159

Carminati D, Giraffa G, Zago M, et al. (2015) Lactic acid bacteria for dairy fermentations: specialized starter cultures to improve dairy products. In: Biotechnol lactic acid bacteria: Nov appl: 191

Chang HN, Kim NJ, Kang J et al (2011) Multi-stage high cell continuous fermentation for high productivity and titer. Bioprocess Biosyst Eng 34(4):419–431

Chang HN, Jung K, Lee JC et al (2014) Multi-stage continuous high cell density culture systems: a review. Biotechnol Adv 32(2):514–525

Chávez B, Ledeboer A (2007) Drying of probiotics: optimization of formulation and process to enhance storage survival. Dry Technol 25(7–8):1193–1201

Chitprasert P, Sudsai P, Rodklongtan A (2012) Aluminum carboxymethyl cellulose–rice bran microcapsules: enhancing survival of Lactobacillus reuteri KUB-AC5. Carbohydr Polym 90(1):78–86

Chun L, Li-bo L, Di S et al (2012) Response of osmotic adjustment of Lactobacillus bulgaricus to NaCl stress. J Northeast Agric Univ (Engl Ed) 19(4):66–74

Cook MT, Tzortzis G, Charalampopoulos D et al (2012) Microencapsulation of probiotics for gastrointestinal delivery. J Control Release 162(1):56–67

de Lara Pedroso D, Thomazini M, Heinemann RJB et al (2012) Protection of Bifidobacterium lactis and Lactobacillus acidophilus by microencapsulation using spray-chilling. Int Dairy J 26(2):127–132

de Souza Oliveira RP, Torres BR, Perego P et al (2012) Co-metabolic models of Streptococcus thermophilus in co-culture with Lactobacillus bulgaricus or Lactobacillus acidophilus. Biochem Eng J 62:62–69

Deb R, Ahmed AB (2013) Pellets and pelletization techniques: a critical review. Int Res J Pharm 4(4):90–95

Dianawati D, Mishra V, Shah NP (2013) Stability of microencapsulated Lactobacillus acidophilus and Lactococcus lactis ssp. cremoris during storage at room temperature at low a_w. Food Res Int 50(1):259–265

Dolly P, Anishaparvin A, Joseph G et al (2011) Microencapsulation of Lactobacillus plantarum (mtcc 5422) by spray-freeze-drying method and evaluation of survival in simulated gastrointestinal conditions. J Microencapsul 28(6):568–574

Foerst P, Kulozik U, Schmitt M et al (2012) Storage stability of vacuum-dried probiotic bacterium Lactobacillus paracasei F19. Food Bioprod Process 90(2):295–300

França M, Panek A, Eleutherio E (2007) Oxidative stress and its effects during dehydration. Comp Biochem Physiol A Mol Integr Physiol 146(4):621–631

Fu N, Chen XD (2011) Towards a maximal cell survival in convective thermal drying processes. Food Res Int 44(5):1127–1149

Gautier J, Passot S, Pénicaud C et al (2013) A low membrane lipid phase transition temperature is associated with a high cryotolerance of Lactobacillus delbrueckii subspecies bulgaricus CFL1. J Dairy Sci 96(9):5591–5602

Ghandi A, Powell I, Chen XD et al (2012) Drying kinetics and survival studies of dairy fermentation bacteria in convective air drying environment using single droplet drying. J Food Eng 110(3):405–417

Golowczyc M, Silva J, Abraham A et al (2010) Preservation of probiotic strains isolated from kefir by spray drying. Lett Appl Microbiol 50(1):7–12

Gong P, Zhang L, Han X et al (2014) Injury mechanisms of lactic acid bacteria starter cultures during spray drying: a review. Dry Technol 32(7):793–800

Jiménez M, Flores-Andrade E, Pascual-Pineda LA et al (2015) Effect of water activity on the stability of Lactobacillus paracasei capsules. LWT-Food Sci Technol 60(1):346–351

Jung I, Lovitt RW (2010) A comparative study of the growth of lactic acid bacteria in a pilot scale membrane bioreactor. J Chem Technol Biotechnol 85(9):1250–1259

Khan NH, Korber DR, Low NH et al (2013) Development of extrusion-based legume protein isolate–alginate capsules for the protection and delivery of the acid sensitive probiotic, Bifidobacterium adolescentis. Food Res Int 54(1):730–737

Kitamura Y, Itoh H, Echizen H et al (2009) Experimental vacuum spray drying of probiotic foods included with lactic acid bacteria. J Food Process Preserv 33(6):714–726

Klemmer KJ, Korber DR, Low NH et al (2011) Pea protein-based capsules for probiotic and prebiotic delivery. Int J Food Sci Technol 46(11):2248–2256

Krzywonos M, Eberhard T (2011) High density process to cultivate Lactobacillus plantarum biomass using wheat stillage and sugar beet molasses. Electron J Biotechnol 14(2):6–6

Lapsiri W, Bhandari B, Wanchaitanawong P (2012) Viability of *Lactobacillus plantarum* TISTR 2075 in different protectants during spray drying and storage. Dry Technol 30(13):1407–1412

Lee JC, Kim DY, Oh DJ et al (2008) Long-term operation of depth filter perfusion systems (DFPS) for monoclonal antibody production using recombinant CHO cells: effect of temperature, pH, and dissolved oxygen. Biotechnol Bioprocess Eng 13(4):401–409

Leroy F, Vuyst L (2004) Lactic acid bacteria as functional starter cultures for the food fermentation industry. Trends Food Sci Technol 15:67–78

Ma C, Ma A, Gong G et al (2015) Cracking *Streptococcus thermophilus* to stimulate the growth of the probiotic *Lactobacillus casei* in co-culture. Int J Food Microbiol 210:42–46

Martín MJ, Lara-Villoslada F, Ruiz MA et al (2015) Microencapsulation of bacteria: a review of different technologies and their impact on the probiotic effects. Innov Food Sci Emerg Technol 27:15–25

Miao S, Mills S, Stanton C et al (2008) Effect of disaccharides on survival during storage of freeze dried probiotics. Dairy Sci Technol 88(1):19–30

Nag A, Das S (2013) Improving ambient temperature stability of probiotics with stress adaptation and fluidized bed drying. J Funct Foods 5(1):170–177

Poddar D, Das S, Jones G et al (2014) Stability of probiotic *Lactobacillus paracasei* during storage as affected by the drying method. Int Dairy J 39(1):1–7

Rajam R, Anandharamakrishnan C (2015) Spray freeze drying method for microencapsulation of *Lactobacillus plantarum*. J Food Eng 166:95–103

Ramchandran L, Sanciolo P, Vasiljevic T et al (2012) Improving cell yield and lactic acid production of *Lactococcus lactis* ssp. *cremoris* by a novel submerged membrane fermentation process. J Membr Sci 403:179–187

Reddy KBPK, Awasthi SP, Madhu AN et al (2009a) Role of cryoprotectants on the viability and functional properties of probiotic lactic acid bacteria during freeze drying. Food Biotechnol 23(3):243–265

Reddy KBPK, Madhu AN, Prapulla SG (2009b) Comparative survival and evaluation of functional probiotic properties of spray-dried lactic acid bacteria. Int J Dairy Technol 62(2):240–248

Riesenberg D, Guthke R (1999) High-cell-density cultivation of microorganisms. Appl Microbiol Biotechnol 51(4):422–430

Santivarangkna C, Wenning M, Foerst P et al (2007) Damage of cell envelope of *Lactobacillus helveticus* during vacuum drying. J Appl Microbiol 102(3):748–756

Savini M, Cecchini C, Verdenelli NC et al (2010) Pilot-scale production and viability analysis of freeze-dried probiotic bacteria using different protective agents. Nutrients 2(3):330–339

Schell D, Beermann C (2014) Fluidized bed microencapsulation of Lactobacillus reuteri with sweet whey and shellac for improved acid resistance and in-vitro gastro-intestinal survival. Food Res Int 62:308–314

Semyonov D, Ramon O, Kaplun Z et al (2010) Microencapsulation of *Lactobacillus paracasei* by spray freeze drying. Food Res Int 43(1):193–202

Semyonov D, Ramon O, Shimoni E (2011) Using ultrasonic vacuum spray dryer to produce highly viable dry probiotics. LWT-Food Sci Technol 44(9):1844–1852

Serrazanetti DI, Guerzoni ME, Corsetti A et al (2009) Metabolic impact and potential exploitation of the stress reactions in lactobacilli. Food Microbiol 26(7):700–711

Shao Y, Gao S, Guo H et al (2014) Influence of culture conditions and preconditioning on survival of *Lactobacillus delbrueckii* subspecies *bulgaricus* ND02 during lyophilization. J Dairy Sci 97(3):1270–1280

Shi LE, Li ZH, Li DT et al (2013a) Encapsulation of probiotic *Lactobacillus bulgaricus* in alginate–milk microspheres and evaluation of the survival in simulated gastrointestinal conditions. J Food Eng 117(1):99–104

Shi LE, Li ZH, Zhang ZL et al (2013b) Encapsulation of *Lactobacillus bulgaricus* in carrageenan-locust bean gum coated milk microspheres with double layer structure. LWT-Food Sci Technol 54(1):147–151

Smid EJ, Lacroix C (2013) Microbe–microbe interactions in mixed culture food fermentations. Curr Opin Biotechnol 24(2):148–154

Sohail A, Turner MS, Coombes A et al (2011) Survivability of probiotics encapsulated in alginate gel microbeads using a novel impinging aerosols method. Int J Food Microbiol 145(1):162–168

Song H, Yu W, Liu X et al (2014) Improved probiotic viability in stress environments with post-culture of alginate–chitosan microencapsulated low density cells. Carbohydr Polym 108:10–16

Strasser S, Neureiter M, Geppl M et al (2009) Influence of lyophilization, fluidized bed drying, addition of protectants, and storage on the viability of lactic acid bacteria. J Appl Microbiol 107(1):167–177

Sung IK, Han NS, Kim BS (2012) Co-production of biomass and metabolites by cell retention culture of *Leuconostoc citreum*. Bioprocess Biosyst Eng 35(5):715–720

Sunny-Roberts E, Knorr D (2009) The protective effect of monosodium glutamate on survival of *Lactobacillus rhamnosus* GG and *Lactobacillus rhamnosus* E-97800 (E800) strains during spray-drying and storage in trehalose-containing powders. Int Dairy J 19(4):209–214

Teixeira P, Castro H, Kirby R (1996) Evidence of membrane lipid oxidation of spray-dried *Lactobacillus bulgaricus* during storage. Lett Appl Microbiol 22(1):34–38

To B, Etzel MR (1997) Spray drying, freeze drying, or freezing of three different lactic acid bacteria species. J Food Sci 62(3):576–578

Tsakalidou E, Papadimitriou K (2011) Stress responses of lactic acid bacteria, Food Microbiology and Food Safety. Springer, New York, pp 23–53

Velly H, Fonseca F, Passot S et al (2014) Cell growth and resistance of *Lactococcus lactis* subsp. *lactis* TOMSC161 following freezing, drying and freeze-dried storage are differentially affected by fermentation conditions. J Appl Microbiol 117(3):729–740

Velly H, Bouix M, Passot S et al (2015) Cyclopropanation of unsaturated fatty acids and membrane rigidification improve the freeze-drying resistance of *Lactococcus lactis* subsp. *lactis* TOMSC161. Appl Microbiol Biotechnol 99(2):907–918

Wang J, Korber DR, Low NH et al (2014) Entrapment, survival and release of *Bifidobacterium adolescentis* within chickpea protein-based microcapsules. Food Res Int 55:20–27

Xu M, Gagné-Bourque F, Dumont MJ et al (2016) Encapsulation of *Lactobacillus casei* ATCC 393 cells and evaluation of their survival after freeze-drying, storage and under gastrointestinal conditions. J Food Eng 168:52–59

Zacharof MP, Lovitt RW (2013) Partially chemically defined liquid medium development for intensive propagation of industrial fermentation *lactobacilli strains*. Ann Microbiol 63(4):1235–1245

Zhang Y, Kumar A, Hardwidge P, Tanaka T et al (2016) D-lactic acid production from renewable lignocellulosic biomass via genetically modified *Lactobacillus plantarum*. Biotechnol Prog 32(2):271–278

Chapter 6
Lactic Acid Bacteria and Fermented Cereals

Bowen Yan and Hao Zhang

6.1 Introduction

6.1.1 Overview of Fermented Cereal Products

It is well-known that cereals are one of the traditional staple foods in many Asian countries. In China, the records of five cereals are reported as early as the Spring and Autumn and Warring States Period in the "Analects of Confucius," including rice, wheat, soybeans, corn, and potatoes. Starch is one of the most important components in cereal; the content accounts for 60% of the total cereals weight and 90% of the total carbohydrate content. In addition, cereals also contain a variety of proteins, lipids, cellulose, minerals, and enzymes, which not only meet the nutritional and metabolic needs of microorganisms but also provide a good substrate for microbial growth (Cho et al. 2013; Waters et al. 2015). Fermentation technology has been known and mastered for thousands of years. As one of the traditional staple foods in northern China (Zhu 2014), steamed bun originated in the Three Kingdoms Period; soy sauce, as an ancient condiment, has a long history of more than 1800 years; fermented bean curd also had historical records in the ancient books of the Wei Dynasty as early as the fifth century AD. However, due to the lack of understanding in fermentation and microorganism at that time, the development of fermented food was limited. With the development of science and technology, we have come to realize that cereal is the natural medium for microbial growth and reproduction. Microorganisms utilize the carbohydrates and amino acids of cereal for fermentation (Oguntoyinbo and Narbad 2015), under a series of physiological and biochemical reactions, which improves the quality, flavor, and nutrition of products. Furthermore, the fermented cereal food also has the effects of regulating

B. Yan · H. Zhang (✉)
Jiangnan University, Wuxi, China
e-mail: zhanghao@jiangnan.edu.cn

© Springer Nature Singapore Pte Ltd. and Science Press 2019
W. Chen (ed.), *Lactic Acid Bacteria*,
https://doi.org/10.1007/978-981-13-7283-4_6

human intestinal health, alleviating constipation, and absorbing heavy metals in the body (Akanbi and Agarry 2014; Brandt 2014; Zhao et al. 2015).

At present, the common fermented cereal-based food on the market can be classified into the following three types:

1) Fermented wheat-based food, such as steamed stuffed bun, steamed bread, bread, rolls, etc.
2) Fermented soybean products, such as soy sauce, black bean, sufu, natto, sour soybean milk, etc.
3) Fermented cereal beverage based on coarse cereal, such as oat fermented milk, barley fermented milk, corn fermented milk, etc.

6.1.2   History and Current Situation of Fermented Cereal Products in China

China has a long history about the cereal fermentation; people have mastered the fermentation skills of fermented cereal products for thousands of years, such as steamed bread, soy sauce, tofu milk, vinegar, etc. The traditional cereal fermented foods are used to natural fermentation, which is one of the oldest methods to extend the shelf life of food due to the complex microbiota. With the development of modern fermentation technology, it is found that the flavor of products is improved after fermentation based on the culture-dependent, gene mutagenesis, and artificially controlled fermentation technology. Therefore, it gradually evolves into a unique food processing method.

6.1.2.1   Fermented Wheat-Based Food

Steamed bread as one of the traditional staple foods in China is a typical wheat-based fermented food. It is used to be made of flour, water, and starter through a series of processes, including mixing, leavening, and steaming (Su 2005). The original steamed bread is called "pastry," which is a kind of unfermented wheat-based food. Until the Wei and Jin dynasties, people began to operate the fermentation skills and applied it to the processing of "pastry," which steamed bread come from this (Su 2009).

In China, about 70% of the wheat flour is used in the processing of steamed bread each year, and the annual consumption of steamed bread is more than 12 million tons. The industrialization of staple food (steamed bread) is not only the trend of market development in the future but also the choice of market consumers. However, the industrialization of steamed bread in China has slowed development; the hand workshop still occupies the majority of the market share. The main reasons are the low profit, added value, and technology content of steamed bread. In addition, the hand workshop still follows the traditional fermentation technology. It takes a long time but improves the flavor and texture of production, which is the important reason that's why the workshop-style processing products are deeply preferred by consumers.

6.1.2.2 Fermented Soybean Products

Soy sauce is a kind of traditional dressing with unique flavor, healthy nutrition, and delicious taste, which promote the appetite of consumers. It is made from cereals with high content of protein and starch and fermented with *Aspergillus* and other microorganisms for a long time (Feng et al. 2015). Soy sauce originated in China, and it evolved from the sauce in the early period. The process of sauce was recorded in the Zhou Dynasty, which has a history of more than 3000 years. The earliest soy sauce was made from fresh meat; the process is similar to the modern processing technology of fish sauce. For the higher demanding of consumer, the raw material of soy sauce evolved by using soybeans for fermentation, which greatly reduced the cost and extended rapidly.

In China, the industrial fermentation technology of soy sauce experienced three phases: natural brewing, traditional biotechnology, and modern bioengineering. A series of advanced technology are applied in the industrial production of soy sauce from strain screening, inoculated fermentation, genetic engineering, and enzyme engineering. Soy sauce industry is in the stage of rapid development and plays an important role in the Chinese traditional food industry. Based on the stable consumer group and high added value of products, soy sauce market has been in full bloom for a long time (Li 2013).

The statistics show that the annual growth rate of soy sauce is more than 10% in China. By the end of 2014, the national output of soy sauce had reached 9.39 million tons, with year-on-year growth of more than 10.63%. China is the birthplace of soy sauce, but it fails to carry forward the soy sauce process technology. Actually, the soy sauce process technology has been improved and innovated in Japan for a long time and forms a unique Japanese fermentation technology. By adding appropriate concentration of fresh yeast during fermentation, the products have the characteristic of intense delicious flavor. Nowadays, the high-end market of soy sauce in the world has been basically monopolized by Japan and Taiwan of China, which is closely related to the serious lack of traditional food culture in China for a long time. Therefore, paying attention on the introduction and application of advanced technology, as well as the continuous innovation and development of new products, has become a matter of urgency to accelerate the domestic soy sauce industry into the leading fermentation industry in the world.

6.1.2.3 Fermented Coarse Cereal Beverage

Fermented coarse cereal beverage is a new kind of beverage product on the market in recent years. However, fermented cereal drinks are not unprecedented product. For example, douzhir (fermented bean drink), the famous snack in old Beijing, has the function of nourishing stomach and interpreting and relieving inflammation or fever, with a history of more than 200 years (Miao et al. 2013). In addition, Gwas, known as "liquid bread," originated from East Slavs before the Principality of Kiev and has a history of 1000 years. At the end of nineteenth century, the declining royal family of Russia introduced the brewing technology of Gwas into China (Sui and Chu 2013). The traditional fermented cereal beverage has strong regional preference, which is the main reason why the popularization of fermented cereal beverage is limited in the market and the innovation has slowed development. With the pressing needs of consumers for the dietary supplement and healthy nutrition of food, fermented coarse cereal-based beverage comes into being. This kind of fermented beverage is mainly made of barley, oat, corn, etc. and fermented with probiotics such as yeast and LAB.

At present, the fermented coarse cereal beverage industry is still in the early stage of development, with fewer product categories and lower market share. However, the loss of national dietary structure balance results in the urgent desire to improve the level of dietary nutrition and health of food. It is an important developing direction of beverage market to develop coarse cereals-based fermented beverage in the future (Cheng et al. 2012; Zhang and Wang 2013).

6.1.3 Development Trend of Fermented Cereal Products in China

Cereal is one of the main sources of basic dietary composition, which plays an important role in the food industry and scientific research. As a large agricultural country, rice, wheat, and corn are the three major cereals in China, among which wheat and corn occupy the first and second place in the total output of the world, respectively. With the transformation of dietary structure of consumers and the understanding of the nutritional value of cereal food, fermented cereal products appear on the stage of global market. Rice as an important cereal is used to produce yellow wine, rice vinegar, and other products. It contains kinds of active ingredients such as threonine, which could effectively prevent memory loss, and a variety of amino acids and minerals also have the positive effects on reducing of blood pressure, blood sugar, and cholesterol (Chen and Xu 2013; Chen et al. 2014a, b). In addition, the total protein content of steamed bread is three to four times higher than that in the noodles and flat cakes, and fermentation also provides better aroma and taste (Rizzello et al. 2012; Katina and Poutanen 2013).

Chinese traditional cereal fermented food with a long history have a wide range of products, which occupy an important position in the dietary habits of consumers and have a wide market prospect. However, there are some disadvantages in the traditional fermented food, such as poor quality of product, imperfect technics and low techniques require, etc. There is still a gap on the process technology as compared with the food industry in the developed foreign countries. Therefore, the urgent affairs to carrying forward Chinese traditional fermented food is developing the modern processing technology to simplify the traditional fermentation process and improving the production efficiency and batch stability of products effectively. With the development of molecular biotechnology method, the combination of culture-independent and culture-dependent analysis provides a more comprehensive and in-depth overview of the microbial communities and changes of fermented food, which plays an important role in the optimal regulation of the final product fermentation process (Minervini et al. 2014). As the interaction of strains during fermentation has been realized gradually, the isolation, identification, cultivation, and application of strain become the research hotspot in the field of fermented food. Modern biological techniques are applied to isolate and screen fermented strains with excellent traits, which are used to ferment food with appropriate ratio. Advances in research technology have greatly simplified the process of the product while increasing the batch stability of the product (Liao et al. 2015; Liu et al. 2015). The development of excellent strain is one of the core technologies in the application of fermented food. With the development of molecular biology, the genetic engineering techniques are used to modify the strains with superior fermentation characteristics and applied in the different fermented foods. However, in order to improve the flavor of product fermented with single strain, the mixed culture fermentation is also used to enhance the quality of fermented food by synergistic fermentation. In brief, the introduction of advanced fermentation equipment, the improvement of current fermentation technology, and the establishment of related technical indexes are the important means to promote the development of Chinese fermentation food industry.

6.2 Sourdough

6.2.1 Microbial Ecosystem of Sourdough

Sourdough is also known as Laomian in China. It is mainly made from wheat flour or miscellaneous cereal flour by mixing with water and spontaneous fermentation with back-slopping for a long time. It is the oldest natural dough starter and also a complex biochemical system, which contains a large number of lactic acid bacteria, yeast, and a small amount of other microorganisms. The total colony number of lactic acid bacteria and yeast is 10^7–10^9 cfu/g and 10^5–10^7 cfu/g, respectively, and the appropriate ratio of lactic acid bacteria to yeast in the "mature" sourdough is about 100:1 (Gobbetti and Gänzle 2012).

6.2.1.1 Yeast Diversity of Sourdough

Due to the acid and osmotic resistance of yeast, it has a good symbiotic relationship with lactic acid bacteria in the sourdough. At present, more than 25 species of yeast isolated from sourdough have been reported, mainly in the genus of *Saccharomyces* and *Candida*. In addition, it also contains a small amount of *Saccharomyces exiguus*,

Issatchenkia orientalis, Pichia anomala, Hansenula subpelliculosa, and *Candida holmii.* In recent years, the yeast diversity of sourdough collected from different regions has been investigated. Iacumin et al. (2009) focused on four sourdough bread samples collected from Northern Italy by culture-dependent and culture-independent method and found that *Saccharomyces cerevisiae* is the dominant yeast in sourdough; Vogelmann et al. (2009) collected from parts of Germany to explore the yeast diversity and found the similar results. In China, Wu Si Ri Gu Leng (2011) isolated and screened 85 strains of yeast from 28 sourdough samples in 7 leagues (cities) of Western Inner Mongolia. According to the 26S rDNA D1/D2 sequence analysis, 43 *Saccharomyces cerevisiae* strains, 22 *Candida mycoderma* strains, 6 *Torulaspora delbrueckii* strains, 3 *Pichia anomala* strains, 2 *Marine yeast* strains, 2 *Pichia kudriavzevii* strains, 2 *Candida glabrata* strains, 2 *Endosporium cerevisiae* strains, 1 *Starch Pichia pastoris* strain, 1 *Meyerozyma guilliermondii* strain and 1 *Rhodotorula mucilaginosa* strain. Liu Tongjie et al. (2014) screened and isolated 60 strains of yeast from 6 sourdough samples in northern China, of which 48 *Saccharomyces cerevisiae* strains were identified. Therefore, *Saccharomyces cerevisiae* is one of the most important yeast in the sourdough. In addition, the yeast diversity of sourdough was also affected by flour substrate, moisture content, fermentation temperature, and other process parameters. Brandt et al. (2004) found that the optimum growth temperature for *Candida* was 27–28 °C. Therefore, when the fermentation temperature is higher than 35 °C, the growth of *Candida* is inhibited, which reduced the competition with *Lactobacillus sanfranciscensis.*

6.2.1.2   Lactic Acid Bacteria Diversity of Sourdough

Sourdough is characterized by high relative abundance and diversity of lactic acid bacteria, which mainly consists of *Lactobacillus* and few amounts of *Weissella, Pediococcus, Streptococcus, Ascococcus,* and *Enterococcus* (Di Cagno et al. 2014). In recent years, various studies made systematic study to explore the LAB diversity of sourdough. More than 60 species of lactic acid bacteria isolated from sourdough have been reported, and heterofermentative LAB is the most representative species in the sourdough. The results showed that obligately heterofermentative LAB has good adaptability of glycometabolism, such as *Lactobacillus fermentum, Lactobacillus sanfranciscensis,* and *Lactobacillus reuteri,* which could metabolize and utilize maltose during sourdough fermentation. The metabolic pathway of arginine deiminase in *Lactobacillus fermentum* and *Lactobacillus reuteri* plays an active role in amino acid assimilation and acid stress regulation.

Scheirlinck et al. (2007) explored the effects of different regions on the LAB diversity of traditional Belgian sourdough. The results showed that 714 LAB strains from 21 different sourdough samples were mainly composed of *Lactobacillus, Pediococcus, Leuconostoc, Weissella,* and *Enterococcus.* Among them, *Lactobacillus sanfranciscensis, Lactobacillus plantarum,* and *Lactobacillus paralimentarius* are the dominant strains in the samples. Zhang et al. (2011a, b) analyzed and compared the LAB diversity of 28 traditional sourdough bread samples from Western China by culture-dependent and culture-independent analysis. The results showed that *Lactobacillus plantarum* was the dominant strain in these sourdough samples. However, *Lactobacillus brevis*, *Lactobacillus plantarum*, *Lactobacillus sanfranciscensis*, *Lactobacillus fermentum*, *Pediococcus pentosaceus*, and *Weissella cibaria* are predominant in the sourdough collected from the Italian region (Minervini et al. 2015). Therefore, regional environment is one of the key factors relating to the LAB diversity of sourdough. Similar to the yeast diversity, many studies confirmed that the substrate, cycle number of fermentation, fermentation temperature, pH, and other process parameters also influence the LAB diversity of sourdough. *Lactobacillus sanfranciscensis* is predominant in the sourdough by spontaneous fermentation due to the lower fermentation temperature and long-term fermentation time. Type II sourdough, which is generally used in industrial production, has a higher fermentation temperature (>30 °C) and dough yield (DY) value during fermentation. *Lactobacillus fermentum*, *Lactobacillus reuteri*, and *Lactobacillus amylovorus* are dominated in the type II sourdough (Gobbetti and Gänzle 2012). Interestingly, *Lactobacillus amylovorus* is predominant in the rye sourdough, but *Lactobacillus fermentum* are dominated in the rye sourdough by increasing the fermentation temperature (Ercolini et al. 2013; Minervini et al. 2014).

6.2.2 Biochemical Activity of LAB During Sourdough Fermentation

A series of biochemical reactions take place during sourdough fermentation, such as acidification, proteolysis, and the generation of exopolysaccharides (EPS) and flavor substrates (Sarfaraz et al. 2014; Wolter et al. 2014). It's a complex interaction, and the component changes greatly influence the sourdough properties and the quality of bread. The effects are associated with the metabolites and enzymes produced by LAB and yeast during fermentation (Corsetti 2013).

6.2.2.1 Acidification

Acidification is one of the important biochemical characteristics of sourdough fermentation (Corsetti 2013; Sarfaraz et al. 2015). Lactic acid bacteria metabolizes the production of organic acids during fermentation and leads to pH drop in the sourdough, which plays an important role in the rheological properties of the dough and the quality of the final product (Marti et al. 2014; Üçok and Hayta 2015).

Organic acids improve the solubility and intermolecular electrostatic repulsion of gluten proteins (Qiu et al. 2013; Robertson et al. 2014), because it is positively charged under acidic systems, which most commonly result in the spread of gluten proteins and expose more hydrophilic groups (Üçok and Hayta 2015).

The intermolecular electrostatic repulsion blocks the formation of new chemical bonds, which leads to dough softening and less time for stirring. The other components, such as starch particles and endogenous cereal protease, are also affected by acid environment. The endogenous protease activity is activated at the pH of around 4.0, which contributes to the proteolysis and the rheological properties of dough. The network structure of gluten greatly affects the physical properties of dough and final product, which is related to the acidification levels. The extension and gas retention capacity of dough are improved by LAB fermentation (Gobbetti et al. 2014). Clarke et al. (2004) focused on the effect of sourdough on the dough microstructure by confocal laser scanning microscope (CLSM). The results showed that the gluten fermented with sourdough was in an amorphous arrangement with aggregate structure, which increased the specific volume and delayed the staling of products. However, the excessive acidification of gluten is one of main barriers for the product-specific volume. Loponen et al. (2007) found that acidification resulted in the hydrolysis of gliadin and glutenin in the dough through quantitative analysis. The degradation of macromolecule glutenin resulted in the decreasing of stability of gluten structure and gas retention capacity of dough. Therefore, the appropriate acidification is one of the important key processes for the high-quality products preparation. In the wheat flour, it mainly consisted of β-amylase but lack of α-amylase. β-amylase has little effect on the starch and has been completely inactivated during starch gelatinization. On the contrary, large amounts of α-amylase are present in rye flour, which leads to the structure of the dough and are generated by the water-binding pentosane (Corsetti et al. 2000). The previous study indicated that the decreasing of pH value of dough contributed to the activation of endogenous xylanase and significantly increased the solubility of arabinoxylan and water-insoluble arabinan, which resulted in the improvement of the dough elasticity and product quality characteristics (Gobbetti et al. 1999).

6.2.2.2 Proteolysis

Mckay and Baldwin (1974) firstly proved that the casein in the milk was hydrolyzed by the proteolytic system of LAB in order to meet the needs of growth and metabolism. The protease produced by LAB is distributed in the cytoplasm and cell wall, which degrades the protein into small molecular peptides and amino acids. In general, the protease activity of *Lactobacillus* is higher than that of *Lactococcus*. The amino acids produced by the *Lactobacillus* metabolism could also be utilized by *Lactococcus* and promote the growth of *Lactococcus*. The nitrogen source is presented in the forms of inorganic compounds in the cereals, which could not be degraded by LAB. Therefore, the protein degradation of LAB is limited in the process of cereal fermentation. As the previous study of Balestra et al. (2014) reported that the proteolysis is mainly dependent on the role of endogenous proteases in cereals during sourdough fermentation. During sourdough fermentation, the acid environment created by LAB activates the endogenous protease activity, which contributes to the hydrolysis of macromolecular protein polymers (Yin et al. 2015). Two steps of proteolysis occur during sourdough fermentation: ① endogenous proteases hydrolyzed macromolecular polymeric proteins to form oligopeptides with low degree of polymerization in the appropriate pH value and the sulfhydryl groups accumulated accompanied by the hydrolysis of gluten and ② proteases produced by LAB transport low-polymerization oligopeptides (4–40 amino acids) into small molecular peptide and amino acids through a series of complex reactions. LAB converts amino acids into ketones, aldehydes, acids, and alcohols, which improve the flavor characteristics of products. In addition, glutathione (GSH) is a kind of important compound with strong reducibility, which reduced the molecular mass and polymerization degree of gluten polymers. The glutathione reductase existed in the heterofermentative LAB, which reduced extracellular oxidized glutathione (GSSG) to GSH and played an active role in dough rheology and product quality (Gänzle 2014).

6.2.2.3 Exopolysaccharides

In recent years, many studies have been reported that LAB produces exopolysaccharides during sourdough fermentation, which improve the tolerance ability of the strain and play a positive effect on the dough rheological property and the texture characteristic of final product (Wolter et al. 2014). Exopolysaccharides are categorized into homopolysaccharides and heteropolysaccharides according to their composition and biosynthesis pathway. Homopolysaccharides are synthesized from one kind of monosaccharide by extracellular glucanase and fructinase, while heteropolysaccharides are composed of one or more monosaccharides by intracellular glycosyltransferase during fermentation. Previous studies have indicated that exopolysaccharides with hydrophilic characteristics are metabolized by LAB, and it plays an important role in the quality of products: ① improving the water-binding capacity and delaying the staling of products and ② the interaction with other components in the dough, such as protein and starch, which improves the stability of network structure and the product quality (Galle and Arendt 2014). The dextran produced by *Leuconostoc* has been proven to extend the shelf life of bread. The optimization of fermentation parameter, such as substrate, dough yield, and fermentation time, which increase the yields of exopolysaccharides, have a great influence on the product quality. Exopolysaccharides are generally used to produce rye bread due to its poor taste and texture property. High molecular weight dextran has a significant effect on the improvement of rye bread quality (Galle and Arendt 2014; Wolter et al. 2014; Di Monaco et al. 2015). However, the screening and isolation of LAB with excellent exopolysaccharide-produced property is still far to work in the further study based on the requirement of fermented food process technology and product quality.

6.2.3 Sourdough Process Technology

6.2.3.1 Traditional Process Technology

In China, the traditional process technology of sourdough is produced by long-term fermentation with daily refreshments. Laomian and jiaozi are the two most representative of Chinese traditional dough starters. Laomian is one kind of type I sourdough, without adding baker's yeast and long-term fermentation with daily refreshment (Hu 2010). However, jiaozi is generally fermented with jiuqu for the sourdough propagation, and the preparation and propagation steps are different from laomian, which the propagation of jiaozi relies on more than three refreshments within 24 h and lasts for several days. Jiaozi is used to dry after fermentation in order to reduce the moisture content for long-term storage. Laomian is used to ferment with wheat, rye, and corn flour, but jiaozi is usually made from corn and rice flour or steamed bread crumbs due to its better dispersibility in the water during steamed bread processing. People in Shandong province used to produce steamed bread fermented with laomian, and jiaozi is usually applied in Henan province. The process technology of sourdough has the region specificity. In Henan province, Shangqiu area is mainly used to produce jiaozi with corn flour as raw material and mixed with Daqu, but rice flour and xiaoqu are used to prepare jiaozi in Nanyang area (Yang and Liu 2007). Han Chanjuan et al. (2010) developed the evaluation method of jiaozi by determining the fermentation and saccharification power. The results indicated that the fermentation and saccharification power of jiaozi are more than 800 mL and 15–300 U/g, respectively, which produced the steamed bread with better texture and flavor properties. Otherwise, there is a negative effect on the quality of steamed bread. In addition, Yang Jingyu et al. (2006) also focused on the changes of the physical and chemical properties of jiaozi. The results showed that the moisture content, protease, and amylase activities of jiaozi gradually increased during fermentation.

6.2.3.2 Modern Process Technology

Type II and type III sourdough are mainly applied in the process of industrialization. The former one is liquid sourdough and dried after preparation and is named type III sourdough. The dough yield of type II sourdough is about 200 and fermented at a controlled temperature that exceeds 30 °C in order to shorten the fermentation time. Long-term fermentation and high dough yield of sourdough resulted in low pH, which the strains with better resistant to low pH are dominated in type II sourdough. Due to the low moisture content of type III sourdough, it's convenient to storage and application. However, according to the preparation of type III sourdough, high temperature is used for sourdough drying, and the stress-resistant strains are dominated in type III sourdough (Brandt 2014). In addition, *Lactobacillus sanfranciscensis*, *Lactobacillus plantarum*, *Lactobacillus fermentum*, and

Lactobacillus brevis are used for type II and type III sourdough fermentation. Unlike type I sourdough, type II and type III sourdough are not suitable for dough leavening. In order to ensure that the dough has better fermentation power, it is necessary to add a proper amount of baker's yeast for dough leavening (Corsetti 2013).

In China, the application of industrial sourdough develops slowly and still has many problems, such as poor fermentation stability of stains, extensive process technology, and simple processing equipment. Therefore, the market and consumer requirements desire us to develop the modern process technology of sourdough rapidly and enhance the added value of products as soon as possible.

6.2.4   Effect of Sourdough on the Quality of Product

Sourdough has been widely used for producing the wheat-based fermented food for a long time, such as steamed bread, bread, and cookie, which have a positive effect on the sensory quality, delay the staling and prolong the shelf life of products (Plessas et al. 2015).

6.2.4.1   Effect of Sourdough on the Sensory Quality of Product

Sourdough fermentation improves the sensory quality of product, such as texture and specific volume, which is related to the enzyme activity and metabolites formation by LAB during fermentation. Proteolysis plays an important role in the texture of bread or steamed bread and is mainly based on the pH-mediated activation of endogenous flour proteases. Katina et al. (2006) found that proteolysis by LAB resulted in better texture of bread compared to the chemically acidified products, which may have been associated with the slow-release acidification during sourdough fermented with LAB. It's indicated that biological acidification plays an important role in improving product sensory quality. In addition, the significant role of proteolysis by LAB and exopolysaccharides produced by LAB in the staling of product also has been confirmed. In addition to the endogenous cereal protease and protease metabolized by LAB, heterofermentative LAB-liberated glutathione reductase also plays a positive effect on the depolymerization of gluten protein. Glutathione reductase reduces GSSG to -GSH, resulting in an increase in the -SH group in gluten. GSH is mainly involved in the thiol-exchange reaction between gluten proteins and reduces cross-linking of disulfide bonds, which results in a decrease in the molecular weight of glutenin macropolymer (Jänsch et al. 2007).

6.2.4.2 Effect of Sourdough on the Flavor of Product

The flavor of steamed bread is one of the most important attributes, which is directly related to the consumers' acceptability. Although the ease of using baker's yeast, sourdough still preferred to apply in the fermented food process due to the improvement of product flavor. Food fermented with sourdough is valued for their odor and taste characteristics. The aroma compounds and taste compounds are generated by LAB and yeast during sourdough fermentation. The volatile compounds are mainly composed of alcohols, esters, aldehydes, and ketones, and taste-active amino acids, amino acid derivatives, and peptides play an important role in the food taste.

6.2.4.2.1 Taste Compounds

The interaction between lactic acid bacteria and exogenous enzymes affects microbial acidification kinetics, the content of acetic acid, and product texture. The generation of flavor compounds has been proven to be related to the activity of enzymes, which are mainly glucose oxidase, lipase, xylanase, amylase, and protease secreted by microorganisms (Pétel et al. 2016). Studies have shown that the addition of enzymes can increase the lactic acid content produced by the metabolism of *Leuconostoc*, *Lactococcus lactis*, and *Lactobacillus hilgardii*. During dough fermentation, moderate acetic acid content has obviously influence on flavor improvement. The synergistic fermentation of *Lactobacillus sanfranciscensis* and yeast can promote the formation of acetic acid, and the symbiosis of *Lactobacillus sanfranciscensis* and *Lactobacillus plantarum* can enhance the acid production capacity. Moreover, the fermentation temperature has a positive effect on the metabolism of the bacteria to produce aroma compounds. Because the optimum fermentation temperature of *Lactobacillus sanfranciscensis* is 32 °C, when it was cultured at 35 °C, the acetic acid content of *Lactobacillus sanfranciscensis* decreased significantly, but the metabolism of lactic acid and ethanol was not affected. In general, the product flavor is preferred with the lactic acid content of 3.11–5.14 g/kg.

6.2.4.2.2 Aroma Compounds

Microbial fermentation has been proved that different strain starters have a significant effect on the variation of product flavor. The aroma compounds produced by yeast fermentation are mainly alcohols, such as 2-methyl-1-propanol and 3-methyl-1-butanol. However, heterofermentative lactic acid bacteria fermentation mainly produced esters, specific alcohols, and aldehyde (Hansen and Schieberle 2005). Lactic acid bacteria play an important role in the improvement of the flavor of the product, and homofermentative lactic acid bacteria and heterofermentative lactic acid bacteria have a synergistic effect with yeast in the generation of aroma compounds. Due to the complex microbial community of sourdough, the flavor of the product is more abundant (Thiele et al. 2002). Free amino acids are important substrates for the formation of volatile aroma compounds. The total amount of free amino acids in environmental system is mainly related to the hydrolysis degree of protein. Furthermore, the acid environment created by the lactic acid bacteria fermentation can increase the production of free amino acids and thus contribute to the formation of aroma compounds such as thiazole, thiophene sulfide, thiophene ketone, pyrrole, and pyrazine.

6.2.4.3 Effect of Sourdough on the Shelf Life of Products

Fermented noodle food is often accompanied by product staling phenomenon during storage, mainly manifested as internal tissue hardening, easy to drop slag, serious water loss, flavor fission, etc. The main causes of this phenomenon are as follows:

① The staling of amylose. Since the amylose is heated and cooled by the product, it is easy to complete its staling process. Therefore, the staling of the product during storage is mainly due to the staling of amylose.
② The migration of water. During the storage process, the water will gradually transfer from the core to the epidermis, resulting in water loss and tissue hardening of the product.

With the development of fermentation technology, it is found that lactic acid bacteria fermentation plays an important role in the anti-staling characteristics of products (Meng 2007). Min Weihong's (Min et al. 2004) research shows that after long-term fermentation of starchy raw materials, the lactic acid produced by the metabolism of lactic acid bacteria can affect the proportion of starch crystallization zone, which leads to the chain scission and debranching of macromolecular amylopectin and the increase of amylose content. In addition, Tieking et al. (2003) also found that some lactic acid bacteria derived from sourdough can be metabolized to produce extracellular polysaccharides, and some exopolysaccharides have good hydrophilic properties, which can replace the use of hydrophilic colloids in fermented foods, in order to reduce the moisture transfer rate of the product in the storage process, effectively improving the anti-staling characteristics of the product.

6.3 Cereal Condiments Fermented with Lactic Acid Bacteria

6.3.1 Microbial Community of Fermented Condiments

6.3.1.1 Vinegar

Vinegar, with complex microbial community and diversity, is one of the traditional fermented condiments in China. The brewing process of vinegar consists of three parts: ethanol fermentation, acetic acid fermentation, and post-fermentation, which is similar to soy sauce fermentation process. The main process of alcohol fermentation is attributed to the fermentation of molds and yeasts. The enzyme produced by microbial metabolism decomposes the macromolecular nutrients to amino acids and sugars, which are utilized by acetic acid bacteria and LAB. Moreover, these metabolites are metabolized and utilized in acetic acid fermentation, and some free amino acids are also the precursor substances of flavor compounds. In the acetic acid fermentation of vinegar brewing, numerous kinds of organic acids such as oxalic acid, tartaric acid, pyruvate, and malic acid are produced by the multi-strain fermentation, which provides amounts of substrates for the improvement of product flavor. In the post-fermentation stage, the color and aroma characteristics of vinegar are improved by microbial fermentation metabolism (Li et al. 2015).

6.3.1.1.1 Fungal Diversity of Vinegar

Different microorganisms play their respective roles in different stages of vinegar brewing. In the process of solid fermentation substrate of vinegar from vinegar, the inoculation of *Aspergillus* causes a large amount of mold to be introduced into the vinegar. Mold mainly acts on the alcohol fermentation stage of vinegar. Because of its good saccharification ability and protein hydrolysis ability, it provides a material basis for the growth and metabolism of yeast, lactic acid bacteria, and acetic acid bacteria in the post-fermentation process of the vinegar. Han Qinghui (2013) analyzed the dynamic structure changes of microbial flora in Liangzhou traditional fumigated vinegar brewing process. It was found that in the early stage of brewing, all kinds of microorganisms showed a tendency of rapid reproduction. However, on the second day of fermentation, the number of mold and yeast showed a rapid decline. On the fourth day, the total number of bacteria peaked and then gradually decreased, eventually gradual. The cultivation of vinegar usually uses *Aspergillus* as the key strain for fermentation. The *Aspergillus* commonly used in vinegar brewing includes *Aspergillus oryzae*, *Aspergillus niger*, and *Monascus*. Haruta et al. (2006) analyzed the molds in the traditional Japanese rice vinegar fermentation process, and they found that *Aspergillus oryzae* was the main fermentation strain in the alcohol fermentation stage. In addition, *Aspergillus oryzae* is the key fermentation strain for the production of traditional fermented condiments in Japan. The study confirms that *Aspergillus oryzae* has a high rate of enzyme secretion and plays a key role in vinegar brewing.

The yeast mainly acts on the ethanol fermentation stage and the acetic acid fermentation stage of the vinegar. In the alcohol fermentation stage, yeast metabolism utilizes monosaccharides to produce ethanol and carbon dioxide. In the acetic acid fermentation stage, the yeast is partially autolyzed, and its intracellular soluble matter can be used by other microorganisms as the nutrient substance. With the deepening of the modern molecular biology technology research, the composition of yeast in the structure of vinegar bacteria has been further understood. Wu et al. (2012) found that the yeast in Shanxi super-mature vinegar is mainly *Saccharomyces cerevisiae*. This is mainly due to the ability of *Saccharomyces cerevisiae* to better adapt to the external environment in the later stage of alcohol fermentation. In addition, Xu Wei et al. (2007) used the molecular biology method to analyze the structure of fungal flora in the fermentation stage of acetic acid in Zhenjiang fragrance vinegar. They found that vinegar was mainly composed of yeasts such as *Saccharomyces cerevisiae*, *Saccharomyces paradoxus*, and *Saccharomyces bayanus*. In the process of brewing super-mature vinegar in China, the yeast is mainly composed of *Dekkera*, *Brettanomyces*, *Oosporium*, *Kluyveromyces*, and *Pichia*. In traditional Italian aromatic vinegar, it is mainly composed of *Zygosaccharomyces* and *Saccharomyces cerevisiae*. Among them, the abundance of *Zygosaccharomyces bailii* accounted for 41%. This is mainly due to the high sugar content in the traditional Italian aromatic vinegar brewing process, and the *Zygosaccharomyces* can still have good growth characteristics under the condition of higher sugar concentration (Solieri and Giudici 2008).

6.3.1.1.2   Bacterial Diversity of Vinegar

The key bacteria for vinegar brewing are acetic acid bacteria, lactic acid bacteria, and *Bacillus*. Acetic acid bacteria are one of the important strains, which oxidize ethanol to acetic acid during vinegar brewing. Acetic acid bacteria are mainly divided into two groups based on their physiological and biochemical characteristics: *Acetobacter* and *Gluconobacter*. However, *Gluconobacter* has a strong ability to oxidize glucose and produce sorbic acid, while *Acetobacter* has higher ethanol oxidation capacity than *Gluconobacter*. Due to the physiological metabolic activity of *Acetobacter pasteurianus*, it is dominated at the end of acetic acid fermentation stage (Li et al. 2015). Lactic acid bacteria produce a large amount of lactic acid during vinegar brewing, which alleviate acetic acid irritation and improve the taste of vinegar. It also produces the fatty acids and reacts with alcohol to generate aromatic esters, resulting in a better mellow and thick aroma of vinegar. Zou et al. (2011) found that *Lactobacillus sanfranciscensis* and *Lactobacillus plantarum* increased the content of lactic acid in the products and play an important role in the flavor compounds formation. *Bacillus* is a class of aerobic bacteria; the organic acids are mainly generated by tricarboxylic acid cycle and improve the sour taste of vinegar. In addition, the highly active protease produced by *Bacillus* hydrolyzes proteins into amino acids, which also play an important role in flavor formation and color improvement.

6.3.1.2   Soy Sauce

6.3.1.2.1   Fungal Diversity of Soy Sauce

Koji-making is an indispensable process in soy sauce production. Mold can fully grow and metabolize in this stage, producing rich enzymes, which contribute to enhance the flavor and quality of soy sauce. The molds commonly used in koji-making of soy sauce are mainly *Aspergillus, Penicillium, Rhizopus, Mucor*, and *Botryosporium*, among which *Aspergillus oryzae* is the dominant species. It is found that mold usually commence to grow at the early stage of soy sauce fermentation. However, the growth of mold was inhibited, and the number decreased gradually with the consumption of oxygen in fermentation process. In addition, molds can generate a large number of enzyme systems, on the one hand, proteins are decomposed to produce small molecular peptide and amino acids by secreted protease, and the amylase can decompose the starch to dextrin and glucose, providing essential nutrients for the growth of lactic acid bacteria and yeasts in the subsequent fermentation; on the other hand, the free glutamine from the raw material was hydrolyzed by glutamate to generate glutamate, which could enhance the umami taste of soy sauce.

Yeast is considered to be another important fermentation strain in brewing of soy sauce. *Zygosaccharomyces rouxii* and *Torulopsis bombicola* are the main yeasts species in the brewing process. *Zygosaccharomyces rouxii* usually contributes to the early stage of fermentation and converts sugars such as glucose and maltose to alcohol. In addition to the formation of aromatic substances, the contents of succinic acid and furfuryl alcohol in soy sauce are increased; these compounds also greatly improved the aroma of soy sauce. Contrarily, *Torulopsis bombicola* has the main advantage in the later fermentation stage, which can reduce the production of sugars and amino acids, and mitigate occurrence of the Maillard reaction. Therefore, the ideal color can be well maintained in the final soy sauce products. In addition, guaiacol, phenylethanol, and other aroma components produced by *Torulopsis bombicola* makes the aroma of soy sauce more mellow. The amount of yeast in the initial stage of fermentation is very low, only 7.6×10^2 cfu/g. However, the growth rate is significantly accelerated when the fermentation is initiating; the highest content can be reached up to 3.2×10^6 cfu/g, then gradually slowed down, and became more balanced, and the number of yeast flora could be finally stabilized at 10^3 cfu/g (Yang et al. 2016). In the modern soy sauce brewing technology, the flavor of soy sauce products is usually improved by adding aroma producing yeast, and the interaction between yeast and *Aspergillus oryzae* can reduce the contents of reducing sugars and increase the content of amino nitrogen.

6.3.1.2.2   Bacterial Diversity of Soy Sauce

The amount of NaCl in soy sauce is as high as 18%, which prevents most bacteria from growing. However, the salt-tolerant lactic acid bacteria mainly include *Pediococcus halophilus* and *Tetracoccus Sojae*, which survive and reproduce in the sauce mash with 18~20% salt content. In addition, it has positive effect on the color and flavor of sauce mash, which makes the product bright, ruddy, and rich in flavor. Lactic acid bacteria with good salt-resistant grow in the early stages of soy sauce brewing. With the decreased of pH during fermentation, yeasts are promoted to grow, and the content of lactic acid bacteria decreased rapidly in the late fermentation stage (Chen et al. 2014a, b).

6.3.1.3   Yellow Wine

6.3.1.3.1   Fungal Diversity of Yellow Wine

Mold existed in wheat koji plays an important saccharification in the brewing process of yellow wine, mainly in the genus *Aspergillus, Rhizopus, Mucor, Penicillium, and Absidia*. *Rhizopus* fungi produces glucoamylase, liquefaction enzyme, and some flavor substances. Moreover, *Aspergillus* fungi can secrete amylase, protease, peptidase, and other enzymes into the environment, which plays an important role in the hydrolysis of residual starch in post-fermentation. *Mucor* has a rich complex enzyme system, which can secrete glucoamylase, α-amylase, and protease. Fang Hua (2006) analyzed the fungi in wheat koji of yellow wine using traditional separation technology. It was found that the main fungi in wheat koji were *Aspergillus oryzae, Rhizopus oryzae, Aspergillus niger, Aspergillus fumigatus, Mucor pusillus*, and *Rhizopus microspores*. However, due to the water evaporation, the fungi were inhibited during process of koji. In addition, Cao et al. (2008) also found that straw and wheat raw materials had a great impact on the dynamic changes of fungal flora during wheat koji fermentation.

The fermented mash of yellow wine is rich in yeasts including *Saccharomyces cerevisiae, Hansenula anomala, Saccharomyces cinerea, Saccharomyces diastaticus*, and *Trichosporon pullulans*. In the early stage of mash fermentation, starch is converted to glucose, and yeast with saccharification ability was dominant in this process (Lv et al. 2013). With the development of fermentation, the alcohol concentration in the wine mash increased, and the *Saccharomyces cerevisiae* with higher alcohol tolerance gradually occupied the dominant position. When the fermentation time exceeds 90 days or more, the alcohol content reaches 19.0% (V/V), and the other yeasts will disappeared in the late fermentation period due to the high alcohol content.

6.3.1.3.2   Bacterial Diversity of Yellow Wine

Lactobacillus is one of the main lactic acid bacteria in yellow wine brewing, including *Lactobacillus pasteurii, Lactobacillus amylophilus, Lactobacillus plantarum, Lactobacillus brevis, Lactobacillus delbrueckii, Lactobacillus delbrueckii* subsp. bulgaricus, and *Lactobacillus thermophilus*. *Lactobacillus pasteurii* decomposes starch, dextrin, and produces exopolysaccharide. The amount of *Lactobacillus* in Lin-fan yeast starter generally is in the range of $0.3 \times 10^9 \sim 1.0 \times 10^9$ cfu/mL. *Lactobacillus pasteurii, Lactobacillus amylophilus, Lactobacillus plantarum, Lactobacillus acidophilus, Lactobacillus casei, Lactobacillus brevis, Lactobacillus debrueckii*, and *Lactobacillus thermophilus* are the most abundant in Lin-fan yeast starter (Fang et al. 2015). In addition, *Lactococcus* is another kind of main lactic acid bacteria in brewing of yellow wine, which stopped growing when pH was lower than 4.5 (Lv et al. 2016).

6.3.2   Biochemical Activity of Lactic Acid Bacteria
During Fermentation

Vinegar and soy sauce are made of cereals through microbial fermentation. The fermentation process is accompanied with a series of biochemical metabolism and reaction, such as proteolysis, starch saccharification, alcohol fermentation, and the formation of organic acids.

6.3.2.1   Proteolysis

The secondary structure of cereal protein can be changed after high-temperature treatments, and protease in the fermentation system hydrolyze proteins into low molecular peptides and amino acids. Numerous studies have demonstrated that most of free amino acids have taste activity and can be used as precursor substances for the formation of flavor substances, which play an important role in the flavor characteristics of the product. During the soy sauce brewing, *Aspergillus oryzae* produce many kinds of proteases and alkaline protease. Therefore, it is necessary to prevent protease activity reduced that caused by low pH in the fermentation process, thus affecting the hydrolysis of protein in the fermentation process and going against the formation of flavor compounds in soy sauce.

6.3.2.2   Starch Saccharification

Cereals are rich in carbohydrates and high temperature of process releases starch granule. Macromolecular dextrin, maltose, and glucose are generated by amylase during fermentation. Glucose is an important carbon source for microbial metabolism in condiment fermentation, and it can also participate in Maillard reaction during high-temperature process. In addition, glucose is also a basic metabolic substrate for alcohol fermentation and organic acid generation. Therefore, the starch saccharification should be strictly controlled in the fermentation process to avoid excessive acidification.

6.3.2.3   Alcohol fermentation

Yeast is an indispensable fermentation strain for alcohol fermentation in the condiment process. The optimum growth temperature of yeast is 28~34 °C. Yeast has a good acid-resistant capability, but its growth can be inhibited on the condition of pH < 3.5. The saccharification of starch in vinegar, soy sauce, and other condiments resulted in the accumulation of a large amount of glucose in the fermentation system, which provides plenty of carbon sources for saccharide metabolism in yeast. In addition to producing alcohols, cereal fermentation also produce aldehydes, acids and esters, and other important aroma components, which plays a crucial role in flavor characteristics of vinegar and soy sauce.

6.3.2.4 Organic Acid Formation

Condiments, such as vinegar and soy sauce, are rich in many organic acids, including lactic acid, acetic acid, succinic acid, malic acid, and citric acid. It was found that different organic acids in condiments had synergistic effects on the flavor characteristics of products. Moreover, the organic acid can also be esterified with alcohols in the later stage of the fermentation, thereby improving the content of ester compounds in the fermented products, which play an active role in the improvement of products flavor.

6.3.3 Processing Technology of Cereal-Based Condiments Fermented by Lactic Acid Bacteria

6.3.3.1 Vinegar

Vinegar is mainly fermented from starchy cereals; the procedure of microbial metabolism and enzymatic conversion converts starch to ethanol and acetic acid. There are significant differences in vinegar brewed in different regions, which is related to the brewing methods and raw materials characteristics. According to fermentation process, it can be divided into three types: solid-state fermentation, liquid state fermentation, and solid-liquid state fermentation.

6.3.3.1.1 Solid-State Fermentation

The traditional vinegar brewing technology in China mostly adopts solid-state fermentation, namely, the fermentation materials is solid-state in acetic acid fermentation stage. Due to acetic acid bacteria which need a certain amount of oxygen to oxidize ethanol to produce acetic acid, it is necessary to add more fluffy food materials, such as millet shell, sorghum shell, and bran. The fermentation environment of traditional solid-state fermentation is open-ended, and the microbial flora structure is complex. So the vinegar produced by this method is rich in flavor compounds and good quality. For example, both of the famous Shanxi aged vinegar and Zhenjiang vinegar are produced by solid-state fermentation. However, this approach has its drawbacks, such as wide coverage, low yield, and low equipments utilization. In recent years, with the improvement of new brewing technology, modern solid-state fermentation technology has been widely used in vinegar processing. Compared with the traditional brewing technology, the production rate has been significantly increased, but the product quality has significantly decreased.

How to keep the quality and flavor of traditional vinegar brewing on the basis of improving technology has become a research problem in the field of vinegar industry (Su 2015).

6.3.3.1.2 Liquid State Fermentation

In China, the common liquid state fermentation technology of vinegar brewing are mainly composed of surface liquid fermentation, quick-brewed vinegar fermentation, and submerged fermentation. Some famous vinegar is also produced by liquid state fermentation technology, such as Jiangsu and Zhejiang rose vinegar, Fujian red koji vinegar, etc.

Surface Liquid Fermentation Process

According to the different operations of the former process, the vinegar produced by surface liquid fermentation can be divided into white vinegar, sweet vinegar, and rice vinegar, and the specific fermentation process is as follows:

1) White Vinegar Process

   White vinegar is produced by placing the vinegar source in a special container, adding ethanol solution and a small amount of nutrients to cover the cylinder head, and fermenting about 20 days at natural temperature or in a greenhouse at 30 °C. The vinegar source is generally the mature vinegar of the previous batch, and the ethanol solution can be prepared by diluting the liquor with water and diluting to 3% ethanol contents. The specific fermentation period depends on the environment temperature. The fermentation period should be extended if the temperature is low. The mature vinegar is clear and colorless with the content of acetic acid of 2.5~39.0 g/100 mL.

2) Sweet Vinegar Process

   Sweet vinegar is made from maltose, which mainly exists in Beijing. After inoculation of acetic acid, the fermentation is carried out in a paper-sealed cylinder at 30 °C. The matured period of fermentation is about 30 days, which is a slightly longer than white vinegar. The acetic acid content of final sweet vinegar is 3~4.59 g/100 mL.

3) Rice Vinegar Process

   Rice vinegar is made from rice by liquid surface fermentation. However, the processing methods have regional divergence. For instance, rice koji can be made by adding *Aspergillus oryzae* to rice, then saccharifying rice by adding water, or saccharifying rice by adding koji. Certainly, wheat flour was inoculated with *Aspergillus oryzae* to make dough and then added with water to saccharify the rice. After the saccharification solution was prepared, yeast was added for ethanol fermentation, and acetic acid bacteria were eventually added to system for surface liquid fermentation. The acetic acid content of rice vinegar is generally 3~5 g/100 mL.

Quick-Brewed Vinegar Process

Quick-brewed vinegar is brewed in a quick brewing tower; raw materials such as liquor are oxidized to acetic acid by acetic acid bacteria and then through aging process. Quick-brewed vinegar is clear, colorless, or yellowish.

Submerged Fermentation Process

Vinegar made by submerged fermentation technology is a new technology developed in modern technology. The period of vinegar brewing is very short with comparatively higher productivity. In addition, most of the production equipments are standard fermentation tank; hence the floor space is small, and no rice husk fillers are used. This new process is a great progress from traditional fermentation to modern mechanized vinegar production. However, the flavor of vinegar by modern process is slightly inferior to that of traditional vinegar.

6.3.3.1.3 Solid-Liquid State Fermentation

The primary difference between solid-liquid state fermentation and other fermentation approaches is that the process of acetic acid fermentation is solid-state, while the saccharification of raw materials and the fermentation of ethanol are carried out in liquid state. Natural ventilation and vinegar reflux combined with enzymatic method are used in solid-liquid state fermentation, which not only reduced the labor force but also improved the utilization ratio of raw materials. In general, 8 kg vinegar can be obtained from 1 kg broken rice material.

6.3.3.2 Soy Sauce

As a traditional fermented condiment in China, soy sauce has a delicious taste and a wide range of soy sauce. Generally speaking, soy sauce is mainly divided into two types: dark soy sauce and light soy sauce. The tastes of dark soy sauce is salty, heavy color, and often used for color improvement, while the light soy sauce has a strong umami taste and is often used for flavor enhancement. At present, the main technology for soy sauce brewing are natural fermentation, high saline diluting fermentation, low saline solid-state fermentation, part-brewing solid-diluted state fermentation, etc. (Zhao 2009).

6.3.3.2.1 Natural Fermentation

Natural fermentation method is the traditional processing technology of soy sauce brewing in China. This method naturally exposed the fermented ingredients to the sun in an open environment. On the one hand, a large number of fungus, yeast, and lactic acid bacteria can be introduced into the fermented system, which is a natural way of koji-making; on the other hand, the moisture in the fermented grains can be evaporated effectively by long-time sunshine. Therefore, the maturation time of fermented soy sauce is longer, and the salt content is higher. However, the higher salt content and lower fermentation temperature can inhibit the enzyme activity in soy sauce cereal.

6.3.3.2.2 High Saline Diluting Fermentation

High saline diluting fermentation is a combinatorial brewing method that combines natural fermentation and modern soy sauce brewing technology. Generally, the fermentation temperature is 10–30 °C; the fermentation time is 6–12 months, and the salt content is 15% or more (Lu and Wei 2006). Compared with the traditional natural fermentation method, modern processing and temperature control technology can effectively eliminate the potential safety problems caused by long-time natural exposure. However, the high saline diluting fermentation cannot avoid the disadvantage of long fermentation period and high investment cost.

6.3.3.2.3 Low-Saline Solid-State Fermentation

Low saline solid-state fermentation is evolved from salt-free solid-state fermentation combined with Chinese brewing technology. It is the most widely used fermentation technology for soy sauce production in China. On account of the salt content in sauce is less than 10%; the inhibitory effect on the enzyme activity is not significant; the ripening period is very short for soy sauce fermentation. Therefore, this brewing approach is suitable for industrial scale production of soy sauce with good quality. However, the flavor taste of this kind of soy sauce is obviously inadequate than the traditional method of natural fermentation and high saline diluting fermentation products (Deng et al. 2015). Moreover, the traditional salt-free solid-state fermentation process is maintained in some rural areas of China, in which the amount of salt added in the brewing process is very little or lacking. In order to ensure that the product is not contaminated by miscellaneous bacteria, the fermentation temperature range is usually 55~60 °C; the increase of temperature can significantly promote the activity of the enzyme, and the fermentation ripening cycle is only 60~72 h, but the products have defects such as serious lack of flavor and insufficient quality.

6.3.3.2.4 Part-Brewing Solid-Diluted State Fermentation

In order to shorten the fermentation period of soy sauce and ensure the flavor of the products, the part-brewing solid-diluted state fermentation was developed as a rapid brewing technology, which is combined with the characteristics of solid and dilute fermentation technology. This method mainly separately prepared sauce mash from protein and starch and optimized the temperature, salt content, and fermentation conditions of solid-liquid mash. Solid-state fermentation is first conducted under a lower salt condition, and then brine is added for diluting fermentation, which effectively reduces the inhibitory effect of salt on enzyme activity. Furthermore, this process can also promote metabolism and growth of fungi and yeast in the early fermentation stage. However, this process is more cumbersome, and the control of segmental fermentation is strict, so that it is not widely used in the manufacturing of soy sauce (Deng et al. 2015).

6.3.3.3   Yellow Wine

Yellow wine is one of the oldest liquors in the world. It is produced by the interaction of different kinds of fungus, yeasts, and bacteria using rice and millet as the main raw materials, and after a series of manufacturing procedure of cooking, saccharifying, fermenting, and squeezing, the low-alcohol brewing yellow wine is obtained, and the alcohol content is usually 14~20%. In addition, yellow wine contains eight kinds of essential amino acids and other nutrients, so it is always called "liquid cake." Due to the differences in raw materials, starter, process technology, and natural conditions yellow wine production, the quality of yellow wines are distinguishable. According to the content of sugar, yellow wines can be divided into dry yellow wine, semidry yellow wine, semisweet yellow wine, and sweet yellow wine. The famous yellow wines include Shaoxing rice wine, Fujian old wine, Jiangxi Jiujiang Sealed Liquor, and Wuxi Huiquan wine. The manufacturing technology of these wines belongs to the traditional rice wine fermentation process, which can be divided into Lin-fan wine, Tan-fan rice wine, and Wei-fan wine.

6.3.3.3.1   Lin-Fan Wine

The name of Lin-fan wine comes from the operation of sprinkling cold water to steaming rice. The process is as follows: the glutinous rice is immersed in water for 48 h, then is steamed and matured into rice, after cooling with cold water to the optimum temperature for saccharification and fermentation of rice, and then mixed with wine, special koji, clear water, etc., and Lin-fan wine produced in fermented condition for 45 days. Lin-fan wine is mainly used in the production of sweet yellow wine. Although the mature Lin-fan wine holds a poor flavor characteristics and monotonous taste, the high yield, cheap price, and fast available market of yellow wine make it popular in China.

6.3.3.3.2   Tan-Fan Wine

Tan-fan wine is made by immersing white glutinous rice with water for 16–20 days, separating the rice water and steaming the raw rice into rice, then spread the rice and cool down to 35 °C, and then mixing it with rice syrup, wheat koji, yeast, and appropriate amount of water for saccharification and fermentation. After 60–80 days, Tan-fan wine can be formed, such as Yuanhong wine. Using the traditional technology of winter brewing, on account of long brewing period, the starch hydrolysis, protein, and fat decomposition of material in the fermentation process are more intensive. Moreover, the degree of formation of various organic acids and flavoring substances are more adequate.

6.3.3.3.3 Wei-Fan Wine

When Wei-fan wine is brewed, the raw rice for brewing is divided into multiple portions; the first one is used to produce yeast starter using Lin-fan method; the first Wei-fan operation is carried out after 27~28 h. Then the materials need to be stir immediately so that the yeast starter of the first fermentation is evenly distributed. Subsequently, add water every 24 h to make the saccharification and fermentation uniform, and control the temperature of the fermentation product. After repeating four times, the Wei-fan wine is produced. Compared with Lin-fan wine and Tan-fan wine, the degree of fermentation is deeper, and the utilization ratio of raw material is higher. There are still many Chinese areas that adopt this traditional technology for yellow wine brewing; a typical representation of this kind of wine is Shaoxing yellow wine.

6.3.4 Effect of Lactic Acid Bacteria Fermentation on the Condiment Quality

6.3.4.1 Effect of Microbial Fermentation on the Vinegar Quality

Vinegar is metabolized and fermented by many microbes and is rich in many organic acids that are mainly classified into volatile acids and nonvolatile acids. Volatile acids are mainly acetic acids produced by the metabolism of acetic acid bacteria, and propionic acid, butyric acid, and valeric acid are present in trace amounts. The smaller the molecular mass of volatile acids makes the stronger flavor irritation. Therefore, the irritative taste of vinegar mainly comes from volatile acetic acid. The nonvolatile acids in vinegar were mainly lactic acid, malic acid, citric acid, and succinic acid, which were also produced by the metabolism of lactic acid bacteria. Such substances can regulate the pH in environmental system, inhibit the growth of spoilage bacteria, and effectively improve the quality of products (Wang et al. 2013). Moreover, nonvolatile acids have a buffer effect on the irritating odor of volatile acetic acid; the products with higher content of nonvolatile acids will weaken flavor irritation of volatile acids. In general, the content of nonvolatile acid in solid-state fermentation vinegar is higher than that in liquid state fermentation vinegar. In addition, esterification reaction occurred in the fermentation process of lactic acid, and bacteria can facilitate the odor of vinegar (Zhao et al. 2014).

The aroma components of vinegar mainly come from acid, aldehyde, alcohol, ester, phenol and furan, and other volatile flavor substances, which are lower in vinegar, but these compounds will give the product a strong acidic flavor in the appropriate proportion. Moreover, the excessive content of aldehydes will lead to vinegar possessing a spicy flavor and strong irritation, while moderate aldehydes concentration can give vinegar rich desirable flavor. Esters are important flavor components that form the unique aroma of vinegar. They are mainly produced by the esterification reaction of organic acids and alcohols. Generally, the high-quality vinegar are rich in ester compounds, such as butyl acetate, ethyl acetate, ethyl lactate, etc. (Zhu et al. 2016). Li Danya's (2008) found that acetic acid and 3-methyl butyric acid were the characteristic aroma components of Zhenjiang fragrant vinegar. In the flavor study of Shanxi aged vinegar, there were significant differences between high-quality vinegar and middling vinegar in the contents of acetaldehyde, ethyl acetate, ethanol, furfural, and propionic acid (Miao et al. 2010). High-quality vinegar needs to go through the aging process, which result in the evaporation of water, so that the content of characteristic flavor substances was improved significantly. Compared with Japanese vinegar, the amino acid level in Chinese vinegar is significantly higher. Taking Japanese rice vinegar as an example, the average amino acid content is 137 mg/100 mL, which is only about one-tenth of Chinese vinegar.

6.3.4.2 Effect of Microbial Fermentation on the Soy Sauce Quality

The flavor formation of soy sauce mainly generate in the later fermentation period. Currently, more than 200 flavor compounds have been reported from soy sauce. Thereinto, there are 20~30 kinds of representative flavoring substances in soy sauce including alcohols, aldehydes, organic acids, phenols, esters etc. The flavor formation mechanism of soy sauce is relatively complex; there are four main generation pathways: ① from the ingredients, different ingredients have a significant impact on the flavor characteristics; ② in the stage of koji-making, metabolism of *Aspergillus* can produce many biochemical enzymes and decompose macromolecular to many flavor compounds, which is beneficial for enhancing the flavor of soy sauce; ③ in the later fermentation stage, acetic acid bacteria, lactic acid bacteria, and yeast produced various flavor compounds by metabolizing the small molecule produced by the decomposition of mold and yeast in the koji-making stage, thus achieving the synergistic effect of enhancing flavor; and ④ the change of nonenzymatic chemical reaction involved in the whole brewing process of soy sauce, which can promote the enhancement of flavor (Singracha et al. 2016).

Lee et al. (2006) compared the volatile flavor compounds of fermented soy sauce and acid hydrolyzed soy sauce. It was demonstrated that alcohols and esters were the major flavor compounds in fermented soy sauce, while heterocyclic compounds were the main components in acid hydrolyzed soy sauce. Alcohol is mainly produced by yeast fermentation metabolism, and ethanol is one of the main substances produced by yeast fermentation with hexacarose; amyl alcohol and isopentanol are important products of yeast decomposing leucine and isoleucine. It was also found that ethanol, pentanol, isoamyl alcohol, phenylethanol, furfuryl alcohol, and other hydroxyl flavor compounds were the main flavor components of low salt solid soy sauce. Esters are the main components of aroma in soy sauce. They are mainly produced by esterification of alcohols and organic acids produced by microbial metabolism (Gao et al. 2010). At present, more than 40 esters were identified from soy sauce, which endows the soy sauce with strong flavor. Aldehydes and acids are also important components of the flavor components of soy sauce. The aldehydes are mainly converted from the organic acid produced by yeast and *Aspergillus* during the koji-making process. Up to now, 18 aldehydes in soy sauce have been reported. The brewing process of soy sauce is based on the growth metabolism of microorganisms. In addition, there is an interaction between different substances, which has the characteristics of synergistic fermentation and flavor enhancement (Cui et al. 2014). Mixed strain fermentation is mainly used in the stage of koji-making and post-fermentation. Multi-strain koji is composed of *Aspergillus oryzae* and mixed with other strains. The main purpose of this method is to improve the utilization rate of raw materials and the productivity of amino acids through the enzymes secreted by strains. The multi-strain fermentation is mainly in the post-fermentation stage, and the flavor characteristics of product are improved by additionally adding yeast and lactic acid bacteria. It was found that mixed fermentation could promote the growth rate of starters; the difference of metabolic pathways between starters led to different metabolites formation, which resulted in the reciprocal growth of microorganisms. It is noteworthy that the later fermentation process is easy to be contaminated by miscellaneous bacteria, and the addition of lactic acid bacteria in mixed fermentation can effectively inhibit the breeding of miscellaneous bacteria (Zhao 2005). In addition, the study also found that *Lactobacillus plantarum* has good salt tolerance, and it has a significant synergistic effect with salt-tolerant yeast under suitable ratio conditions. Therefore, it plays an active role in improving the flavor of soy sauce prepared by low salt solid-state quick-brewing process.

6.4   Other Cereal Products Fermented with Lactic Acid Bacteria

6.4.1   Fermented Rice Flour

6.4.1.1   Microbial Diversity of Fermented Rice Flour

Natural fermentation is usually used in the process of fermented rice flour. Tong Litao et al. (2013) have found that the microbial diversity of rice flour is changing during the process of fermentation, in which lactic acid bacteria are the main bacteria with dominant microflora in the process of fermentation. The growth rate of yeast was faster during the first 24 h of fermentation and then tended to be stable. Due to the large amount of oxygen needed for the growth of mold, most of them are grown on the surface of fermentation broth in the process of rice flour fermentation, which has a limited effect on the fermentation process of rice flour. In recent years, many scholars have studied on the diversity of microbial flora in fermented rice flour. Lu Zhanhui et al. (2006) found that rice flour in Changde (Hunan Province) were mainly composed of *Lactobacillus*, *Streptococcus*, and *Saccharomyces cerevisiae*. The analysis of microbial community showed that *Lactobacillus plantarum* and *Saccharomyces cerevisiae* were the dominant fermentation strains for rice flour according to the 14 samples from 3 regions of China and South Asia.

6.4.1.2   Process Technology of Fermented Rice Flour

Rice flour, which is similar to the noodles in northern China with long strips form, is an important dietary component in Southern China and Southeast Asia. The processing technology of common rice flour consists three types: cutting powder, pressing powder, and the other kinds. Among them, the cutting-powder type of rice flour is prepared by gelatinizing part of the rice milk in advance, then mixing it with raw rice milk, and coating on the conveyor belt, which is heated by the steam tunnel furnace, air-cooled, and sliced into shape. The pressing-powder type of rice flour is mainly processed by extrusion molding, that is, the rice is soaked, wet-milled, filtered, and dehydrated until the moisture content is 38%~42%, and then extruding inside the machine after boiling or steaming. In addition, rice flour is divided into fermented and non-fermented. Compared with the non-fermented type, the fermented rice flour has special texture characteristics and chewiness. Ma Xia et al. (2015) found that soaking time and temperature are two important factors to determine the quality of fermented rice flour.

6.4.1.3   Effect of Lactic Acid Bacteria on the Rice Flour Quality

Lactic acid bacteria play an indispensable role in the natural fermentation of rice flour, which has a positive effect on the texture, nutritional, and functional characteristics of rice flour. Starch, the main component of rice, which makes rice flour can be regarded as a kind of starch gel. The previous studies found that the rice flour produced by *Lactobacillus* fermentation can produce organic acids such as extracellular enzyme and lactic acid during the fermentation process, and the content and average degree of polymerization of amylose are significantly increased after fermentation. The elasticity of starch gel was improved by increasing the average polymerization degree of amylose. Therefore, compared with the non-fermented rice flour, the product has better edible quality, such as flexibility, chewiness, and tensile properties. Li Lite et al. (2001) further confirmed that the main reason for the increase of amylose content in rice flour was that lactic acid bacteria metabolized during the fermentation process could affect the proportion of starch crystalline region; thus, the amylopectin of macromolecular was broken and debranched, and the content of amylose was increased. In addition, the fermentation of lactic acid bacteria can metabolize many bacteriocins with the inhibition of spoilage bacteria and effectively prolong the shelf life of products.

6.4.2 Fermented Cereal Beverage

6.4.2.1 Microbial Diversity of Fermented Cereal Beverage

The bacteria commonly used in traditional cereals fermented drinks are *Leuconostoc mesenteroides*, *Lactobacillus*, *Streptococcus*, and *Pediococcus*, the fungus include *Aspergillus*, *Paecilomyces*, *Cladosporium*, *Fusarium*, *Penicillium*, and *Trichothecium*. Boza in Turkey is a mixture of wheat, rye, corn, and other cereals with sucrose and fermented for a long time at 30 °C. Based on the culture-dependent and culture-independent analysis. Boza is mainly composed of *Lactobacillus sanfranciscensis*, *Leuconostoc mesenteroides*, *Lactobacillus coryniformis*, *Leuconostoc dxtranicum*, *Lactobacillus fermenti*, *Leuconostoc oenos*, *Saccharomyces uvarum*, and *Saccharomyces cerevisiae*. Mahewu is a traditional fermented coarse grain beverage in Zimbabwe, which is mainly made from corn and sorghum by 16 h fermentation at 45 °C. It has been proved that the main fermentation microorganism is *Lactococcus*. Due to *Lactococcus lactis* which can metabolize *nisin*, the products could be resistance to *Salmonella*, *Campylobacter jejunii*, and *Escherichia coli*.

6.4.2.2 Process Technology of Fermented Cereal Beverage

There are many kinds of fermented coarse grain beverages reported in China, which are mainly processed from oats, corn, barley, etc. Yan Haiyan et al. (2008) mixed corn with water according to 1:10 (w/V), inoculated with lactic acid bacteria by 12 h fermentation at 37 °C, and then added 0.1% xanthan gum and 0.15% sodium carboxymethyl cellulose after fermentation to produce a lactic acid corn fermented beverage. Jia Jianbo (2002) focused on a mixture of skim milk powders and oats, which was fermented by *Lactobacillus rhamnosus*. Firstly, the oats were hydrolyzed by two enzymes, then add skim milk powders, sterilized and cooled, inoculated and fermented to make the biological milk with active ingredients and healthcare function. Its viable count reached 10^{10} cfu/ml; acidity was 108 °T, and β-glucan content was 236 mg/L. It can be seen that fermented coarse grain beverage has a good market prospect not only due to the effectively enhancement of nutritional value of grain itself but also has the same healthcare function as fermented products such as lactic acid bacteria.

6.4.2.3   Effect of Lactic Acid Bacteria on the Cereal Beverage Quality

After fermentation with lactic acid bacteria, the contents and types of trace nutrients such as vitamins, amino acids, and minerals in cereals were significantly increased, and organic acids, alcohols, aldehydes, and ketones were produced by metabolism, which endow the products with more intense aroma. Onyango et al. (2004b) used corn and millet milk as raw materials to produce Uji, a traditional food in East Africa. By comparing the amino acid content in the mixture before and after fermentation, it was found that the contents of Lys, Tyr, Met, Cys, Gly, and Asp were significantly increased after fermentation of Uji. Wang Fengling and Liu Aiguo (2003) also found that based on the combined action of microbes and enzymes, the protein, starch, fat, and other macromolecular nutrients decompose into small molecular substances such as amino acids, polysaccharides, monosaccharides, and fatty acids, which improve the absorption of body. Onyango et al. (2004a) processed the fermented mixture of corn and millet into a kind of food for nourishing and weaning, which is more suitable for children because of lower content of fiber and tannin after fermentation. Nanson and Fields (1984) reported that the content of lysine and tryptophan increased greatly after natural fermentation of corn flour.

6.4.3   Fermented Soya Beans

6.4.3.1   Microbial Diversity of Fermented Soya Beans

As one of the traditional fermented products in China, the fermentation process of fermented soya beans is coordinated by a variety of complex microorganisms. Traditional fermented soya beans fermentation is mainly divided into koji-making stage and brine stage. The main fermentation bacteria were filamentous fungi, such as *Aspergillus*, *Mucor*, and *Rhizopus*. The types of filamentous fungi used in koji-making stage are mainly affected by local environment and climate due to the natural fermentation processing. When the fermentation temperature is about 15 °C, the main fermentation bacteria are low-temperature type of *Mucor*, and when the temperature is 20~30 °C, the main fermentation bacteria is medium-temperature type of *Rhizopus*. Chen et al. (2011) used modern molecular biological techniques to analyze the diversity of microbial communities in the koji-making stage of "Daoxiangyuan fermented soya beans." It was found that *Aspergillus oryzae* and *Bacillus subtilis* were the main fermentation bacteria in the stage. In addition, they found a variety of microbes known as *Bacillus amyloliquefaciens*, *Bacillus brevis*, *Aspergillus Niger*, *Staphylococcus saprophyticus*, and *Saccharomyces cerevisiae*. But the brine stage is mainly the anaerobic fermentation process with high osmotic pressure, which needs long time and plays an important role in the formation of flavor and quality improvement of fermented soybeans. The mold is useless in the process because of the anaerobic environment. However, due to the metabolism of many enzymes in the stage of koji-making, sufficient material basis was provided by the mold for the brine stage of fermentation process. The selection of salt concentration in brine stage plays an important role in microbial growth. High concentration of salt can inhibit the growth of bacteria and prolong the ripening process. In contrast, low salt concentration lead to the rapid growth of lactic acid bacteria and excessive acidification of the product, which is not conducive to the sensory quality of the product. Therefore, suitable selection of salt concentration determines the growth and metabolism of microorganisms and affects the quality characteristics of the product. Chen et al. (2011) used traditional separation techniques combined with PCR-DGGE to investigate the microbial community composition of the fermented soya beans meal brine phase. The study found *Lactococcus lactis* and *Staphylococcus aureus* were the main bacteria in the post-fermentation process, and a variety of microbes, such as *Bacillus subtilis and Enterobacter*, has been found in the process.

6.4.3.2   Process Technology of Fermented Soya Beans

Fermented soya beans are mainly made of beans, processed by raw material handling, stewing, koji-making, fermentation, and drying. As one of the four traditional fermented soybean products in China, these not only have many types but also use the natural fermentation method, which is made by using the microorganism in the environment and raw materials combined with the local climate fermentation. However, the open fermentation methods are often accompanied by serious safety risks and poor batch stability, which is not conducive to industrial standardized production. Fermented soya beans have only a limited presence in marketplace due to the long fermentation cycle, high storage cost, and low market value. In addition, in order to prevent the spoilage of the products, high concentration of salt was used to extend the shelf life, which further discourages the consumers. In recent years, with the development of modern fermentation technology, directional fermentation with specific dominant strains has become the direction of industrial processing of modern fermented foods. Therefore, the strains can be divided into *Aspergillus*, *Mucor*, *Rhizopus*, *Bacteria*, and *Neurospora* according to the different strains used in the stage of koji-making. Among them, *Aspergillus*-type of fermented soya beans are mostly used in China, and tempeh of Indonesia and natto of Japan are representatives of *Rhizopus*-type and Bacterial-type, respectively. Therefore, there were significant differences in sensory and texture properties of fermented soya beans by different strains.

6.4.3.3   Effect of Lactic Acid Bacteria on Fermented Soya Bean Quality

Lactic acid bacteria are not the key fermentation bacteria in the process of fermented soya beans production, but they play an important role in the formation of flavor, enhancements to the functionality, and nutritional characteristics of fermented soybeans during the post-fermentation stage. Similar to other fermented soybean products, the lactic acid bacteria with good salt-tolerant characteristics, such as *Tetragenococcus halophilus*, can use small molecular sugar to produce a variety of organic acids, which can form esters with unique aroma with the alcohols produced by yeast fermentation. At the same time, a large number of small molecules of saccharides in the system are metabolized and utilized by microorganisms in the post-fermentation process so as to reduce the Maillard reaction in the post-ripening process and effectively maintain the colors of the products. In addition, it was found that *Lactobacillus plantarum* and *Lactococcus lactis* contained in fermented soya beans can metabolically produce bacteriocins, which can inhibit the growth of spoilage bacteria, to effectively extend shelf life to meet current consumer demand for green and chemical-free food market.

References

Akanbi B, Agarry O (2014) Hypocholesterolemic and growth promoting effects of *Lactobacillus plantarum* AK isolated from a Nigerian fermented cereal product on rats fed high fat diet. Adv Microbiol 4(3):160–166

Balestra F, Gianotti A, Saa DT et al (2014) Durum wheat and Kamut® bread characteristics: influence of chemical acidification. 7th International Congress Flour-Bread'13 and 9th Croatian Congress of Cereal Technologists

Brandt MJ (2014) Starter cultures for cereal based foods. Food Microbiol 37:41–43

Brandt MJ, Hammes WP, Gänzle MG et al (2004) Effects of process parameters on growth and metabolism of *Lactobacillus sanfranciscensis* and *Candida humilis* during rye sourdough fermentation. Eur Food Res Technol 218(4):333–338

Cao Y, Lu J, Fang H et al (2008) Fungal diversity of wheat Qu of shaoxing rice wine. Food Sci 29(3):277–282

Chen S, Xu Y (2013) Effect of 'wheat Qu'on the fermentation processes and volatile flavour-active compounds of Chinese rice wine (Huangjiu). J Inst Brew 119(1-2):71–77

Chen T, Wang M, Jiang S et al (2011) Investigation of the microbial changes during koji-making process of Douchi by culture-dependent techniques and PCR-DGGE. Int J Food Sci Technol 46(9):1878–1883

Chen HX, Zhao LQ, Yun TT et al (2014a) Study on manufacturing process and nutritional value of brown rice beverage by probiotic fermentation. Food and Fermentation Industries 11:269–275

Chen L, Xu S, Pan Y et al (2014b) Diversity of lactic acid bacteria in Chinese traditional fermented foods. In: Beneficial microbes in fermented and functional foods. CRC Press, Boca Raton, p 1

Cheng ZY, Mo SP, Bai JL et al (2012) A survey of research progress and production of cereal beverages in China. Beverage Ind 15(6):6–10

Cho SS, Qi L, Fahey GC et al (2013) Consumption of cereal fiber, mixtures of whole cereals and bran, and whole cereals and risk reduction in type 2 diabetes, obesity, and cardiovascular disease. Am J Clin Nutr 98(2):594–619

Clarke CI, Schober TJ, Dockery P et al (2004) Wheat sourdough fermentation: effects of time and acidification on fundamental rheological properties. Cereal Chem 81(3):409–417

Corsetti A (2013) Technology of sourdough fermentation and sourdough applications. In: Handbook on sourdough biotechnology. Springer, New York, pp 85–103

Corsetti A, Gobbetti M, De Marco B et al (2000) Combined effect of sourdough lactic acid bacteria and additives on bread firmness and staling. J Agric Food Chem 48(7):3044–3051

Cui RY, Zheng J, Wu CD et al (2014) Effect of different halophilic microbial fermentation patterns on the volatile compound profiles and sensory properties of soy sauce moromi. Eur Food Res Technol 239(2):321–331

Deng YJ, Liu S, Liu K et al (2015) Exploration of improving the utilization and flavor of low-salt solid-state soy sauce. China Condiment (11):57–58, 63

Di Cagno R, Pontonio E, Buchin S et al (2014) Diversity of the lactic acid bacteria and yeast microbiota switching from firm- to liquid- sourdough fermentation. Appl Environ Microbiol AEM 80(10):3161–3172

Di Monaco R, Torrieri E, Pepe O et al (2015) Effect of sourdough with exopolysaccharide (EPS)-producing lactic acid bacteria (LAB) on sensory quality of bread during shelf life. Food Bioprocess Technol 8(3):691–701

Ercolini D, Pontonio E, De Filippis F et al (2013) Microbial ecology dynamics during rye and wheat sourdough preparation. Appl Environ Microbiol 79(24):7827–7836

Fang H (2006) Primary study of microorganism on wheat Qu of Shaoxing rice wine. Master dissertation, Jiangnan University, Wuxi

Fang RS, Dong YC, Chen F et al (2015) Bacterial diversity analysis during the fermentation processing of traditional Chinese yellow rice wine revealed by 16S rDNA 454 pyrosequencing. J Food Sci 80(10):M2265–M2271

Feng Y, Su G, Zhao H et al (2015) Characterisation of aroma profiles of commercial soy sauce by odour activity value and omission test. Food Chem 167:220–228

Galle S, Arendt EK (2014) Exopolysaccharides from sourdough lactic acid bacteria. Crit Rev Food Sci Nutr 54(7):891–901

Gänzle MG (2014) Enzymatic and bacterial conversions during sourdough fermentation. Food Microbiol 37:2–10

Gao XL, Cui C, Zhao HF et al (2010) Changes in volatile aroma compounds of traditional Chinese-type soy sauce during moromi fermentation and heat treatment. Food Sci Biotechnol 19(4):889–898

Gobbetti M, Gänzle M (2012) Handbook on sourdough biotechnology. Springer, New York

Gobbetti M, De Angelis M, Arnaut P et al (1999) Added pentosans in breadmaking: fermentations of derived pentoses by sourdough lactic acid bacteria. Food Microbiol 16(4):409–418

Gobbetti M, Rizzello CG, Di Cagno R et al (2014) How the sourdough may affect the functional features of leavened baked goods. Food Microbiol 37:30–40

Han QH (2013) Study on the relationship between microbial community and vinegar flavor in the traditional brewing process of Liangzhou Fumigated Vinegar. Master dissertation, Gansu Agricultural University, Gansu

Han CJ, Liu CH, Zhou X (2010) Judgement indicators and measures of Jiaozi quality. Sci Technol Food Ind (5):107–113

Hansen A, Schieberle P (2005) Generation of aroma compounds during sourdough fermentation: applied and fundamental aspects. Trends Food Sci Technol 16(1):85–94

Haruta S, Ueno S, Egawa I et al (2006) Succession of bacterial and fungal communities during a traditional pot fermentation of rice vinegar assessed by PCR-mediated denaturing gradient gel electrophoresis. Int J Food Microbiol 109(1):79–87

Hu LH (2010) Screening of microorganisms from traditional starter cultures and their effects on the quality of Mantou. Master dissertation, Henan University of Technology, Zhengzhou

Iacumin L, Cecchini F, Manzano M et al (2009) Description of the microflora of sourdoughs by culture-dependent and culture-independent methods. Food Microbiol 26(2):128–135

Jänsch A, Korakli M, Vogel RF et al (2007) Glutathione reductase from *Lactobacillus sanfranciscensis DSM20451T*: contribution to oxygen tolerance and thiol exchange reactions in wheat sourdoughs. Appl Environ Microbiol 73(14):4469–4476

Jia JB (2002) Development of oat probiotics milk of *lactobacillus rhamnosus*. Sci Technol Food Ind 23(10):40–42

Katina K, Poutanen K (2013) Nutritional aspects of cereal fermentation with lactic acid bacteria and yeast. In: Handbook on sourdough biotechnology. Springer, New York, pp 229–244

Katina K, Salmenkallio-Marttila M, Partanen R et al (2006) Effects of sourdough and enzymes on staling of high-fibre wheat bread. LWT-Food Sci Technol 39(5):479–491

Lee S, Seo B, Kim YS (2006) Volatile compounds in fermented and acid-hydrolyzed soy sauces. J Food Sci 71(3):C146–C156

Li DY (2008) The variation of the flavors and functional factors during the production of Zhenjiang Vinegar, Master dissertation, Jiangnan University, Wuxi

Li WY (2013) Analysis on current situation and development trends of soy sauce industry. Jiangsu Condiment Subsidiary Food (1):1–3

Li LT, Lu ZH, Min WH (2001) Influence of natural fermentation on the physicochemical characteristics of rice and gelation mechanism of rice noodle. Food Ferment Ind 27(12):1–6

Li S, Li P, Feng F et al (2015) Microbial diversity and their roles in the vinegar fermentation process. Appl Microbiol Biotechnol 99(12):4997–5024

Liao YT, Wu J, Long M et al (2015) Screening of dominant lactic acid bacteria from naturally fermented yak milk in Tibetan pastoral areas and optimization of fermentation conditions for yak yogurt production. Food Sci (11):140–144

Liu TJ, Li Y, Wu SR et al (2014) Isolation and identification of bacteria and yeast from Chinese traditional sourdough. Mod Food Sci Technol 30(9):114–120

Liu Y, Hu MF, Liu SC (2015) Effect on volatile flavor compounds in broad bean sauce fermented in four different ways. Mod Food Sci Technol 31(3):190–196

Loponen J, Sontag-Strohm T, Venäläinen T et al (2007) Prolamin hydrolysis in wheat sourdoughs with differing proteolytic activities. J Agric Food Chem 55(3):978–984

Lu ZY, Wei KQ (2006) Discuss on the high salt liquid state fermentation of the soy sauce. China Condiment (1):28–31

Lu ZH, Peng HH, Li LT (2006) Isolating and identifying microbes in fermented rice noodles of Changde. J Chin Cereal Oils Assoc 21(3):23–26

Lv XC, Huang XL, Zhang W et al (2013) Yeast diversity of traditional alcohol fermentation starters for Hong Qu glutinous rice wine brewing, revealed by culture-dependent and culture-independent methods. Food Control 34(1):183–190

Lv XC, Jia RB, Li Y et al (2016) Characterization of the dominant bacterial communities of traditional fermentation starters for Hong Qu glutinous rice wine by means of MALDI-TOF mass spectrometry fingerprinting, 16S rRNA gene sequencing and species-specific PCRs. Food Control 67:292–302

Ma X, Zhang MM, He Y et al (2015) Research development of effect of fermentation on the quality of fresh rice noodle. China Brew 34(4):5–7

Marti A, Torri L, Casiraghi MC et al (2014) Wheat germ stabilization by heat-treatment or sourdough fermentation: effects on dough rheology and bread properties. LWT-Food Sci Technol 59(2):1100–1106

McKay L, Baldwin K (1974) Simultaneous loss of proteinase-and lactose-utilizing enzyme activities in *Streptococcus lactis* and reversal of loss by transduction. Appl Microbiol 28(3):342–346

Meng XY (2007) Study on retrogradation mechanism and influencing factors of starch retrogradation. Food Eng (2):60–63

Miao ZW, Liu YP, Chen HT et al (2010) Analysis of volatile components in Shanxi overmature vinegar with different staling periods. Food Sci (24):380–384

Miao ZW, Liu YP, Huang MQ et al (2013) The change of volatile aroma components of Douzhi in the heating process. J Chin Inst Food Sci Technol (2):199–204

Min WH, Li LT, Wang CH (2004) Effects of lactic acid bacteria fermentation of rice starch on physical properties. Food Sci 25(10):73–76

Minervini F, De Angelis M, Di Cagno R et al (2014) Ecological parameters influencing microbial diversity and stability of traditional sourdough. Int J Food Microbiol 171:136–146

Minervini F, Lattanzi A, De Angelis M et al (2015) House microbiotas as sources of lactic acid bacteria and yeasts in traditional Italian sourdoughs. Food Microbiol 52:66–76

Nanson NJ, Fields ML (1984) Influence of temperature of fermentation on the nutritive value of lactic acid fermented cornmeal. J Food Sci 49(3):958–959

Oguntoyinbo FA, Narbad A (2015) Multifunctional properties of *Lactobacillus plantarum* strains isolated from fermented cereal foods. J Funct Foods 17:621–631

Onyango C, Henle T, Ziems A et al (2004a) Effect of extrusion variables on fermented maize–finger millet blend in the production of uji. LWT-Food Sci Technol 37(4):409–415

Onyango C, Noetzold H, Bley T et al (2004b) Proximate composition and digestibility of fermented and extruded uji from maize–finger millet blend. LWT-Food Sci Technol 37(8):827–832

Pétel C, Onno B, Prost C (2016) Sourdough volatile compounds and their contribution to bread: a review. Trends Food Sci Technol 59:105–123

Plessas S, Mantzourani I, Bekatorou A et al (2015) New biotechnological approaches in sourdough bread production regarding starter culture applications. Advances in Food Biotechnology, Hoboken, pp 277–285

Qiu C, Sun W, Zhao Q et al (2013) Emulsifying and surface properties of citric acid deamidated wheat gliadin. J Cereal Sci 58(1):68–75

Rizzello CG, Coda R, Mazzacane F et al (2012) Micronized by-products from debranned durum wheat and sourdough fermentation enhanced the nutritional, textural and sensory features of bread. Food Res Int 46(1):304–313

Robertson GH, Cao TK, Gregorski KS et al (2014) Modification of vital wheat gluten with phosphoric acid to produce high free swelling capacity. J Appl Polym Sci 131(2):39440

Sarfaraz A, Azizi M, Esfahani H et al (2014) Evaluation of some variables affecting the acidification characteristics of liquid sourdough. J Food Sci Technol 12(46):65–74

Sarfaraz A, Azizi M, Hamidi EZ et al (2015) Evaluation of some variables affecting the acidification characteristics of liquid sourdough. Iran J Food Sci Technol 13(60):115–124

Scheirlinck I, van der Meulen R, van SA et al (2007) Influence of geographical origin and flour type on diversity of lactic acid bacteria in traditional Belgian sourdoughs. Appl Environ Microbiol 73(19):6262–6269

Singracha P, Niamsiri N, Visessanguan W et al (2016) Application of lactic acid bacteria and yeasts as starter cultures for reduced-salt soy sauce (moromi) fermentation. LWT-Food Sci Technol 78:181–188

Solieri L, Giudici P (2008) Yeasts associated to traditional balsamic vinegar: ecological and technological features. Int J Food Microbiol 125(1):36–45

Su DM (2005) Studies on classification and quality evaluation of staple Chinese steamed bread. PhD dissertation, China Agriculture University, Beijing

Su DM (2009) Probe into the origin of the steamed bun and its historical development. J Henan Univ Technol (Soc Sci Ed) 5(2):14–18

Su YH (2015) Microorganisms and flavor formation in vinegar production with solid-state fermentation. China Brew 34(3):137–140

Sui CG, Chu YY (2013) Study on the production technology of Gwas. Packag Food Mach 31(3):60–62

Thiele C, Gänzle M, Vogel R (2002) Contribution of sourdough lactobacilli, yeast, and cereal enzymes to the generation of amino acids in dough relevant for bread flavor. Cereal Chem 79(1):45–51

Tieking M, Korakli M, Ehrmann MA et al (2003) In situ production of exopolysaccharides during sourdough fermentation by cereal and intestinal isolates of lactic acid bacteria. Appl Environ Microbiol 69(2):945–952

Tong LT, Zhou SM, Lin LZ et al (2013) Changes of main microflora in Changde fresh wet rice noodles. Mod Food Sci Technol 29(11):2616–2620

Üçok G, Hayta M (2015) Effect of sourdough addition on rice based gluten-free formulation: rheological properties of dough and bread quality. Qual Assur Saf Crops Foods 7(5):643–649

Vogelmann SA, Seitter M, Singer U et al (2009) Adaptability of lactic acid bacteria and yeasts to sourdoughs prepared from cereals, pseudocereals and cassava and use of competitive strains as starters. Int J Food Microbiol 130(3):205–212

Wang FL, Liu AG (2003) Study on the fermentation of Mimi. Sci Technol Food Ind 24(5):47–49

Wang WG, Cao W, Zhu XS (2013) Determination of organic acids in vinegar and difference analysis. Sichuan Food Ferment 49(2):81–84

Waters DM, Mauch A, Coffey A et al (2015) Lactic acid bacteria as a cell factory for the delivery of functional biomolecules and ingredients in cereal-based beverages: a review. Crit Rev Food Sci Nutr 55(4):503–520

Wolter A, Hager AS, Zannini E et al (2014) Evaluation of exopolysaccharide producing Weissella cibaria MG1 strain for the production of sourdough from various flours. Food Microbiol 37:44–50

Wu SRGL (2011) Identification and biodiversity of yeast and LAB isolated from sourdoughs collected from western region of inner Mongolia. Master dissertation, Agricultural University of the Inner Mongol, Hohehot

Wu JJ, Ma YK, Zhang FF et al (2012) Biodiversity of yeasts, lactic acid bacteria and acetic acid bacteria in the fermentation of "Shanxi aged vinegar", a traditional Chinese vinegar. Food Microbiol 30(1):289–297

Xu W, Zhang XJ, Xu HY et al (2007) Analysis of bacterial communities in aerobic solid-fermentation culture of Zhenjiang Hengshun vinegar. Microbiology 34(4):646–649

Yan HY, Zhan P, Liu YD et al (2008) Study on production technologies of the corn fermented beverage. Cereal Oils Process 6:117–119

Yang JY, Liu CH (2007) Industrialization of Chinese traditional Jiaozi. Food Res Dev 28(2):164–166

Yang JY, Liu CH, Niu L et al (2006) Research on the change of physical and chemical index of traditional Jiaozi during the making procedure. Cereal Oils Process (10):70–72

Yang Y, Deng Y, Jin Y et al (2016) Dynamics of microbial community during the extremely long-term fermentation process of a traditional soy sauce. J Sci Food Agric 97(10):3220–3227

Yin Y, Wang J, Yang S et al (2015) Protein degradation in wheat sourdough fermentation with *Lactobacillus plantarum M616*. Interdiscip Sci Comput Life Sci 7(2):205–210

Zhang Z, Wang QY (2013) Study on current situation of cereal-based beverages: a review. Beverage Ind 16(8):45–50

Zhang J, Liu W, Sun Z et al (2011a) Diversity of lactic acid bacteria and yeasts in traditional sourdoughs collected from western region in Inner Mongolia of China. Food Control 22(5):767–774

Zhang ZL, Xiong L, Zhao YL et al (2011b) Study on effect of amylose content and pasting properties on rice noodles gels texture. J Qingdao Agric Univ 28(1):60–64

Zhao DA (2005) Mixed ferment and pure-blood ferment. China Condiment 3:3–8

Zhao DA (2009) Evolution and development of soy sauce production technology in China. China Brew 28(9):15–17

Zhao GZ, Sun FY, Yao YP et al (2014) Screening of lactic acid bacteria in the fermentation of mature vinegar and its effects on flavors. Sci Technol Food Ind 35(24):159–163

Zhao CJ, Kinner M, Wismer W et al (2015) Effect of glutamate accumulation during sourdough fermentation with *lactobacillus reuteri* on the taste of bread and sodium-reduced bread. Cereal Chem 92(2):224–230

Zhu F (2014) Influence of ingredients and chemical components on the quality of Chinese steamed bread. Food Chem 163:154–162

Zhu H, Zhu J, Wang L et al (2016) Development of a SPME-GC-MS method for the determination of volatile compounds in Shanxi aged vinegar and its analytical characterization by aroma wheel. J Food Sci Technol 53(1):171–183

Zou X, Chen Z, Shi J et al (2011) Near infrared modeling of total acid content in vinegars based on LS-SVM. China Brew 3:63–65

Chapter 7
Lactic Acid Bacteria and Fermented Fruits and Vegetables

Bingyong Mao and Shuang Yan

7.1 Overview

China has a vast territory and abundant resources with an annual output of hundreds of millions of tons of fruits and vegetables. The per capita consumption and export volume are among the highest in the world. However, China's annual losses in fruit and vegetable products amount to tens of billions of yuan due to the lag of processing technology and industrialization and the lack of storage and processing methods, and the cost of losses is also at the top in the world. As common microbes in fermented food, lactic acid bacteria not only can improve the flavor and storage of the product but also have certain health functions. Therefore, it can not only facilitate the preservation but also increase the added value and meet the market demand for product diversity by applying lactic acid bacteria to the processing of fruit and vegetables.

Fermentation technology is one of the oldest food preservation technologies in the world, and fermented foods are widely distributed all over the world, especially in some underdeveloped countries. Fermented fruit may be the first fermented food in human history. Hunter gatherers often eat fresh fruit, but they will also eat rotten or fermented fruit when food is scarce. Repeated consumption of fermented fruit may prompt our ancestors to consciously improve the flavor of the fermented fruit. Studies have shown that fermented beverages appeared in Babylon 7000 years ago (Stanton 1998).

The production of traditional fermented foods is often considered as unsanitary or unsafe. Although it does occur at some point, but it is often overstated. Many fermented foods can inhibit the growth of spoilage bacteria due to lower moisture or higher acidity and salt concentration. So the fermented foods have a certain degree of security. In addition, fermentation is a very appropriate food processing and

B. Mao (✉) · S. Yan
Jiangnan University, Wuxi, China
e-mail: maobingyong@jiangnan.edu.cn

© Springer Nature Singapore Pte Ltd. and Science Press 2019
W. Chen (ed.), *Lactic Acid Bacteria*,
https://doi.org/10.1007/978-981-13-7283-4_7

181

preservation technology for developing or remote areas because fermentation technology is less dependent on refrigeration or other food preservation technologies. Therefore, natural foods or raw materials can be more effectively used, and waste materials can be even converted through microbial fermentation.

For the fermented foods like cheese, bread, beer, and wine, the production process is mostly in commercial scale, and the microbial fermentation processes had been studied in-depth. However, the relevant research and theory are still scarce for traditional fermented foods in Asia, Africa, and Latin America. Although the basic theory has broad consistency, the production conditions are quite different due to the diversity of fermentation products in different regions. Research on these production processes can be helpful to guide the product manufacture to improve the yield and quality of the fermented food.

Traditionally the fermented food is a natural fermentation process without sterilization. The microorganisms are selected in special environmental conditions and could produce unique end products. Lactic acid bacteria are the most widely used in fermented vegetables, and the fermentation process is shown in Fig. 7.1. Lactic acid

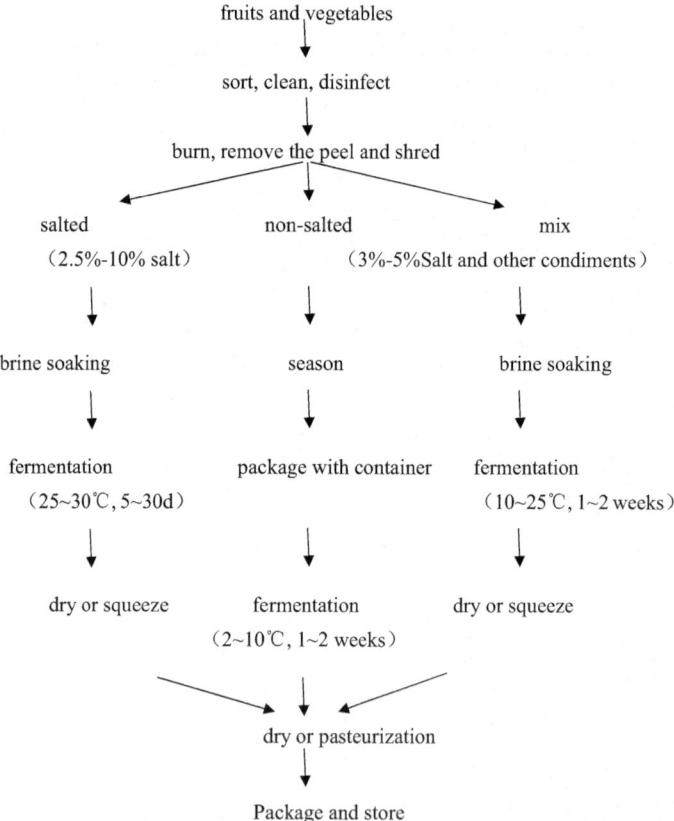

Fig. 7.1 General process of fermentation of fruits and vegetables (Swain et al. 2014)

bacteria have long been used in vegetable fermentation, and China is one of the first countries to adopt this method for vegetable preservation. About 3000 years ago, Chinese people invented the production process of pickles, and the technology was transmitted to North Korea 1300 years ago which developed Korean kimchi that is popular in the world today. In Europe, lactic acid bacteria-fermented vegetables also have a long history. As early as the first century AD, people first reported the acid kimchi made from cabbage. In modern times, lactic acid bacteria-fermented vegetables are still very popular among the consumers at home and abroad due to its unique flavor and rich nutrition. Vegetables are rich in nutrients, including vitamins, cellulose, and minerals, but these ingredients are easily lost during the processing. However, the "cold processing" method of lactic acid bacteria fermentation is extremely beneficial to the maintenance of vegetable nutrients and color. At the same time, kimchi contains a large amount of lactic acid bacteria; after entering the digestive tract of the human body, it can relieve constipation by promoting gastrointestinal motility, lowering blood fat, and enhancing the host immunity. With the increasing in-depth research and development of lactic acid bacteria-fermented vegetables, traditional foods have become an emerging modern food industry.

Besides fermented vegetables, lactic acid bacteria-fermented fruit and vegetable juice is a new type of beverage developed in recent years. The research and development of lactic acid bacteria-fermented fruit and vegetable juice has broad prospects in enriching the beverage market and increasing the vegetable value due to the good flavor and nutritional health value.

7.1.1 The Main Products of Fermented Fruit and Vegetables

7.1.1.1 Fermented Vegetables

Fermented vegetables are obtained under certain process conditions using beneficial microorganisms, including lactic acid bacteria, yeast, and acetobacteria, of which lactic acid bacteria are the most important microorganisms. Many vegetables can be used as raw materials for fermentation, such as cabbage, radish, kale, and green head. Common fermented vegetable products mainly include kimchi, sauerkraut, mustard, snow, fermented radish, and fermented olive oil.

7.1.1.2 Fermented Fruit and Vegetable Juice Drink

7.1.1.2.1 Fermented Fruit and Vegetable Juice Mix

Among the current probiotic products on the market, fermented dairy products still occupy a dominant position, and 78% of probiotic products are produced and sold in the form of yogurt (Abu-Ghannam and Rajauria 2015). In recent years, vegetarian food has become more and more popular worldwide due to the high cholesterol content in animal-derived products, and there are different proportions of patients with lactose intolerance in many countries. Nondairy probiotic products have become a new research hotspot. A variety of fruit and vegetable juices and cereal products are likely to be suitable carriers for lactic acid bacteria. Fruit and vegetable resources are extremely rich and inexpensive in China, but they depend on the seasons and can't be supplied all the year round. Fruit and vegetable juice are fermented by lactic acid bacteria and can be supplied all year round to meet the market needs. Moreover, the juices fermented by lactic acid bacteria such as *Lactobacillus plantarum* and *Lactobacillus acidophilus* are natural, nutritious, and healthy. Therefore, there are a large number of researches on the fruit and vegetable juices fermentation by lactic acid bacteria, including tomatoes, carrots, kale, pumpkins, and apples.

7.1.1.2.2 Fermented Fruit and Vegetable Juice Milk Drink

Fresh and delicious fermented fruit and vegetable juice milk drinks can be produced by adding a small amount of milk powder and lactose. The strains used in the production process are the same as that used for producing yogurt, except that the amount of raw milk is reduced and 10–30% of the juice is added. The resulting beverage has both the frankincense of yogurt and the scent of fruits and vegetables and is more nutritious than yogurt.

7.1.2 Lactic Acid Bacteria Commonly Used in Fruits and Vegetables Fermentation

7.1.2.1 Homofermentative Lactic Acid Bacteria

7.1.2.1.1 *Lactobacillus plantarum*

The cells are nonmotile rods without flagella, occurring singlely, in pairs or in short chains. Strains are usually facultative anaerobic and Gram positive. The optimum growth temperature range is 30–35 °C, and sugars are fermented to produce lactic acid. It plays a major role in the fermentation of fruits and vegetables due to its strong tolerance to acid and salt.

7.1.2.1.2 *Streptococcus faecalis*

The cells are spherical, occurring in pairs or in chains. The strains are facultative anaerobic and Gram positive, and they do not produce spores. The growth temperature range is 25–45 °C, and the optimum temperature is 37 °C. The strains usually have poor tolerance to acid, and some strains are pathogens to humans and animals.

7.1.2.1.3 *Pediococcus pentosaceus*

The cells are spherical occurring usually in pairs and not in chains. The strains are facultative anaerobic and Gram positive, and they don't produce spores. The optimum growth temperature range is 25–40 °C.

7.1.2.1.4 *Lactobacillus casei*

The cells are rod without flagella. The strains are facultative anaerobic and Gram positive, and they don't produce spores. The optimum growth temperature is 37 °C, and glucose is fermented to produce acid. The strains have good tolerance to salt.

7.1.2.2 Heterofermentative Lactic Acid Bacteria

7.1.2.2.1 *Leuconostoc mesenteroides*

The cells are spherical or elliptical occurring in pairs or in chains, usually in the form of short chains. The strains are facultative anaerobic and Gram positive, and they don't produce spores. It can grow at 10–32 °C, and the optimum growth temperature is 20–30 °C. It can ferment sucrose to produce a characteristic and viscous dextran capsule. It grows faster in the early stage of fermentation and plays an important role in the start-up stage of fruit and vegetable fermentation. The pH drops rapidly and inhibits the growth of spoilage bacteria. The acid, alcohol, and other substances produced during the fermentation could be combined with other metabolites and play an important role in the formation of flavor of the product. However, the strains have poor tolerance to the acid and the growth is slow in the later stage. *Leuconostoc mesogenes* can convert excess sugar into mannitol or dextran, while mannitol and dextran can only be used by lactic acid bacteria and cannot be used by other microorganisms or react with amino acids to form aldehyde or ketone groups. Therefore, it does not cause browning of food. *Leuconostoc mesenteroides* relies entirely on the pentose phosphate pathway to degrade glucose due to the lack of aldolase and isomerase.

7.1.2.2.2 *Lactobacillus brevis*

The cells are straight rods with rounded ends occurring singlely or in short chains. The strains are facultative anaerobic and Gram positive, and they don't produce spores. The growth temperature is 15–40 °C, and the optimum growth temperature is 30 °C. The strains can ferment pentose and form a unique flavor.

7.1.2.2.3   *Lactobacillus fermentum*

The cells are usually short rods occurring in pairs or in chains. The strains are facultative anaerobic and Gram positive, and they don't produce spores. The growth temperature is 30–35 °C. It can grow above 45 °C and can't grow below 15 °C. The strains usually distribute on the surface of plants and in the gastrointestinal tract of animals and can survive in low pH and high bile acid intestinal environment. It has a promoting effect on regulating the balance of host intestinal flora and host health. The strains can ferment glucose to produce lactic acid, acetic acid, and ethanol, which promote the formation of flavor of the product (Zhang Yan-chao 2008).

7.1.3   Nutritional Value and Functional Characteristics of Fermented Fruits and Vegetables

7.1.3.1   Effects of Lactic Acid Bacteria on Nutritional and Physicochemical Properties of Fermented Fruits and Vegetables

7.1.3.1.1   Improve the Nutrition of Products

Lactic acid fermentation is a cold processing method, and the fermentation process does not reduce the nutritional value of vegetables. On the contrary, lactic acid bacteria can produce a variety of amino acids, vitamins, and enzymes by using soluble substances in raw materials. Lactic acid bacteria also have a weak decomposition ability for fat, which can increase the amount of free fatty acids in food. Wen-hui and Bin (1995) used *Lactobacillus bulgaricus* and *Streptococcus thermophilus* to ferment carrot juice and tomato juice and the content of vitamin B1, vitamin B2, and pyridoxine increased greatly after fermentation. Moreover, the content of glutamic acid and aspartic acid in the fermented product is increased, and the content of sulfur-containing amino acids like methionine and histidine is decreased. The calcium and phosphorus are in the form of ions after fermentation, which can be easily absorbed and promote the development of bones and can also prevent anemia and rickets caused by calcium deficiency and iron deficiency.

7.1.3.1.2   Improve the Flavor of the Products

The fermentation products of lactic acid bacteria are mainly organic acids including lactic acid, acetic acid, and propionic acid, and these flavor substances can bring a soft sour taste. Ethyl acetate or ethyl lactate can be formed by combining lactic acid and acetic acid with an alcohol, imparting a fruity aroma to the product. Meanwhile, 2-heptanone and 2-nonanone can impart a certain fragrance and refreshing after fermentation (Shin et al. 2003). In addition, the sour taste of lactic acid can mask the odor (green odor) of vegetable products (Ju-hua et al. 2003).

7.1.3.1.3 Improve the Safety of Products

Improve the Preservation of Products

Fermentation is a low-cost, energy-saving food preservation technology. Fruit and vegetable products are prone to spoilage after harvest, especially in humid tropical regions, where special environments can accelerate the process of corruption. In order to prevent the corruption of food, there are drying, freezing, canning, pickling, and other processing methods. Among these methods, the freezing method has higher requirements on equipment, and the cost is also higher. The fermentation of fruits and vegetables does not require complicated equipment, and the low-acid environment after fermentation is also beneficial to product preservation. In addition, lactobacillin, hydrogen peroxide, and diacetyl formed during the fermentation have been proven to inhibit the growth of some spoilage bacteria and pathogenic bacteria. In recent years, lactobacillin has become a hot topic all over the world and is used as a natural preservative in the food industry. At present, nisin has been used as a natural preservative to extend the shelf life of foods, which has a broad-spectrum inhibitory effect on Gram-positive bacteria. Micrococcin can also inhibit the growth of Gram-positive bacteria, but it has not yet been approved by the food safety department. Thus both lactic acid bacteria and their metabolites can effectively prevent food spoilage (Martínez-Castellanos et al. 2011; Trias et al. 2008).

Remove Anti-nutritional Factors or Toxins

Some fruits and vegetables contain naturally occurring toxins and anti-nutritional compounds that can be detoxified or degraded by the fermentation of microorganisms. For example, cassava naturally contains cyanogenic glycosides, and it will release deadly cyanide in the body without proper treatment. After cassava is peeled and fermented for a certain period, toxic substances can be degraded by 90–95% (Battcock 1998).

Many fruits and vegetables generally contain a certain amount of nitrate, and some bacteria can convert nitrate to nitrite during the fermentation. Nitrite can be further converted to nitrosamine after intake, which is harmful to the human body. Lactic acid bacteria in fermented fruits and vegetables can reduce the formation of nitrite in fruit and vegetable products by inhibiting the proliferation of harmful bacteria and degrading nitrite. Lactic acid bacteria such as *Lactobacillus plantarum*, *Lactobacillus brevis*, *Lactobacillus fermentum*, and *Leuconostoc mesophila* were isolated from Chinese kimchi by Yan et al. (2008) and can degrade nitrite by more than 97%. Guang-yan et al. (2006) found that the content of nitrite in kimchi inoculated with *Lactobacillus plantarum* was significantly lower than that of naturally fermented kimchi.

7.1.3.2 Functional Properties of Fermented Fruits and Vegetables

7.1.3.2.1   Relieve Constipation and Diarrhea

Constipation is caused by abnormal mechanical movement of the large intestine, mainly manifested by difficulty in defecation, rectal swelling, incomplete emptying, and prolonged emptying time. Diet is one of the common causes of constipation, and non-amyloid polysaccharides in the large intestine can restore the balance of intestinal flora and help relieve constipation. Kimchi contains a large amount of lactic acid bacteria and can relieve symptoms of constipation by producing short-chain fatty acids and regulating the intestinal flora. Kimchi is an essential dish in the homes of people in southwestern China, and kimchi water is a common prescription for treating chronic diarrhea in rural areas. Kimchi water is beneficial for spleen and dampness. Ping-chun et al. (2010) found that kimchi water had significantly benefi-cial effects in the treatment of diarrhea for children, with an effectiveness of 87%, which was better than Pfeiffer with an effectiveness of 68%.

7.1.3.2.2   Lower Cholesterol

Cholesterol is closely related to the occurrence of diseases such as hyperlipidemia, diabetes, and coronary heart disease. Studies have shown that lactic acid bacteria can absorb cholesterol into their own cell membranes. The bile salt hydrolase pro-duced by the lactic acid bacteria can decompose the bile salt and precipitate together with the lactic acid bacteria cells, thus lowering the blood cholesterol level (Li-na and Chang-qing 2007). The mechanism of lowering blood lipids by lactic acid bacteria may mainly include the following aspects. Some salts formed by organic acids such as acetate, propionate, and lactate can regulate the fat metabo-lism by lowering plasma total cholesterol (TC) and low-density lipoprotein (LDL) and increasing high-density lipoprotein (HDL). The enzymes produced by lactic acid bacteria can inhibit the synthesis of endogenous cholesterol. Some lactic acid bacteria can colonize in the intestinal mucosa, and its metabolism can reduce the absorption of cholesterol in the intestine, which may be related to the assimilation of cholesterol by lactic acid bacteria. Lactic acid bacteria can convert absorbed cholesterol into cholate and promote its excretion (Ling-yan et al. 2004; Yong 2003). Choi et al. (2013) studied the fasting blood glucose and serum cholesterol levels in two groups of volunteers with different intakes of kimchi per day. They found that eating kimchi every day can reduce the fasting blood glucose and serum cholesterol levels in both groups and improve the antioxidant capacity. Moreover, the effect was dose dependent and better effects were observed in the group of the volunteers with higher intake of kimchi (210 g/day) than that with lower intake of kimchi (15 g/day).

7.1.3.2.3   Enhance Immunity

Xiao-ran et al. (2010) studied the effects of fermented fruit and vegetable juices on the immune functions in mice. Three kinds of fermented juices were prepared by mixing orange juice, pear juice, apple juice, and carrot juice with the ratio of 1:1:1:1 and then inoculated with *Bifidobacterium bifidum*, *Lactobacillus bulgaricus*, or *Streptococcus thermophiles*. The results showed that the three fermented juices significantly increased the thymus index and spleen index of the mice, indicating that they enhanced the body's immunity. Moreover, it can promote the production of serum hemolysin antibody in mice and enhance the degree of delayed-type hypersensitivity reaction and the phagocytic ability of macrophages, thereby improving the body fluid and cellular immune function of the body.

7.1.3.2.4   Anticancer

In recent years, the anticancer effect of kimchi has attracted the attention of many scholars. Choi et al. (Choi and Park 1999) extracted the effective healthcare ingredients from fermented cabbage with methanol and then fed the extract to mice with gastric cancer. The extract was found to inhibit the growth of 60.7% of gastric cancer cells and prolong the life of the mice in the experimental group. Studies have also shown that the weight of tumors in mice with transplanted cancer cells was reduced by half and 40% of cancer cells are inhibited to metastasize after 3 weeks of feeding kimchi extract (Wen-bin et al. 2006).

7.1.3.2.5   Lose Weight

Kimchi is a high-fiber, low-calorie food. Regular consumption of kimchi can stimulate the secretion of adrenal hormones, accelerate the burning of body fat, reduce the accumulation of fat in the blood and liver, and enhance the function of spleen immune cells. Korean scientists have shown that the mice fed kimchi had only 145–149 mg/g fat, while the fat content of mice without kimchi was 167–169 mg/g and the total fat content in the blood of mice fed kimchi was 170–200 mg/kg, while the total fat content in the blood of mice without kimchi was 246.1 mg/kg. Subcutaneous fat test also demonstrated that mice fed kimchi showed significant weight loss.

7.1.3.2.6   Antiaging

Lactic acid bacteria and their metabolites have strong antioxidant activity because lactic acid bacteria can produce superoxide dismutase (SOD). SOD is a kind of enzyme containing metal elements, which can remove excess superoxide anion radicals generated during metabolism in the body and improve immunity, delay aging, and reduce fatigue. In addition, lactic acid produced by lactic acid bacteria can inhibit the growth of spoilage bacteria and pathogenic bacteria in the intestinal tract. To a certain extent, the carcinogens and other toxic substances such as H_2S and sputum matrix produced by these harmful bacteria are reduced, and the aging process of the body was slow down (Zhang and Cao 2005).

7.2   Fermented Vegetables by Lactic Acid Bacteria

7.2.1   Main Types of Fermented Vegetable Products

7.2.1.1   Fermented Pickles by Lactic Acid Bacteria

Fermented pickles are a kind of food with unique flavors by natural fermentation or artificial inoculation, which enables lactic acid bacteria to ferment the soluble nutrients in vegetables. During the fermentation, lactic acid bacteria can ferment the soluble components of plants and some saline extracts to produce flavor components such as lactic acid. In addition, the lactobacillin produced during the growth and metabolism can also plays an antiseptic role.

7.2.1.2   Fermented Sauerkraut by Lactic Acid Bacteria

Sauerkraut is mainly distributed in the vast areas of northern China and is a kind of fermented vegetable formed by brewing fresh cabbage in an airtight container with brine for some period. Traditional sauerkraut is produced by natural fermentation, and lactic acid bacteria play an important role in the natural fermentation. Lactic acid produced during the fermentation can give the sauerkraut unique flavors. In the process of pickling, inoculation of mixed lactic acid bacteria can effectively improve the flavor and product quality of pickles.

7.2.1.3   Fermented Olive by Lactic Acid Bacteria

Olive is a subtropical woody oil fruit with fleshy stone fruit native to west Asia and later expanded to Mediterranean coastal countries. Olive is rich in oleic acid, linoleic acid, linolenic acid, and other unsaturated fatty acids, and they are mainly n-3 unsaturated fatty acids, the proportion of which is most suitable for human needs. It has certain medical and health functions for the digestive system, cardiovascular disease, bone, and nervous system development. The salted olive fermented by lactic acid bacteria was table olive, which was widely consumed in Europe (Gang 2007). According to the process method, there are two kinds of fermented olive including Greek-style and Spanish-style. Lactic acid bacteria fermentation can change the pH and total acidity of the olive and improve the palatability of the product. Compared with naturally fermented olive, the bitter taste of fermented olive by lactic acid bacteria inoculation was decreased, and the aromatic taste was increased with increasing concentration of various flavor compounds (Sabatini et al. 2008). The results of isolation and culture of lactobacillus in fermented olive indicated that *Lactobacillus coryniformis* and *Lactobacillus pentosus* were identified as the main lactic acid bacteria during the fermentation of olive (Aponte et al. 2012).

7.2.1.4 Fermented Pickled Mustard by Lactic Acid Bacteria

Pickled mustard was a well-known Chinese vegetable product, which was called "pickled mustard" because water was extracted from vegetables by pressing method during the processing (Zeng-san 2005). The pickled mustard was the main pickling vegetables, along with French pickles and German *Brassica oleracea*. Pickled mustard was fresh, crisp, delicious, and nutritious, with a good taste of sour and special salty taste. Compared with traditional pickled mustard, those fermented by lactic acid bacteria had higher yield of lactic acid, more flavor substances, stronger aroma and taste, and better sensory characteristics (Yun-bin et al. 2012).

7.2.1.5 Fermented Potherb Mustard by Lactic Acid Bacteria

Potherb mustard is cultivated in south and north of China. Potherb mustard is very nutritious containing 7.3 mg of carotene, 0.35 mg of vitamin B1, 0.7 mg of vitamin B2, 4 mg of niacin, and 400 mg of ascorbic acid every 500 g fresh potherb mustard. In addition to rich nutrition, potherb mustard also has a unique delicious flavor. Therefore, whether fresh or pickled, potherb mustard attracts many dinners, especially pickled potherb mustard which has a long history of production. Fermentation of pickled potherb mustard by lactic acid bacteria is an effective biochemical preservation method. The main lactic acid bacteria during the fermentation of potherb mustard were *Streptococcus thermophilus*, *Lactobacillus bulgarian*, *Lactobacillus acidophilus*, and *Lactobacillus plantarum* (Gang 2007).

7.2.1.6 Fermented Dried Turnip by Lactic Acid Bacteria

The traditional dried turnip is mainly fermented by lactic acid bacteria, which has a higher salinity in the curing environment, forming a hyperosmotic dehydration effect, thus inhibiting the growth of spoilage microorganisms. In the process of curing, a series of biochemical changes caused by the metabolism of beneficial microorganisms attached to the surface of radish strip can promote its fermentation and maturation, forming unique taste and nutrition. Compared with the natural fermentation, the dried turnip inoculated with lactic acid bacteria could accelerate the decline rate of pH in the fermentation process, shorten the fermentation period, inhibit the growth of mixed bacteria, and improve the flavor of the product (Xiang-yang et al. 2016). In addition, some researchers isolated one *Lactobacillus sakei* from the dried turnip, which showed good cholesterol-lowering functions (Chang-jian et al. 2011).

7.2.2 The Microecology of Fermented Vegetable Products

The production of traditional Chinese fermented pickles mainly relies on natural fermentation, which mainly relies on the microorganisms attached to the surface of vegetables. In the brine immersion environment, various microorganisms naturally attached to vegetables grow rapidly and use the dissolved carbohydrates and other components of raw materials for metabolic activities. During the fermentation, the species and quantity of microorganisms continue to change, and metabolites like lactic acid, acetic acid, and ethanol and a variety of volatile flavor substances are produced, so as to obtain mature pickles with good flavors. Lactic acid bacteria play a major role on flavor formation, while other microorganisms also exist including yeast and molds.

Tao et al. (2015) studied the changes in the amount of lactic acid bacteria, yeast, acetic acid bacteria, and molds in the fermentation of traditional pickles. The number of lactic acid bacteria increased rapidly in the first 2 days after the start of pickles fermentation and then remained stable until the end of fermentation. Yeast, acetic acid bacteria, and molds increased in the initial stage of fermentation and then decreased gradually until disappearing. As can be seen from the numbers, lactic acid bacteria are the dominant microorganisms in the fermentation of pickles and have good acid toler-ance. The anaerobic environment is formed by the consumption of oxygen in the mid-fermentation vessel and the carbon dioxide production by the metabolism of heterofermentation. Acetic acid bacteria and molds are aerobic microorganisms, so their growth is inhibited. In addition, the decrease of pH by the production of acids can also inhibit the growth of yeast and molds. Further analysis of the dynamic change of lactic acid bacteria revealed that *Leuconostoc mesenteroides* subsp. *mesenteroides* started the fermentation, followed by *Enterococcus faecalis*, *Lactococcus lactis* subsp. *lactis*, and *Lactobacillus zeae*, and finally *Lactobacillus plantarum* and *Lactobacillus casei* terminated the fermentation process (Tao et al. 2012).

Traditional analysis method revealed that *Leuconostoc mesenteroides* and *Lactobacillus plantarum* were the dominant bacteria during the fermentation of Korean pickles. However, more bacteria belonging to *Leuconostoc* and *Lactobacillus* in pickles were identified with the use of molecular identification techniques, such as *Leuconostoc citreum*, *Leuconostoc gasicomitatum*, *Leuconostoc gelidum*, *Lactobacillus sakei*, *Lactobacillus brevis*, and so on. However, some researchers analyzed the microbial composition of commercial pickles by non-culture method and found that *Weissella koreensis* was also the dominant bacteria in pickles (Kim and Chun 2005). Lee et al. (2005) found that *Weissella confusa*, *Leuconostoc citreum*, *Lactobacillus sakei*, and *Lactobacillus curvatus* were the dominant bacteria in the fermentation of pickles using the molecular method. These results suggest that the fermentation of pickles involved a variety of lactic acid bacteria, including *Leuconostoc*, *Lactobacillus*, *Lactococcus*, *Pediococcus*, and *Weissella*.

Park et al. (2014) analyzed the composition of lactic acid bacteria of 13 Korean pickles (with a pH of 4.2–4.4) by pyrosequencing and found that *Weissella koreensis* accounted for the highest proportion (27.2%) in the lactic acid bacteria of pickles, followed by *Lactobacillus sakei* (14.7%), *Weissella cibaria* (8.7%), *Leuconostoc mesenteroides* (7.8%), *Lactobacillus gelidum* (6.3%), *Leuconostoc inhae* (1.2%), *Leuconostoc gasicomitatum* (1.2%), *Weissella confusa* (0.3%), and *Leuconostoc kimchii* (0.3%). At the genus level, *Weissella* had the highest proportion of lactic acid bacteria (44.4%), followed by *Lactobacillus* (38.1%) and *Leuconostoc* (17.3%). *Weissella*, especially *Weissella koreensis*, were the dominant bacteria in pickles.

Wei (2013) studied the microbial community of Sichuan pickles and industrial pickles by building 16S rRNA gene library and found that the microorganisms were mainly *Lactobacillus* and *Pediococcus*, accounting for 88.4% and 10.1%, respectively. The dominant bacteria were *Lactobacillus pentosus*, *Lactobacillus plantarum*, and *Pediococcus damnosus*, which accounted for 50.4%, 16.3%, and 10.1%, respectively. In addition, *Lactobacillus paralimentarius*, *Lactobacillus sunkii*, *Lactobacillus brevis*, *Lactobacillus kisonensis*, *Lactobacillus acetotolerans*, and *Lactobacillus namurensis* were also detected. The main microorganisms in Sichuan industrial pickles were *Halomonas*, *Lactobacillus*, and *Vibrio*, accounting for 13.5%, 32.4%, and 12.1%, respectively.

7.2.3 Lactic Acid Bacteria in Fermented Vegetable Products

7.2.3.1 Species of Lactic Acid Bacteria Commonly Found in Fermented Vegetable Products

The common lactic acid bacteria in fermented vegetables mainly include *Lactobacillus*, *Lactococcus*, *Pediococcus*, *Leuconostoc*, and *Weissella*. The dominant lactic acid bacteria in pickles are mainly *Lactobacillus* and *Leuconostoc*, in which *Leuconostoc* dominates in the early stage of fermentation and *Lactobacillus* rapidly grow and replace its dominant position in the later stage of fermentation (Tao et al. 2012). Specifically, *Lactobacillus plantarum*, *Lactobacillus casei*, *Lactobacillus brevis*, *Lactobacillus pentose*, *Lactobacillus curvatus*, and *Leuconostoc mesenteroides* were the common lactic acid bacteria in the fermentation of pickles (Xiao-lin et al. 2011; Yuan-feng and Li-gen 2007). The species and quantities of lactic acid bacteria in the fermentation have a crucial influence on the flavor and the qualities of the product.

7.2.3.2 Effects of Lactic Acid Fermentation on the Flavor Qualities of Fermented Vegetable Products

Generally, the volatile flavor components detected in fermented pickles with pure inoculation are less than that in traditional naturally fermented pickles. In the fermentation of pickles by pure inoculation, acid was rapidly produced with short fermentation period, and thus fewer flavor substances were accumulated. As a result, the taste and flavor were inferior to traditional naturally fermented pickles. Chun-yan et al. (2015) studied the differences of flavor substances in natural or inoculated fermentation of pickles and found that there was no significant difference in flavor substances, but the content of organic acids was different. However, Dan-ping et al. (2015) found that the main flavor components of pickles were different from each other and only nonaldehyde was the shared flavor component of all kinds of pickles. Bong et al. (2013) found that pickles fermented by a mixture of *Leuconostoc mesenteroides* and *Lactobacillus plantarum* had better sensory properties and higher free radical scavenging abilities than the naturally fermented pickles.

Zu-fang et al. (2008) made the pickled mustard by artificially inoculating *Lactobacillus plantarum* lact-8 and *Leuconostoc mesenteroides* lact-2 and optimized the fermentation process to obtain the optimal fermentation conditions. The conditions were shown as follows: salt addition 8%, fermentation temperature 25 °C, and inoculation amount 2% (the ratio was 1:1 for the two strains). The pickled mustard prepared under this condition had excellent physical, chemical, and sensory properties and was significantly superior to pickled mustard by natural fermentation in terms of nitrites, free amino acids, lactic acid bacteria content, acidity, pH, fermentation time, and sensory quality (Table 7.1).

Table 7.1 Comparison of qualities between natural fermentation and artificially inoculated fermentation pickled mustard (Zu-fang et al. 2008)

Quality indicators	Optimization of fermentation	Natural fermentation
Fermentation time/day	15	17
Terminal pH	4.20	5.15
Terminal acidity/%	0.432	0.315
Lactic acid bacteria content/ (cfu/ml)	2.127×10^7	3.733×10^5
Free amino acid/(mg/kg)	22.076	21.765
Nitrite/(mg/kg)	5.63	22.54
Sensory quality	Light yellow, strong aroma, crisp and tender, stable quality	Yellow, aromatic, crisp, and difficult to control

7.2.4   Technologies and Characteristics of Fermented Vegetable Products by Lactic Acid Bacteria

Fermented vegetables are prepared by placing washed vegetables in containers and adding ingredients, so that soluble ingredients in vegetables and some brine extracts can be used by lactic acid bacteria for fermentation. Brined vegetables usually contain high concentration of salt up to 12–15%, while pickles contain low concentration of salt. Fermenting vegetables with lactic acid bacteria will not reduce the nutritional value of vegetables but can improve their nutritional value. Lactic acid bacteria do not destroy plant cells and tissues, nor do they break down proteins because of their low activities of cellulolytic enzymes and proteases. The nutrients for lactic acid bacteria are mainly soluble substances of plants and some saline extracts. Lactic acid produced during the fermentation can not only act as a preservative but also improve product flavors.

7.2.4.1   Fermented Pickles by Lactic Acid Bacteria

According to the raw materials and processing techniques, fermented pickles can be divided into Chinese pickles (Pao cai), Korean pickles, Japanese pickles, and Sauerkraut.

7.2.4.1.1   Chinese Pickles

Material Selection

The basic requirement of material selection is high content of solid, such as Chinese cabbage, cabbage, or radish, which can exceed 20% in some materials. Physical impact resistance is a direct advantage for materials of high content of solids. It should be noted that the nutrients of materials can meet the requirements of lactic acid bacteria.

Garlic, ginger, chili, onion, and cloves can be added during the pickle fermentation, which not only help to improve the flavors but also exert antibacterial effects. For example, the addition of clove can significantly prevent the growth of bacteria and delay the pH decrease and inhibit the decline of hardness of pickles during the storage. In addition, clove also contains a large amount of flavone and flavonoids with good antioxidant effects (Xiang-yang et al. 2015; Xue-ping et al. 2011).

Crafting Process

Classification

Check the cabbage for pests, yellow leaves, and rotting and inedible vegetables, and then the roots are repaired.

Segmentation

Cabbage is usually segmented into two or four lengthwise or into small pieces. Cabbage and radish are usually cut into small pieces, while garlic, ginger, and chili are cut into shreds.

Add Salt and Accessories

Generally, the amount of salt is 2.0–2.5%, and the amount of sugar is 2–3%.

Inoculation

Fermented vegetables are traditionally inoculated with old pickle saline, while modern fermented vegetables are inoculated with pure cultures of lactic acid bacteria. Lactic acid bacteria at the end of the logarithmic phase are inoculated into the processed vegetables, which helps to better control the product quality and shorten fermentation period. The cultures are generally *Lactobacillus plantarum*, *Leuconostoc mesenteroides*, or mixed culture of *Lactobacillus delbrueckii* subsp. *bulgaricus* and *Streptococcus thermophilus*.

Fermentation

After inoculation and blending, the vegetables are divided into containers, which are pressed tightly and sealed tightly with water. The fermentation temperature is usually 20–25 °C, and the fermentation time is around 2 days to decrease the pH to 4.2.

Preservation

The fermented vegetables are placed at 1–5 °C cold storage for preservation and sailed through the cold chain transportation.

7.2.4.1.2 Japanese and Korean Pickles

Kimchi is the main food dish in Korea. Korean kimchi is spicy, refreshing, and delicious, with Chinese cabbage as the main raw material and red pepper as the main ingredients. Fish sauce, spices, mushrooms, seaweed sauce, and other ingredients are often added to increase the flavors. Some researchers also added mustard and Korean parasitic dendrobium extract to develop kimchi products with special healthcare effects (Fig. 7.2) (Park et al. 2014). Chinese cabbage are salted at 1–5 °C overnight and then mixed with chili paste for natural fermentation. After fermentation, the mature kimchi are packed for sale. Korean kimchi was mainly made by the hypertonic effect of salt and the fermentation by lactic acid bacteria, which not only keeps the vegetables fresh but also improves the nutritional value and healthcare functions (Lim et al. 2013; Yun-rong 2002).

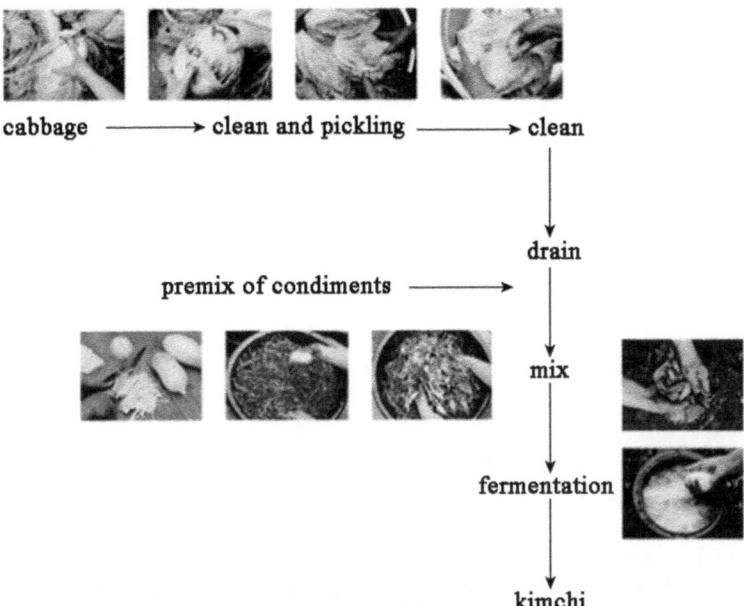

cabbage ⟶ clean and pickling ⟶ clean

premix of condiments ⟶ | drain

mix

fermentation

kimchi

Fig. 7.2 Making of Korean fermented vegetables

Japanese kimchi is developed from Korean kimchi and has formed a unique style, which is different from Chinese kimchi and Korean kimchi in taste and production process. It was mainly made of natural pigments and soy sauce without the fermentation of lactic acid bacteria. Thus, Japanese kimchi belongs to a low-salt and low-acid and non-fermented product (Yan-gang and Quan 2011).

7.2.4.1.3 Sauerkraut

Sauerkraut is a traditional fermented cabbage in Germany and is now commonly found in Germany and Wisconsin State of America. The fermentation of Sauerkraut is a natural fermentation process dominated by the microorganisms on cabbage leaves. After fermentation, the plant tissue of cabbage is damaged, increasing the surface area and releasing nutrients for microbial growth under anaerobic conditions. The ion concentration and osmotic pressure are increased with the addition of salt in the juice, releasing nutrients from plant tissues and promoting the growth of required microorganisms. In the initial stage, the fermentation was mainly conducted by heterofermentative *Leuconostoc mesenteroides* and followed by the homofermentation of *Lactobacillus plantarum* in the late stage. During the fermentation, sucrose, fructose, glucose, and other carbon sources were transformed into lactic acid, acetic acid, ethanol, carbon dioxide, or mannitol, and the final fermented cabbage can be kept in a fermentation vessel for up to 1 year (Dijk et al. 2000).

7.2.4.2 Fermented Cucumber by Lactic Acid Bacteria

Cucumber is fresh and tender, fragrant, and delicious, containing protein, fat, sugar, cellulose, a variety of vitamins and calcium, phosphorus, iron, potassium, sodium, magnesium, and other rich mineral ingredients. In particular, cucumber contains special fine cellulose, which can reduce blood cholesterol and triglycerides, promote intestinal peristalsis, accelerate waste discharge, and improve human metabolism. Fermented cucumber is a popular fermented vegetable because of its crisp, sweet, and sour taste. The main process of its production was shown as follows: fresh cucumber → selection → sorting → cleaning → cutting → draining → bottling → adding salt water, white sugar, etc. → inoculation → brewing → bottling → seasoning → packaging → sterilization → finished products (Ying et al. 2015).

7.2.4.3 Fermented Sauerkraut by Lactic Acid Bacteria

Sauerkraut is a kind of fermented vegetable product with a long history. Sauerkraut is very fresh and delicious, increasing the appetite. The sauerkraut is rich in lactic acid bacteria with nutritional and health functions (Xiao-hui et al. 2009). The main process of its production was shown as follows: Chinese cabbage → drying and finishing → cleaning →entering barrel → crushing stone → filling water → fermentation → finished product.

7.2.4.4 Fermented Olive by Lactic Acid Bacteria

7.2.4.4.1 Lactic Acid Bacteria in the Olive Fermentation

The lactic acid bacteria on the surface of olive include *Lactococcus lactis*, *Pediococcus pentosaceus*, *Leuconostoc mesenteroides*, *Lactobacillus plantarum*, *Lactobacillus delbrueckii*, and *Lactobacillus brevis*. In the past, natural fermentation was used by controlling the temperature, humidity, and concentration of salt solution. Now the fermentation is mainly inoculation by pure cultures. However, the olive fruit contains oleoresin (mainly glucoside), which inhibits the growth of lactic acid bacteria. So the variety characteristics of olive should also be considered during fermentation. Generally, *Lactobacillus plantarum* or micrococcus or mixed strains are mainly used for fermentation, and the effect of heterofermentation alone is not good (Gang, 2007).

7.2.4.4.2 Process Flow

Olive → selection grading → water washing → removing → water washing →acid neutralization →water washing → heat → inoculation → fermentation → finished product

7.2.4.4.3   Key Points of Olive Fermentation

A small amount of sugar can be added during the fermentation. Fermentation temperature should be kept at 20 °C with regular air sterilization. The fermentation can be ended while the pH arrived at 3.8–4.0 and the yield of lactic acid is above 0.5%. The fermented olive is yellow and moderate in hardness and has no bitter taste and salty taste.

7.2.4.5   Fermented Pickled Mustard by Lactic Acid Bacteria

At present, high-salt pickling is commonly used in the production of pickle, and the batch salting process is mainly to prevent excessive salt solution from causing severe infiltration of pickling, resulting in sudden water loss and shrinkage of pickling tissue.

At the same time, vigorous fermentation can produce enough lactic acid, which inhibits the activities of other harmful microorganisms, which is not only conducive to the preservation of vitamins but also can shorten the time needed to reach the permeation balance and improve the curing effect (Yao-guang and Cheng-dong 1994). Natural fermentation adopts the method of high-salt pickling, while artificial fermentation adopts the method of low-salt pickling and inoculation with specific lactic acid bacteria starters. Some studies had shown that the artificial inoculation fermentation was significantly superior to the traditional natural fermentation, including reducing the use of salt, reducing the content of nitrite in products, improving the acidity of products, and inhibiting the contamination of miscellaneous bacteria (Shi-yang et al. 2013).

At present, the pickle pickling process is mainly divided into two kinds, namely, air dehydration and salt dehydration. Most of the pickle production in Sichuan and Chongqing adopts the way of wind dehydration, and the curing process is as follows: material selecting → stripping → pruning → wind dehydration → first curing → first pressing → second curing → second pressing → panning → mixing → loading and ripening. Most of the pickled vegetables in Zhejiang province are salted and dehydrated, and the curing process is as follows: selecting materials → stripping vegetables → cleaning → entering into the pond → adding salt → first application → turning the pond and adding salt → second pickling → ripening.

7.2.4.6   Fermented Potherb Mustard by Lactic Acid Bacteria

The pickling process of potherb mustard is as follows: fresh cabbage → pretreating (selection, cleaning, etc.) → Pao cai → pickling → stepping vegetable → inverting cylinder or pond → sealing cylinder or pond.

Traditional potherb mustard preservation is a natural fermentation method; to join a high concentration of salt, most of the growth of microorganisms in the process of curing is restrained, but there is a small amount of microorganisms in growth;

some of the beneficial microbes has certain contribution to the flavor of potherb mustard; potherb mustard produced by the traditional curing process has good flavor. However, the fermentation cycle of traditional potherb mustard is long, and the operation process is complicated, which often makes the product quality difficult to reach the unified standard. For this reason, people adopted methods to improve the growth conditions of lactobacillus and add lactobacillus liquid and, combined with modern industrial production technology, greatly shortened the production cycle, simplified the production process, and increased the output (Wei 2005).

7.2.4.7 Fermented Dried Turnip by Lactic Acid Bacteria

Dried turnip is a traditional fermented vegetable and very popular in China, among which Xiaoshan dried turnip is the most famous. Its processing technology has a history of several thousand years, and it is the specialty of Xiaoshan district of Zhejiang province. Xiaoshan dried turnip is yellow and bright in color, crisp and refreshing, and nutritious and has anti-inflammatory, anti-heat, and appetizing effects. The processing method of dried radish in Xiaoshan is commonly known as air dehydration method, and the process could be summarized as follows: selection of raw materials → washing and cutting strips → white strips (fresh cut radish strips) → drying in the sun → salting and curing → loading altar with mixed materials (Zhi-qun 1986). At present, the production of dried turnip in China mainly adopts the traditional natural fermentation method with long fermentation period, high content of nitrite, and short shelf life. Therefore, the application of artificial inoculation of lactic acid bacteria has a broad prospect for dried turnip.

7.3 Fermented Fruit and Vegetable Juice Drinks by Lactic Acid Bacteria

7.3.1 Main Types of Fermented Fruit and Vegetable Juice

7.3.1.1 Fermented Fruit and Vegetable Juice

Fermented fruit and vegetable juice is prepared by various fruit and vegetable juices as main raw materials and fermented by lactic acid bacteria or yeast alone or in combination. As a functional food, fermented fruit and vegetable juice by lactic acid bacteria has been studied in some developed countries including Japan, Korea, Germany, etc. It also has formed a great market development trend. The development of research on fermented fruit and vegetable juice is late and slow in China. There are few products in the market.

Fruit and vegetable juices are rich in nutrients, which will also increase after fermentation by lactic acid bacteria. Besides, the flavor in fermented fruit and vegetable juices will be improved. For example, carrot has rich nutrient because of carotene, vitamin B1, vitamin B2, and vitamin C. Tomatoes are nutritious and have a cool effect that promotes appetite. A variety of amino acids and polysaccharides are obtained in pumpkin, which can effectively prevent diabetes and high blood pressure. The germinated soybeans are rich in essential amino acids and minerals such as calcium, iron, zinc, and phosphorus. The health-promoting ingredients such as B vitamins and carotene are also multiplied. The fruit and vegetable pulp mixed by different fruit and vegetable juices is rich in nutrients and bright in color. After fermentation, its health function is greatly increased. So fermented fruit and vegetable juice is an ideal health drink. After acclimation, different strains of lactic acid bacteria can gradually adapt to the physical and chemical environment of compound vegetable pulp. They can use the nutrients in vegetable pulp to grow and produce flavor substances such as lactic acid. According to the needs of different groups of people, various fruits and vegetables are fermented separately, or several kinds of fruit and vegetable juices are combined to produce a variety of fermented fruit and vegetable juices, which can meet the demands and preferences of different consumers.

7.3.1.2   Fermented Fruit and Vegetable Juice Milk Drink

The active lactic acid bacteria milk beverage is a new type of milk beverage that integrates nutrition and health functions developed in recent years. The product contains rich minerals, vitamins, proteins, and carbohydrates. The production process of fermented fruit and vegetable juice milk drink is similar to that of yogurt. Generally, *Lactobacillus delbrueckii* subsp. *bulgaricus* and *Streptococcus thermophilus* are used for fermentation. The amount of raw milk is reduced instead of 10–30% of vegetable juice. The beverages have lactic acid and milky aromas of yogurt, as well as the scent of vegetables, and are more nutritious than yogurt.

7.3.2   Microecology of Fermented Fruit and Vegetable Juice

The microbial composition during fermentation is relatively simple because artificial inoculation is adopted by the fermentation of fruit and vegetable juice. Microbial community structure in natural fermentation enzymes was studied by Fang (Fang 2016), who believed that the microorganisms in the fermentation process of the enzyme mainly included bacteria and yeasts. The dominant bacteria mainly included *Lactobacillus harbinensis*, *Lactobacillus acetotolerans*, *Lactobacillus kefiri*, *Acetobacter pasteurianus*, and *Bacteroides thetaiotaomicron*. *Pichia, Issatchenkia,* and *Saccharomyces* are the dominant yeast.

7.3.3   Lactic Acid Bacteria in Fermented Fruit and Vegetable Juices

7.3.3.1   The Common Lactic Acid Bacteria in Fermented Fruit and Vegetable Juice

Recently, the mainly lactic acid bacteria used in fermented fruit and vegetable juice are *Lactobacillus* and *Bifidobacterium* including *Lactobacillus plantarum*, *Lactobacillus casei*, *Lactobacillus delbrueckii* subsp. *bulgaricus*, *Lactobacillus acidophilus*, *Lactobacillus salivarius*, *Lactobacillus brevis*, *Bifidobacterium longum*, and *Bifidobacterium animalis* subsp. *lactis*. The common fruit and vegetable juice products fermented by lactic acid bacteria are shown in Table 7.2.

7.3.3.2   Effects of Lactic Acid Bacteria Fermentation on Flavor Quality of Fermented Fruit and Vegetable Juice Beverage

The effects of lactic acid bacteria in the processing of fermented fruit and vegetable juices can be summarized as follows: improving the nutrients in vegetable juices, improving the flavor of vegetable juice, preventing the corruption of fruit and vegetable juices, extending their shelf life, and increasing the health benefits.

Table 7.2 Examples of probiotics fermented fruit and vegetable juice

Material	Lactic acid bacteria	Reference
Tomato juice	*L. acidophilus, L. plantarum, L. casei, L. delbrueckii*	Yoon et al. (2004)
Carrot juice	*L. rhamnosus, L. bulgaricus*	Nazzaro et al.2010)
Cabbage juice	*L. casei, L. delbrueckii, L. plantarum*	Yoon et al. (2006)
Beet juice	*L. acidophilus, L. casei, L. delbrueckii, L. plantarum Y*	Yoon et al. (2005)
Pumpkin juice	*L. plantarum, Saccharomyces cerevisiae*	Chun-li et al. (2014)
Ginger juice	*B. longum, L. casei* subsp. *casei, L. acidophilus*	Chen et al. (2009)
Litchi juice	*L. casei*	Zheng et al. (2014)
Apple juice	*L. paracasei* ssp. *paracasei*	Pimentel et al. (2015)
Coconut milk	*L. plantarum*	Pimentel et al. (2015)
Pineapple juice	*L. casei*	Costa et al. (2013)
Lemon juice	*L. casei, L. delbrueckii, L. plantarum, L. helveticus*	Islam et al. (2015)
Grenadine juice	*L. plantarum, L. delbrueckii*	Dogahe et al. (2015)
Banana juice	*L. plantarum, L. delbrueckii*	Tsen et al. (2009)
Cranberries, pineapples, orange juice	*L. salivarius, L. casei, L. rhamnosus, B. lactis, L. paracasei*	Sheehan et al. (2007)
Black raspberry juice	*L. brevis*	Kim et al. (2009)
Blackcurrant juice	*L. plantarum*	Luckow and Delahunty (2004)

For example, the flavor of tomato juice changed after lactic acid bacteria fermentation. The content of glutamic acid and aspartic acid was increased, which can increase the umami taste of the product. Some new flavor components were found by GC-MS after fermentation, which had a positive effect on increasing the fragrance and freshness of the product, including diacetyl, ethyl acetate, 2-heptanone, and 2-ketone. Low concentrations of diacetyl can present a creamy aroma, while esters of lower saturated fatty acids and fatty alcohols have a fruity aroma. Therefore, the flavor of fermented tomato juice is the combination of various components. The new components such as lactic acid, malonic acid, and succinic acid were found, and lactic acid was the highest organic acid in the fermented tomato juice (Li-hua and Jie-bin 1993). The relative content of carbonyl compounds in the volatile flavor components decreased by 61.6%, while the relative content of alcohol compounds increased by 2.4 times after the fermentation of pumpkin juice with lactic acid bacteria and *Saccharomyces cerevisiae*. The main flavor substances were ethanol, isoamyl alcohol, and phenylethyl alcohol. The content of esters and organic acids in the pumpkin juice after fermentation was significantly increased. It demonstrated that fermentation process had a significant effect on pumpkin juice.

7.3.4   Process and Characteristics of Fermented Fruit and Vegetable Juice

7.3.4.1   Lactic Acid Bacteria-Fermented Fruit and Vegetable Juice

Typically, a variety of fruit and vegetable juice can be fermented by lactic acid bacteria, such as apple juice, tomato juice, carrot juice, and a variety of other vegetable juice.

7.3.4.1.1   Fermented Tomato Juice by Lactic Acid Bacteria

Preparation of Tomato Juice

Tomatoes with the maturity of more than 90% and dark red color are selected as raw material. After washing, they are blanched in hot water at 90–95 °C for 3 min, so that the outer skin is soft. The main effect is killing the microorganisms on the surface of the tomato, destroying the activity of the enzyme, and increasing the juice yield. Then tomatoes are put into the juicer for juice. Tomato juice is not appropriate for the growth of lactic acid bacteria because of its high acidity. Therefore, the acidity is adjusted with sodium carbonate before the colloid mill is homogenized, so that the pH is about 6.5, and 3–4% of sucrose is added.

Preparation of Starters

The species used for fermentation should be based on actual conditions. *Streptococcus thermophilus* and *Lactobacillus delbrueckii* subsp. *bulgaricus* or *Lactobacillus plantarum*, *Lactobacillus acidophilus*, and other lactic acid bacteria may be used in mixed fermented vegetable juice. Generally, the starters are first activated 2–3 times, and then the compound vegetable juice and skim milk powder are used as a medium for inoculum enlargement. A seed starter of 10^8 cfu/ml is used as a working starter, and the inoculum amount is usually 3–4%.

Fermentation

After the fermenter was thoroughly sterilized, the treated tomato juice was placed, sterilized at a temperature of 90–95 °C for 20 min, and cooled to 43 °C for inoculation. The fermentation temperature is 40–43 °C until the pH is reduced to 4.0–4.5, and the lactic acid concentration reaches 0.85–1.00%. The taste is best and the fermentation can be ended.

Termination of Fermentation

The temperature in the fermenter is rapidly raised to 70 °C or higher to kill the lactic acid bacteria and then to fill and cool. Another choice is that the fermented tomato juices are rapidly filled in bottles after fermentation and stored at a temperature of 1–5 °C. The fermented tomato juice is red and turbid with tomato flavor and lactic acid bacteria fermentation flavor and sweet and sour taste. The pH is 4.0–4.5 and the lactic acid concentration is 0.85–1.0%. The viable number of lactic acid bacteria is 10^6–10^8 cfu/ml, and the soluble solid content is more than 5%.

7.3.4.1.2   Fermented Carrot Juice by Lactic Acid Bacteria

Carrot is one of the main vegetables all over the country. Its price is low and it is rich in nutrients, minerals, and vitamins. The fermented carrot juice not only maintains the nutrient composition of carrot but also has the characteristics of lactic acid fermentation, which can improve the flavor and taste of carrot juice and improve the intestinal functions and immunities.

The process is as follows: carrot → cleaning → slicing → squeezing → filtering → sterilization → inoculation → fermentation → post-cooking → mixing → packing → inspection → finished product (Rui and He 2002).

7.3.4.1.3   Fermented Purple Potato Juice by Lactic Acid Bacteria

Purple sweet potato is rich in anthocyanins, dietary fiber, selenium, iodine, zinc, and other minerals. In addition to the ingredients and functions of ordinary sweet potato, it also has a variety of special physiological health functions and is an important raw material in the fields of food, medicine, and cosmetics. China has rich purple sweet potato resources, but at present the domestic purple sweet potatoes are mainly used for fresh food and processing purple sweet potato red pigment and purple sweet potato total powder. Although the nutrient content of purple potato is more abundant, but its flavor is not as good as ordinary sweet potato, and the processing variety is single. Therefore, the purple potato food developed for the majority of consumers has a good market prospect. In order to retain the nutrients and bioactive substances in purple sweet potato as much as possible and increase its flavor, artificial inoculation of lactic acid bacteria is adopted for fermentation, and the production of purple sweet potato series of nutritious food, which is of great significance to improve the utilization rate of raw materials, enriches product varieties and enhances product added value. Xing-zhuang et al. (2013) prepared the fermented purple potato juices using *Lactobacillus plantarum*, *Lactobacillus* Reitman, and *Lactobacillus brevis* as starters. The process conditions were optimized by sensory evaluation and pH as indicators. The most ideal fermented purple potato juice could be obtained at the conditions of the inoculum of 2.0% with the ratio of three *Lactobacillus* 1:1:1, fermentation temperature 18–28 °C, and fermentation time 7–15 days.

7.3.4.1.4 Fermented Strawberry Juice by Lactic Acid Bacteria

Strawberry is a kind of berry, which is not easy to store at room temperature and easy to rot. However, after fermentation by lactic acid bacteria, the fermented beverage of strawberry juice not only has the unique fragrance of fermented fruit juice but also has the fragrance of fruit. The product is sweet and sour and at the same time is beneficial for storage, which has considerable economic benefits. The technological process is as follows: strawberry juice → pH adjustment to 6.5 → heat sterilization (95 °C, 15mim) → cooling, inoculation (*Lactobacillus acidophilus*, *Lactobacillus delbrueckii* subsp. *bulgarian*, *Streptococcus thermophiles*) → fermentation → blending → heat sterilization → canning → finished product (Chun-bao et al. 2001).

7.3.4.2 Fermented Mixed Fruit and Vegetable Juice Beverage by Lactic Acid Bacteria

7.3.4.2.1 Purple Sweet Potato Milk Beverage by Lactic Acid Bacteria

He-sheng (He-sheng and Hai-ping 2015) developed purple sweet potato milk beverage with purple sweet potato and fresh milk as the main raw materials. The manufacturing process was as follows: fresh milk → adding purple potato, white sugar, and stabilizer → homogenization → sterilization → cooling → inoculation → aseptic filling → heat preservation fermentation → cold fermentation leave → finished product. Through orthogonal test, the optimal process conditions were obtained as fermentation time 6.5 h, citric acid addition 0.4%, white sugar addition 8%, and purple potato addition 20%. The product was uniform in texture and delicate in taste, sweet and sour, with purple potato and lactic acid fermented aroma and rich milky aroma.

7.3.4.2.2   Pear Milk Beverage by Lactic Acid Bacteria

Xu-guang et al. (2015) developed a pear milk beverage; the manufacturing process was as follows: mixing raw materials (fermented milk, pear juice, stabilizer, sweetener) → adding citric acid to adjust acid → preheating (50–55 °C) → sterilization → canning → after cooking → finished product. The optimum parameters were 20% pear juice, 10% sugar, 0.15% citric acid, and 0.4% stabilizer.

7.3.4.2.3   Longan Mixed Fruit and Vegetable Juice Drink by Lactic
Acid Bacteria

An-shu et al. (2012, 2013) studied the preparation process of longan carrot, tomato, fruit, and vegetable juice milk beverage. *Lactobacillus delbrueckii* subsp. *bulgaricus* and *Streptococcus thermophilus* were used as starters. Longan juice, carrot juice, and tomato juice were mixed at 3.5:5:1.5; seed solution was inoculated to 4% and fermented at 37 °C for 24 h. The fermented drink was orange-red, had longan and lactic acid fermentation flavor, tasted sweet and sour, and was refreshing and soft. On this basis, the bactericidal beverage was developed. It was found that the sterilization conditions of longan fruit and vegetable juice mixed directly could affect the stability of the final drink. The best sterilization conditions were at 100 °C for 30 min.

References

Abu-Ghannam N, Rajauria G (2015) Non-dairy probiotic products. Adv Probiotic Technol 1:356–374. Foerst P, Santivarangkna C

An-shu X, Chun-mei T, Wei L (2012) Production process of Longan mixed fruit and vegetable juice lactic acid bacteria beverage. Food Ferment Ind 38(6):128–132

An-shu X, Hong-li Z, Wei L (2013) Stability of mixed juice lactobacillus drink for longan carrot and tomato. Food Sci technol 38(6):125–130

Aponte M, Blaiotta G, La CF, Mazzaglia A, Farina V, Settanni L, Moschetti G (2012) Use of selected autochthonous lactic acid bacteria for Spanish-style table olive fermentation. Food Microbiol 30(1):8–16

Battcock M (1998) Fermented fruits and vegetables: a global perspective. Food & Agriculture Orginasation, Rome

Bong Y-J, Jeong J-K, Park K-Y (2013) Fermentation properties and increased health functionality of kimchi by kimchi lactic acid bacteria starters. J Korean Soc Food Sc Nutr 42(11):1717–1726

Chang-jian L, Ben-guo J, Bo J, Qiu L, Yu-bo C (2011) Screening, identification and cholesterol-lowering probiotic characteristics of lactic acid bacteria MR25. J Chin Inst Food Sci Technol 11(8):42–46

Chen IN, Ng CC, Wang CY, Chang TL (2009) Lactic fermentation and antioxidant activity of Zingiberaceae plants in Taiwan. Int J Food Sci Nutr 60(sup2):57–66

Choi W-Y, Park K-Y (1999) Anticancer effects of organic Chinese cabbage kimchi. J Food Sci Nutr 4(2):113–116

Choi IH, Noh JS, Han J-S, Kim HJ, Han E-S, Song YO (2013) Kimchi, a fermented vegetable, improves serum lipid profiles in healthy young adults: randomized clinical trial. J Med Food 16(3):223–229

Chun-bao G, Wan-zhong Z, Qiu-lan L, Run-guang Z, Cui-ling G, Gui-yun L (2001) Development of lactic acid fermented strawberry juice. Food Sci 22(9):52–55

Chun-li Z, Wei L, Hui L, Jing Z, Chi Y, Quan-hong L (2014) Mixed culture fermentation of pumpkin juice and its aroma analysis. Mod Food Sci Technol 30(5):301–310

Chun-yan L, Ming-fu D, Jiao X, Ting Z, Li-wen S, Biao P (2015) Study on the flavor of different lactic acid bacteria inoculated fermentation pickle. Sci Technol Food Ind 36(7):154–158

Costa MGM, Fonteles TV, de Jesus ALT, Rodrigues S (2013) Sonicated pineapple juice as substrate for L. casei cultivation for probiotic beverage development: process optimisation and product stability. Food Chem 139(1–4):261–266

Dan-ping X, Biao P, Shu-liang L, Zhi-hang Z, Nan Z (2015) Analysis of volatile components in pickles fermented with different starter cultures. Food Sci 36(16):94–100

Dijk CV, Ebbenhorstselles T, Ruisch H, Stollesmits T, Schijvens E, And WVD, Boeriu C (2000) Product and redox potential analysis of sauerkraut fermentation. J Agric Food Chem 48(2):132

Dogahe MK, Khosravi-Darani K, Tofighi A, Dadgar M, Mortazavian AM (2015) Effect of process variables on survival of bacteria in probiotics enriched pomegranate juice. Br Biotechnol J 5(1):37

Fang Y (2016) Analysis of microbial community construction in different self-made enzyme samples by PCR-DGGE technique. Da Li University. Dali, China

Gang, Z. (2007). Lactic acid bacteria: foundations, techniques and applications. Chemical Industry Press. Beijing, China

Guang-yan Z, Xiao-ping Z, Kai Z, Gui-dan S (2006) Effect of different lactic acid bacteria inoculation on the nitrite concentration and the quality of pickle. South West China J Agric Sci 19(2):290–293

He-sheng H, Hai-ping W (2015) Study on the technology of purple sweet potato yogurt. Food Res Dev 36(20):60–63

Islam MK, Hasan MS, Al Mamun MA, Kudrat-E-Zahan M, Al-Bari MAA (2015) Lemon juice synergistically preserved with lactobacilli ameliorates inflammation in shigellosis mice. Adv Pharmacol Pharm 3(1):11–21

Ju-hua Z, Yang S, Gao-yang L (2003) Research progress on lactic acid bacteria fermented vegetable juice. Beverage Ind 6(6):27–31

Kim M, Chun J (2005) Bacterial community structure in kimchi, a Korean fermented vegetable food, as revealed by 16S rRNA gene analysis. Int J Food Microbiol 103(1):91–96

Kim JY, Lee MY, Ji GE, Lee YS, Hwang KT (2009) Production of γ-aminobutyric acid in black raspberry juice during fermentation by lactobacillus brevis GABA100. Int J Food Microbiol 130(1):12–16

Lee JS, Heo GY, Lee JW, Oh YJ, Park JA, Park YH, Pyun YR, Ahn JS (2005) Analysis of kimchi microflora using denaturing gradient gel electrophoresis. Int J Food Microbiol 102(2):143–150

Li-hua F, Jie-bin Y (1993) Study on flavoring substances of lactic acid fermented tomato juice and carrot juice. Food Ferment Ind (2):18–24

Lim J-H, Park S-S, Jeong J-W, Park K-J, Seo K-H, Sung J-M (2013) Quality characteristics of kimchi fermented with abalone or sea tangle extracts. J Korean Soc Food Sci Nutr 42(3):450–456

Li-na Z, Chang-qing Y (2007) Cholesterol-lowering effect of lactic acid bacteria and its application in fermented products. Sci Technol Food Ind 28(7):228–231

Ling-yan Z, Fang-ming D, Fu-lin Y (2004) Characters of lactobacillus and its application in fermented fruits and vegetables. China Food Addit (5):77–80

Luckow T, Delahunty C (2004) Which juice is 'healthier'? A consumer study of probiotic non-dairy juice drinks. Food Qual Prefer 15(7–8):751–759

Martínez-Castellanos G, Pelayo-Zaldívar C, Pérez-Flores LJ, López-Luna A, Gimeno M, Bárzana E, Shirai K (2011) Postharvest litchi (Litchi chinensis Sonn.) quality preservation by lactobacillus plantarum. Postharvest Biol Technol 59(2):172–178

Nazzaro F, Fratianni F, Sada A, Orlando P (2010) Synbiotic potential of carrot juice supplemented with lactobacillus spp. and inulin or fructooligosaccharides. J Sci Food Agric 88(13):2271–2276

Park KY, Jeong JK, Lee YE (2014) Health benefits of kimchi (Korean fermented vegetables) as a probiotic food. J Med Food 17(1):6–20

Pimentel TC, Madrona GS, Garcia S, Prudencio SH (2015) Probiotic viability, physicochemical characteristics and acceptability during refrigerated storage of clarified apple juice supplemented with Lactobacillus paracasei ssp. paracasei and oligofructose in different package type. LWT Food Sci Technol 63(1):415–422

Ping-chun C, Zong-xiang Y, Xue-feng M (2010) Treatment of 60 cases of infantile diarrhea with oral kimchi water. Mod J Integr Tradit Chin West Med 19(18):2282–2283

Rui X, He W (2002) Development of lactic acid fermented carrot juice. Breverage Ind 23(2):21–23

Sabatini N, Mucciarella MR, Marsilio V (2008) Volatile compounds in uninoculated and inoculated table olives with lactobacillus plantarum (*Olea europaea* L., cv. Moresca and Kalamata). LWT-Food Sci Technol 41(10):2017–2022

Sheehan VM, Ross P, Fitzgerald GF (2007) Assessing the acid tolerance and the technological robustness of probiotic cultures for fortification in fruit juices. Innovative Food Sci Emerg Technol 8(2):279–284

Shin D, Park Y, Kim Y (2003) Study on the optimum fermentation condition for making lactic acid beverage by u-sing mixed-vegetable juice. Food Sci Technol 5(13):1256–1260

Shi-yang G, Zhi-dong S, Xin-yong D, Guo-qing H (2013) Physicochemical properties and flavor components of low-salt pickle inoculated with lactic acid bacteria. Mod Food Sci Technol 29(11):2663–2668

Stanton WR (1998) Food fermentation in the tropics. In: Wood BJB (ed) Microbiology of fermented foods. Springer, New York

Swain MR, Anandharaj M, Ray RC, Parveen Rani R (2014) Fermented fruits and vegetables of Asia: a potential source of probiotics. Biotechnol Res Int 2014:250424

Tao X, Song S, Hao M, Xie M (2012) Dynamic changes of lactic acid bacteria flora during Chinese sauerkraut fermentation. Food Control 26(1):178–181

Tao X, Fei P, Xiao L, Jun-bo L, Qian-qian G (2015) Changes and metabolic characteristics of Main microorganisms during Chinese sauerkraut fermentation. Food Sci 36(3):158–161

Trias R, Bañeras L, Badosa E, Montesinos E (2008) Bioprotection of Golden delicious apples and iceberg lettuce against foodborne bacterial pathogens by lactic acid bacteria. Int J Food Microbiol 123(1):50–60

Tsen JH, Lin YP, King VAE (2009) Response surface methodology optimisation of immobilised lactobacillus acidophilus banana puree fermentation. Int J Food Sci Technol 44(1):120–127

Wei Z (2005) Study on the semifinished product and the flavor of processing pickled potherb mustard. Hunan Agricultural University. Changsha, China

Wei T (2013) Analysis of microorganism community structure in the fermentation process of Sichuan tradition pickles and industrial pickles. Xihua University. Chengdu, China

Wen-bin L, Zhong-wei T, Min-li S (2006) The up to date returns of nutritive value and health protection of corea pickled vegetable. Acad Period Farm Prod Process (8):83–84

Wen-hui W, Bin B (1995) Study on active lactic acid bacteria beverage. J Inn Mong Inst Agric Anim Husbandry 16(1):54–60

Xiang-yang W, Bing D, Xing-wei Y, Ling Y (2015) The effect of clove on the storage property of pickled cabbage. China Condiment (11):7–9

Xiang-yang W, Zhou T, Long J (2016) Effect of salt and lactobacillus additive amount on the processing technology and preservation of pickled radish. China Condiment 41(11):28–31

Xiao-hui L, Shun C, Xiao-mei G, Yan Z, Rui C (2009) Isolation and identification of lactic acid bacteria in pickled cabbage. China Brew 28(2):62–64

Xiao-lin A, Xiao-ping Z, Ling S, Xian-qin Z (2011) Identification of two lactic acid bacterial strains isolated from Sichuan pickles and effect of fermentation conditions on the quality of pickles co-fermented by them. Food Sci 32(11):152–156

Xiao-ran G, De-chun Z, Jin-ling L, Qing X (2010) Effect of Bifidobacterium fermented mixed fruit and vegetable juice on immunological function of mice. Chin J Microecol 22(2):110–113

Xing-zhuang, W., Hua, Z., Xiao-li, Z., Xin, F., Chang-yi, L., Chao, C. (2013). Study on the optimum fermentation conditions of lactic acid bacteria fermented purple potato. Agric Sci Technol Equip (9):55–58

Xue-ping S, Liang-liang W, Peng G, Wei-ming Z (2011) Ethanol extracts from twenty edible spices: antioxidant activity and its correlations with total flavonoids and total phenols contents. Food Sci 32(5):83–86

Xu-guang Z, Chun-guang L, Jin-chao M (2015) Study on pear lactic acid bacteria beverage. Food Res Dev 36(3):47–49

Yan P-M, Xue W-T, Tan S-S, Zhang H, Chang X-H (2008) Effect of inoculating lactic acid bacteria starter cultures on the nitrite concentration of fermenting Chinese paocai. Food Control 19(1):50–55

Yan-gang L, Quan D (2011) The comparison of Chinese pickle, Japanese pickle and kimchi processing technology. Food Ferment Technol 47(4):5–9

Yao-guang Z, Cheng-dong C (1994) Mechanism and influencing factors of vegetable pickling. Food Ind Technol (5):23–26

Ying C, Jin-hui W, Wen-ling-zi Z, Xin-yi H (2015) Study on quality cucumber pickles fermented by lactic acid bacteria. Food Mach (4):208–211

Yong N (2003) Assimilation of cholesterol by lactic acid bacteria and bifidobacteria. Ind Microorganism 33(2):58–58

Yoon KY, Woodams EE, Hang YD (2004) Probiotication of tomato juice by lactic acid bacteria. J Microbiol 42(4):315–318

Yoon KY, Woodams EE, Hang YD (2005) Fermentation of beet juice by beneficial lactic acid bacteria. LWT Food Sci Technol 38(1):73–75

Yoon KY, Woodams EE, Hang YD (2006) Production of probiotic cabbage juice by lactic acid bacteria. Bioresour Technol 97(12):1427–1430

Yuan-feng W, Li-gen Z (2007) Isolation, identification and fermentation performance of lactic acid bacteria from pickles. J Chin Inst Food Sci Technol (5):42–46

Yun-bin Z, Yuan G, Ye S (2012) Comparative study on sensory properties between lactobacillus fermented and traditionally salted mustard tubers. J Shanghai Inst Technol (Nat Sci) 12(3):175–181

Yun-rong Z (2002) Kimchi – a representative traditional fermented food in Korea. Food Ferment Technol (3):73–73

Zeng-san C (2005) Sichuan fuling mustard production method. China Condiment (10):42–46

Zhang J-W, Cao Y-S (2005) Antioxidative activities of lactic acid bacteria. Zhonggue Rupin Gongye 1:010

Zhang Yan-chao BY, Wang S-q, Dong X-m, Zhang H-P (2008) Review on biological characteristics, physiological functions and application prospects of *Lactobacillus Fermentum*. Agric Food Prod Sci Technol 2(4):33–36

Zheng X, Yu Y, Xiao G, Xu Y, Wu J, Tang D, Zhang Y (2014) Comparing product stability of probiotic beverages using litchi juice treated by high hydrostatic pressure and heat as substrates. Innovative Food Sci Emerg Technol 23(3):61–67

Zhi-qun Y (1986) Xiaoshan radish. Chin Vegetables 1(1):11–10

Zu-fang, W., Pu, L., Pei-fang, W. (2008). Study on application of lactic acid bacteria technology to the processing of traditional pickled mustard tuber. Food Ind Technol (2):101–103

Chapter 8
Lactic Acid Bacteria and Fermented Meat Products

Shumao Cui and Zhexin Fan

8.1 Introduction

8.1.1 The Types of Fermented Meat Products

Fermented meat products are a kind of meat products with special flavor, color, texture, nutrition, and prolonged shelf life, which are produced from livestock or poultry meat with a series of natural or artificially controlled processing methods (such as curing, fermentation, drying, or smoking). Fermented meat products are an important branch of traditional Chinese meat products. They have a long history of production and consumption. Due to their unique flavor and rich nutrition, they are deeply loved by consumers. The types of fermented meat products mainly include fermented sausage, fermented ham, cured products and smoked meat, etc.

At present, there has been no inherent standard for the classification of fermented meat products, and different countries and regions have different standards. Fermented meat products in China mainly refer to fermented sausages and fermented hams, while foreign ones are mainly filled meat products (sausages), such as Lebanese big sausages, Cervelat sausages, and salami sausages.

According to the degree of fermentation, meat products can be classified into low-acid fermented meat products and high-acid fermented meat products. Low-acid fermented meat products refer to the fermented meat products with a pH of >5.5, such as Spanish ham, salami sausage, and other dry fermented sausages, which are produced by curing, fermentation, drying, etc. at a low temperature of 0–25 °C. High-acid fermented meat products refer to the products with a pH of <5.5, of which the production generally needs the addition of starters and fermentation at >25 °C.

S. Cui (✉) · Z. Fan
Jiangnan University, Wuxi, China
e-mail: cuishumao@jiangnan.edu.cn

© Springer Nature Singapore Pte Ltd. and Science Press 2019
W. Chen (ed.), *Lactic Acid Bacteria*,
https://doi.org/10.1007/978-981-13-7283-4_8

8.1.2 Lactic Acid Bacteria Commonly Used in Fermented Meat Products

In 1919, Cesari discovered the presence of yeast in naturally fermented meat products and began the study of microorganisms in fermented meat products. In 1955, Niven et al. isolated *Pediococcus acidilactici* from fermented meat products and successfully applied it to Summer sausages. In modern technology, commercial starters are generally added to ensure product stability. In the production of traditional fermented meat products, the bacteria from the environment compete with the microorganisms in the raw materials for growth, completing the fermentation process. Lactic acid bacteria become the main microflora in the later stage of fermentation. Different kinds of microorganisms produce various substances in the fermentation process. These by-products react to form unique flavors and nutrients. To some extent, the sensory properties of fermented meat products can be considered to be determined by fermenting microorganisms (Cocolin et al. 2011).

The microorganisms in the fermented meat product mainly include bacteria, molds, yeasts, and so on. In general, lactic acid bacteria are an important class of microbial species required for production and play a leading role in the fermentation stage. In addition, *Micrococcus* and *Staphylococcus* in *Micrococcus* family, although not belonging to the lactic acid bacteria, have a strong ability to decompose nitrous acid and have an important influence on the color formation of the fermented meat product. Yeast is generally resistant to high salt and has a strong fermentation capacity and grows on the surface of fermented meat. In the fermentation production of meat products, yeast is rarely used alone, and most of them are combined with lactic acid bacteria, microspheres, and the like to complete the fermentation. Some of the commonly used strains in fermented meat products are shown in Table 8.1.

Table 8.1 Commonly used strains in fermented meat products

Genus	Name
Lactobacillus	Lactobacillus plantarum
	Lactobacillus sakei
	Lactobacillus casei
	Lactobacillus curvatus
Mold	Penicillium chrysogenum
	Penicillium nalgiovense
Pediococcus	Pediococcus lactis
	Pediococcus pentosaceus
Streptococcus	Streptococcus thermophilus
	Streptococcus lactis
	Streptococcus diacetilactis
Yeast	Debaryomyces hansenii
	Candida famata
Micrococci	Staphylococcus xylosus
	Staphylococcus carnosus

8.1.3 The Nutritional Value and Functional Properties of Fermented Meat Products

Fermentation of raw meat has a long history and has formed distinctive products based on regional climatic conditions and consumer preferences. Typical representatives of fermented meat products in China are fermented sausages and fermented ham. In a broad sense, some salted products (such as sausage, bacon, and sour meat) also belong to fermented meat products. Fermented meat products undergo a series of biochemical reactions during fermentation. Some properties of the raw meat are changed. For example, the decomposition of protein will make the fermented meat more nutritious than the raw meat (Steinkraus 1994), because proteins are degraded into amino acids and peptides by the enzymes during the fermentation process, improving the digestibility of the product and increasing the nutritional value of the product (Guo et al. 2009). Some characteristics of fermented meat products have the following aspects.

1. Beautiful color

Color is an index to evaluate the quality of meat products. Chromogenic agents (such as nitrite) are often used in the processing to improve the color of meat products, but residual nitrite is carcinogenic. Studies have shown that lactic acid produced by lactic acid bacteria can low the pH of meat products in the fermentation process, resulting in the production of free nitrite, which is decomposed into NO. NO combines with myoglobin in meat to form nitroso-myoglobin, which remains stable under thermal conditions, giving the product a bright red color. The contaminated bacteria in the meat will produce hydrogen peroxide, which can form biliary myoglobin with myoglobin, making the meat green.

2. Unique flavor

On the one hand, organic acids such as lactic acid, acetic acid, and propionic acid produced by *lactic acid bacteria* during the fermentation process give food a mild acidity and form flavor substances, giving fermented meat products a unique flavor, by interacting with other fermented substances (such as alcohols, ketones, aldehydes, etc.). On the other hand, lactic acid Fe can also eliminate some peculiar smell in raw materials.

3. High nutritional value

The proteins in meat are decomposed to peptides and amino acids by protease from the metabolism of lactic acid bacteria, which greatly increases the digestibility of fermented meat (Steinkraus 1994). And some essential amino acids, vitamins, and bifidoxin in the process of fermentation can be produced, which further enhance the nutritional value of the product.

4. Long shelf life

Lactic acid bacteria produce a large amount of acids in the fermentation process, making the pH lowered. Under these conditions, some spoilage microorganisms cannot grow and reproduce, so that the shelf life of the products has been greatly improved. Other lactic acid bacteria may produce bacteriocin, which can also effectively inhibit the growth of spoilage bacteria and pathogenic bacteria in meat.

8.2 Fermented Sausage

Fermented sausages are prepared under natural or manual conditions. Fresh raw meat is chopped, then added with accessories (such as spices), mixed evenly, and then poured into the casings, which are then fermented and air-dried at a low temperature for a long period. Fermented sausages, with good preservation, typical fermentation flavor, beautiful color, and unique flavor, are the largest category of fermented meat products in China. The types of fermented sausages are complex. According to the shape of stuffing, they are divided into coarse sausages and finely twisted sausages. They are divided into semidry fermented sausages and dried fermented sausages based on the moisture content of the products. The water content of the former is 40–45%, while one of the latter is 25–40%. According to the degree of fermentation, they are divided into low-acid fermented sausages and high-acid fermented sausages. Low-acid fermented sausages refer to sausage products with pH >5.5 and prepared at 0–25 °C, such as Salami sausages in France, Italy, and Hungary. High-acid fermented sausages refer to fermented sausages with pH <5.5, most of which are prepared by inoculating an external starter. The final water content and water activity of some fermented sausages are shown in Table 8.2.

8.2.1 Microecology of Fermented Sausages

The microecosystem of traditional fermented sausages is complex, mainly including bacteria, yeast, and mold. Among them, bacteria mainly refer to lactic acid bacteria. The number of LAB in chilled fresh meat is relatively small. LAB gradually

Table 8.2 Water content and water activity of fermented sausage

Name	Processing cycle	Moisture content	Water activity	Typical products
Smear type	3–5 days	34–42%	0.95–0.96	Germany
				Teewurst frisch, Mettwurst
Sliced type (short-term fermentation)	1–4 weeks	30–40%	0.92–0.94	America Summer sausage
				Germany Thuringer
Sliced type (long-term fermentation)	12–14 weeks	20–30%	0.82–0.86	Salami of Germany, Denmark, Hungary
				Genoa in Italy
				The French Saucisson

become dominant bacteria with fermentation. This is because the low oxygen content and low pH in meat stuffing are not conducive to the growth of some enterobacterium in raw meat but beneficial for the growth and reproduction of LAB, *Staphylococcus* and *Micrococcus*. The LAB in traditional fermented sausages mainly include *Lactobacillus sake, Lactobacillus campylobacter* and *Lactobacillus plantarum*, etc., as well as some *Streptococcus enterococcus, Staphylococcus*, and *Kocuria* (Fontana et al. 2016).

8.2.1.1 Diversity of Yeasts and Molds in Fermented Sausages

The flavor of some fermented sausages in Southern Europe is affected by yeast and mold. Yeast plays an irreplaceable role in the process of sausage fermentation, which can inhibit rancidity by utilizing residual oxygen in minced meat. Hydrogen peroxide produced by yeast fermentation can prevent oxidative discoloration of meat. Yeast can also decompose fat and protein in meat and produce flavor substances (Li and Meng 2010). *Debaryomyces hansenii* is a kind of common yeast, which is salt-tolerant and gas-tolerant and can grow on the surface of sausage.

Mold is a fungus commonly used in dry fermented sausages. Most of these molds belong to the genus *Penicillium* and *Scopulariopsis*. The two common fungi are *Penicillium flavus* and *Penicillium natripenicillium*. They grow on the surface of sausages and form a film on the epidermis that prevents oxygen from entering and prevents rancidity. In addition, the aroma of fermented sausage is partly dependent on the production of aromatic substances by lipase and protease secreted by molds that degrade fats and proteins. However, it should be noted that many molds have the ability to produce toxins, so the molds used in meat products must be strictly screened (Talon and Leroy 2011).

8.2.1.2 Diversity of Bacteria in Fermented Sausages

The main bacteria in fermented sausages are *Lactobacillus, Streptococcus, Micrococcus*, and *Staphylococcus. Lactobacillus* in fermented sausages include *Lactobacillus plantarum, Lactobacillus campylobacter*, and *Lactobacillus sake*. The genus *Planococcus* mainly includes *Pediococcus pentosaceus, Pediococcus bacillus, Pediococcus lactici*, and so on. *Micrococcus* mainly includes *Micrococcus* and *Staphylococcus* (Nan 2008). *Staphylococci* commonly found in fermented sausages include *Staphylococcus carinii, Staphylococcus xylose, Staphylococcus amber*, and *Staphylococcus equi. Staphylococcus carnosus* is the key bacteria for flavor formation of fermented sausage. *Staphylococcus xylosus* does not exist throughout the fermentation period. *Staphylococcus succinus* and *Staphylococcus equorum* play important roles in the maturation process (Simonová et al. 2006). In fermented sausages, there are also a small amount of *Leuconostoc* (such as

Leuconostoc gelidum), *Weiss* (such as *Weissella viridescens*), and *Enterococcus* (such as *Enterococcus faecium* and *Enterococcus faecalis*). Rare lactobacillus such as *Lactobacillus brevis* and *Lactobacillus rhamnosus* were isolated from fermented sausages (Rebucci et al. 2007; Chen et al. 2015). Lactic acid bacteria mainly form acidic environment, which can inhibit the growth of pathogenic bacteria and spoilage bacteria and accelerate the formation of color and luster. The main function of cocci is to form bright color, remove excess nitrate, and form fragrance and special flavor. Lactic acid bacteria ensure product stability, while coccus determines the product color and flavor. Some studies have shown that *Staphylococcus aureus* can improve the color and aroma characteristics of the product (Stahnke et al. 2002).

8.2.2 Lactic Acid Bacteria (LAB) in Fermented Sausages

The content of LAB in raw meat was relatively low, which is 10^2–10^4 cfu/g. They grow rapidly in the fermentation process and soon become the dominant strain of fermentation in the environment of anoxic and high-salt cured meat. LAB in fermented sausages mainly include *Lactobacillus sake*, *Lactobacillus campylobacter*, and *Lactobacillus plantarum* (Urso et al. 2006). Generally, LAB selected for sausage fermentation are homofermentative LAB, which ferment the carbohydrates in meat stuffing to produce lactic acid, while heterofermentative LAB produce lactic acid, acetic acid, carbon dioxide, and peroxide. These metabolites have side effects on the color and flavor formation of sausage. *Lactobacillus sake* is the dominant LAB in fermented sausages, because of its special metabolic system, the arginine deiminase pathway (ADI), which has strong competitiveness and adaptability (Ravyts et al. 2012). *Flaccoccus* is a facultative anaerobic LAB, which can ferment glucose to produce L-lactic acid and D-lactic acid by the Embden-Meyerhof-Parnas pathway (EMP). *Lactococcus* lactic acid and *Streptococcus* pentose are commonly used.

8.2.3 The Role of LAB in Sausage Fermentation

LAB play two main roles in sausage fermentation: one is favorable to the formation of flavor substances and the other is conducive to reducing the acidity of meat products and prolonging the shelf life. LAB can produce large amounts of lactic acid in fermented sausages. Meat proteins and fats are more likely to undergo a series of physical, chemical, and biological reactions under acidic conditions, increasing the content of free amino acids and fatty acids while improving the digestibility of protein.

8.2.3.1 Fermentation

8.2.3.1.1 Reduce the Content of Nitrosamine

In the processing of meat products, the added nitrite interacts with the dimethylamine in raw meat to produce dimethyl nitrosamine, which is carcinogenic. LAB produce a variety of organic acids by fermentation to form a low-acid environment, which makes nitrite reduction and reduces the formation of dimethylnitrosamine, improving the safety of fermented sausages and being beneficial to health.

8.2.3.1.2 Increase Nutritional Value

During fermentation, LAB metabolize to produce different hydrolytic enzymes, which degrade the proteins, fats, and other macromolecules in muscle into small molecular substances (such as amino acids, peptides, etc.) and improve the digestibility and absorptivity of the products. In addition, proteins in muscle gradually produce colloids in acidic environment, promote meat elasticity, and improve meat structure.

8.2.3.2 Physiologic Function

8.2.3.2.1 Improve the Color and Flavor of the Product

In the meat processing and manufacturing, chromogenic reagent, such as nitrate and nitrite, is often added in order to make meat products rose red and increase consumers' desire to buy, chromophore. Fermentation by LAB can not only give bright color to meat products but also reduce the formation of dimethylnitrosamine. LAB can produce organic acids by fermenting carbohydrates, such as propionic acid and acetic acid. The organic acids interact with ketones and alcohols produced in fermentation process and produce various flavor substances which improve the flavor of fermented sausages. In addition, organic acids improve the sensory evaluation of the product, of which the sour taste is soft and not pungent (Xiong et al. 2013).

8.2.3.2.2 Reduce Cholesterol and Delay Senility

The probiotic LAB in fermented sausages can reduce cholesterol levels and prevent heart disease and atherosclerosis to some extent after entering the digestive tract (Schiffrin et al. 1995). LAB can produce a superoxide dismutase that removes excess free radicals from the body and delays senility.

8.2.4   Processing Technology and Characteristics of Fermented Sausage

The production of fermented sausages originated in the Mediterranean. The Romans began to prepare fermented sausages using ground meat more than 2000 years ago. Through development of 2000 years, the production process of fermented sausage is gradually mature. The process methods are similar in different regions, and the process is similar. Selection of raw meat → Mince and mix → Filling → Fermentation → Dry and ripening → Inspection and packaging → Finished product. Traditional fermented sausages are mainly matured by natural fermentation. Drying is accomplished by natural weather. Due to the uncontrollable factors of fermenting seeds and weather during the fermentation process, fermented sausages vary in quality. With the development of modern meat processing technology and the comprehensive study of fermentation microorganisms, some European countries have developed non-dry sausages that are fermented but do not need to be ripe (Wang 2006).

8.2.4.1   Pretreatment of Filling

Pig, beef, and mutton are generally chosen to make fermented sausages. The best choice is a site with more muscle and less connective tissue (such as rump or leg meat). The back fat is chosen as the auxiliary adipose tissue. Lean meat and adipose tissue are cut separately. The process of meat grinding is that lean meat is twisted into relatively large particles at 4 °C and the fat part is cut up at 8 °C. After the meat is minced, they are mixed in a certain proportion according to the processing requirements. Salt, sugar, cooking wine, and spices are added, and the mixture is chopped at low temperature. A small amount of nitrates and sodium ascorbate are added to industrial products. Chopping directly determines the quality of fermented sausage. On the one hand, chopping can make salt and other spices evenly distributed in meat fillings; on the other hand, it can also exclude part of oxygen.

8.2.4.2   Filling

Filling refers to stuffing mixed ground meat into the prepared natural or artificial casing. The casing can affect the maturity and quality of the sausage. The temperature of meat filling should be less than 2 °C.

8.2.4.3   Fermentation

Fermentation refers to the stage in which under natural or artificial conditions, LAB in sausage grow and metabolize vigorously with rapid drop in pH and the formation of sausage flavor and color under the action of other microorganisms. In traditional processing, sausage fermentation is mainly completed by LAB naturally existing in raw meat. Although the content of LAB in raw meat is not high, it will reach a high level after 2–5 days under normal circumstances. However, if the fermentation of LAB is delayed and the pH decreases slowly, *Staphylococcus aureus* will grow in large amounts and produce enterotoxin as well as other miscellaneous bacteria, which will lead to worse flavor of sausage (Li and Meng 2010). Because of the uncontrollability of traditional fermentation methods, starter cultures are gradually used to control the fermentation process in modern processing technology. Commercially available starter cultures generally conclude LAB, *Micrococcus*, and mold.

8.2.4.4 Drying and Maturation

Drying is the process wherein water in fermented sausages evaporates under natural or artificial conditions. Traditional drying methods generally use natural environmental conditions for drying, such as air-drying, sun drying, or shade drying. So it is also called natural drying. In modern production, artificial technology is often used to adjust the temperature and humidity of the drying chamber to effectively dry fermented sausages. The final moisture content of different types of fermented sausage products varies greatly. The final characteristics and flavor of fermented sausage products largely depend on the changes of its properties during drying period. All fermented sausages need to control the rate of dehydration during the drying stage, so that the rate of water transfer from the inside of the sausages to the surface is equal to the rate of water loss from the surface, thus ensuring that the fermented sausages are stable and dry and the surface will not form dry skin (Xu 2011). The changes in chemical properties of fermented sausages during drying and molds or yeasts growing on the surface even go on to the consumption stage and are all a process of continuous maturation. The maturation of fermented sausages affects the final sensory properties of the product, especially flavor and aroma.

In recent years, some domestic and foreign scholars began to study the effects of enzymes on fermentation and maturation of fermented meat products. These enzymes are mainly lipases derived from microorganisms. Fat degradation in fermented sausage is the main source of its special flavor. Lipase secreted by fermentation microorganisms plays an important role in the production of fatty acids through lipase hydrolysis and lowering pH. Adding extraneous enzymes can shorten the fermentation time and reduce production cost. Endogenous lipase in meat also has the effect of degrading fat (Hierro et al. 1997). It has been reported that adding exogenous fat enzymes can promote the formation of flavor substances in the fermentation of sausages, but it does not shorten the fermentation time (Zalacain et al. 1997; Liu et al. 2014). Studies have found that compared with the control group, the sensory characteristics of fermented sausages with exogenous streptomycin protease or papain have no obvious difference, and some proteases could improve the sensory characteristics of sausages (Diaz et al. 1997; Benito et al. 2004).

8.3   Fermented Ham

8.3.1   Types of Fermented Ham

The fermented ham is divided into traditional dry-cured ham and Western ham. The traditional dry-cured ham is made from the pork's fresh hind or front legs (with skin, bone, and claw), belonging to a kind of fermented meat product that is dry-cured at low temperature (0–4 °C) and dried, fermented, and matured at higher temperature (15–20 °C). The production technology of fermented ham in most countries and regions of the world is similar. The representative of dry-cured ham in China are Xuanwei ham in Yunnan, Xianning ham, Jinhua ham in Zhejiang, Enshi ham and Rugao ham in Jiangsu, and so on. Western fermented ham is also divided into boned ham and boneless ham. Typical representatives are country-cured ham in the United States, Iberian ham in Spain, Westphalian ham in Germany, and Parma ham in Italy. Parma ham is especially famous among Western hams.

8.3.2   The Microecology of Fermented Ham

Traditional fermented ham has a long history in China. Due to the differences in climatic conditions, raw meat and processing techniques and the number and species of microorganisms in different hams are obviously different. The microorganisms in different kinds of ham produced by different factories in the same area are also different. Even if the same type of ham is produced by the same factory, the microbial species and quantity will change greatly at different stages of production. The microorganisms in traditional fermented ham mainly depend on the extraneous bacteria existing in the natural environment. The extraneous bacteria compete with the microorganisms of raw meat, and the fermentation microorganisms gradually occupy the dominant position. Under natural conditions, the fermented ham is formed after a long period of fermentation. In modern production, the starter cultures with known ingredients are used to the product stable and easy to control. The main components of the starter cultures are LAB, *Staphylococcus*, yeast, and other microorganisms, which form special flavor through complex biochemical reaction during fermentation process.

   The species and quantity of microorganisms in different positions of the same kind of ham in the same period are different, such as muscle tissue and ham surface. Anaerobic microorganisms easily reproduce in muscle tissue, while aerobic microorganisms easily grow on the surface of ham.

   The molds in fermented ham are a kind of important microorganisms in the fermentation ecosystem, mainly distribute on the surface of the ham and the lower part of the adjacent surface. During the ripening period of ham, mold grows rapidly on the surface of meat products and forms a protective film, which endows the product with a unique appearance, due to that the temperature and humidity of the environment are suitable for the growth of mold. The protective film has the function of blocking oxygen, avoiding light, and preventing rancidity, which is conducive to the formation of the product's unique flavor. Most of the molds on the surface of ham are *Penicillium, Aspergillus*, and *Eurotium*. Some studies have found that the dominant mold in fermented ham at mature stage is *Eurotium, Penicillium*, and *Aspergillus*. The quantity and species of molds are mainly determined by the temperature and relative humidity of mature environment, but some molds are also found to be capable of producing toxins (Comi et al. 2004). For example, the predominant molds in the early stage of maturation of Iberian ham are mainly *Penicillium* and in the later stage are *Microcystis* (Núñez et al. 1996).

In the process of ham processing, *Micrococcus* bacteria is an important part. *Staphylococcus* is the dominant species, which mainly include *Staphylococcus xylose, Staphylococcus equinus, Staphylococcus saprophyticus*, and *Staphylococcus squirrel*. Ham fermentation is a process in which microorganisms of *Micrococcidae* gradually grow from the surface of ham to the muscle (Zhu 2009). At the late stage of maturation, the number of micrococci and staphylococci decreased, while the number of molds and yeasts increased. Yeast mainly distributes on or near the surface of ham. During ham processing, cleaning can reduce the amount of yeast on the surface of muscle and fat, but the number of unskinned muscle tissue increases sharply during maturation. Yeast has a strong ability to hydrolyze protein and fat and has an influence on the sensory properties of ham, especially the formation of volatile substances. The common yeasts in the fermented ham include *Debali* yeast, *Hansen* yeast, *Candida, Rhodotorula, Pichia pastoris*, etc. Simoncini et al. (2007) screened and identified 261 strains of yeast from 40 mature Italian hams, of which *Debaryomyces hansenii, Candida zeylanoides*, and *Debaryomyces maramus* were the dominant bacteria. The amount and species of yeast in different fermented hams were not exactly the same due to the influences of processing operation, production environment, and external factors (Gu and Lian 2007).

8.3.3 LAB in Fermented Ham

The amount of LAB in fermented ham is significantly less than that of yeasts and molds. In the production of ham, the main LAB are *Pediococcus acidilactici, Pediococcus pentosaceus, Pediococcus cerevisiae*, and *Lactobacillus. Pediococcus acidilactici* is the first LAB used in meat fermentation. They play an important role in the process of ham fermentation. *Pediococcus* can rapidly ferment glucose to produce lactic acid (Bartholomew and Blumer 1977).

8.3.4   The Processing Technology and Characteristics of Traditional Fermented Ham

Ham has a long history in China. Jinhua ham, Rugao ham, Xuanwei ham, etc. are famous ham products in China. Ham is mainly made from the front or back legs of pigs and is processed by curing, washing, drying, hanging, fermentation, etc. The process of making ham in different regions is slightly different, but the main operations are similar. The processing technology of traditional fermented ham mainly includes raw material selection, dressing, curing, washing, drying, shaping, hanging, fermentation, dressing and finished products, etc. Jinhua ham is briefly introduced as an example.

8.3.4.1   Raw Material Selection

Different types of ham generally choose different types of pork raw materials, such as Jinhua ham choose two black pork in Jinhua City. Raw material legs should be fresh, undamaged, full of muscles, fat and thin and white, and suitable size. They should be fully cooled, spread, or suspended for natural cooling at least. The selected pig legs are also required to meet hygienic standards.

8.3.4.2   Raw Material Dressing

Raw material dressing mainly includes removing hair, repairing bone, trimming legs surface, and removing blood stasis.

8.3.4.3   Curing

Curing is an important operation in ham processing technology. The curing process has certain requirements on ambient temperature and humidity, which is an important factor affecting product quality. If the curing period is not well controlled, the ham will deteriorate. The curing process of Jinhua ham is the key step. Although the amount and intervals of salt used are not identical, the basic principles are the same (Han 2007). Different hams differ slightly in times and duration of salting. Traditionally, the production of Jinhua ham is usually cured between the beginning of winter and the beginning of spring. The winter natural climate in Jinhua area is used to control the curing temperature. Generally, it needs to be salted about seven times, and the curing time is controlled about 35 days.

8.3.4.4 Washing and Drying

The cured ham is immersed in water, soaked, and cleaned, and the excess salt of meat and skin was removed. The ham was cleaned with bamboo brush. The water temperature is controlled at 5–10 °C and the soaking time is 4–6 h. After the main washing is finished, the ham should be re soaked in clean water for 18 h and then dried after washed for two times.

8.3.4.5 Shaping and Hanging

Shaping means that the meat is extruded to the middle with the help of tools to make the muscles bulge. After the shaping is finished, the ham is hung in pairs.

8.3.4.6 Fermentation

The ham is fermented for about 25 days. The mold begins to grow on the muscle surface. In general, the surface of normally fermented ham is mainly yellow and green, also known as "oil flower," due to the dominant green mold. This indicates that fermentation temperature, salt content, and water content are normal. If the salt content is too high, no mold grows on the surface of the meat, which is called "salt flower" (Huang 2009).

The fermentation stage is an important period for the formation of ham flavor. It is generally required that the temperature in the early stage of fermentation should be low (15–25 °C) and in the late stage should be raised to 30–37 °C. The relative humidity in the fermentation chamber should be between 60% and 70%. During the whole fermentation period, muscle and fat tissues are degraded and oxidized under the action of enzymes, resulting in the formation of small molecular substances (such as polypeptide, amino acid, and fatty acid), thus forming the unique smell of ham (Yang and Liu 2008).

8.3.4.7 Dressing

During fermentation, the evaporation of water makes the ham muscle dry and shrink, which affects the appearance of the ham. Therefore, it is necessary to trim the ham, to flatten the bones that protrude from the meat surface, to cut off the excess fat and skin, and to trim the meat surface, so as to make it flat to the two sides of the meat surface and arc, reaching the standard shape of Jinhua ham. After dressing, which usually begins in early April, is finished, the fermentation will continue until mid-August.

8.3.4.8 Finished Products

After the fermentation, the ham muscles are dried and hard. Coating a layer of vegetable oil on the surface of the meat, on the one hand, softens the muscle and, on the other hand, prevents the fat from oxidizing. Then the ham is transferred to the product store for stacking and ripening.

References

Bartholomew D, Blumer T (1977) Microbial interactions in country-style hams. J Food Sci 42(2):498–502

Benito M a J, Rodríguez M, Martín A, Aranda E, Córdoba JJ (2004) Effect of the fungal protease EPg222 on the sensory characteristics of dry fermented sausage "salchichón" ripened with commercial starter cultures. Meat Sci 67(3):497–505

Chen Q, Kong B, Sun Q, Dong F, Liu Q (2015) Antioxidant potential of a unique LAB culture isolated from Harbin dry sausage: in vitro and in a sausage model. Meat Sci 110:180–188

Cocolin L, Dolci P, Rantsiou K (2011) Biodiversity and dynamics of meat fermentations: the contribution of molecular methods for a better comprehension of a complex ecosystem. Meat Sci 89(3):296–302

Comi G, Orlic S, Redzepovic S, Urso R, Iacumin L (2004) Moulds isolated from Istrian dried ham at the pre-ripening and ripening level. Int J Food Microbiol 96(1):29–34

Diaz O, Fernandez M, De Fernando GDG, de la Hoz L, Ordoñez JA (1997) Proteolysis in dry fermented sausages: the effect of selected exogenous proteases. Meat Sci 46(1):115–128

Fontana C, Bassi D, López C, Pisacane V, Otero MC, Puglisi E, Rebecchi A, Cocconcelli PS, Vignolo G (2016) Microbial ecology involved in the ripening of naturally fermented llama meat sausages. A focus on lactobacilli diversity. Int J Food Microbiol 236:17–25

Gu Y, Lian D (2007) Microbe and its variety of traditional fermented meat. J Yinbin Univ 7(6):61–65

Guo X, Zhang Y, Zhang Q (2009) Nutrition processing characteristics and research progress of fermented meat. Meat Ind 5:47–50

Han S (2007) Processing technology of ham. Farm Technol 2:36–36

Hierro E, de la Hoz L, Ordóñez JA (1997) Contribution of microbial and meat endogenous enzymes to the lipolysis of dry fermented sausages. J Agric Food Chem 45(8):2989–2995

Huang X (2009) Value-added processing technology of meat products. Henan Science and Technology Publishing House, Zhengzhou

Li Z, Meng L (2010) Fermented food technology. China Metrology Publishing House, Beijing

Liu E, Wu Y, Zhang J, Li Y, Song H (2014) Shortening fermentation period of fermented sausage by adding exogenous enzymes. Adv Mater Res 941–944:1146–1150

Nan Q (2008) Meat industry handbook. China Light Industry Press, Beijing

Núñez F, Rodríguez M, Bermúdez M, Córdoba J, Asensio M (1996) Composition and toxigenic potential of the mould population on dry-cured Iberian ham. Int J Food Microbiol 32(1–2):185–197

Ravyts F, Vuyst LD, Leroy F (2012) Bacterial diversity and functionalities in food fermentations. Eng Life Sci 12(4):356–367

Rebucci R, Sangalli L, Fava M, Bersani C, Cantoni C, Baldi A (2007) Evaluation of functional aspects in Lactobacillus strains isolated from dry fermented sausages. J Food Qual 30(2):187–201

Schiffrin E, Rochat F, Link-Amster H, Aeschlimann J, Donnet-Hughes A (1995) Immunomodulation of human blood cells following the ingestion of lactic acid bacteria. J Dairy Sci 78(3):491–497

Simoncini N, Rotelli D, Virgili R, Quintavalla S (2007) Dynamics and characterization of yeasts during ripening of typical Italian dry-cured ham. Food Microbiol 24(6):577–584

Simonová M, Strompfová V, Marciňáková M, Lauková A, Vesterlund S, Moratalla ML, Bover-Cid S, Vidal-Carou C (2006) Characterization of Staphylococcus xylosus and Staphylococcus carnosus isolated from Slovak meat products. Meat Sci 73(4):559–564

Stahnke LH, Holck A, Jensen A, Nilsen A, Zanardi E (2002) Maturity acceleration of Italian dried sausage by Staphylococcus carnosus—relationship between maturity and flavor compounds. J Food Sci 67(5):1914–1921

Steinkraus KH (1994) Nutritional significance of fermented foods. Food Res Int 27(3):259–267

Talon R, Leroy S (2011) Diversity and safety hazards of bacteria involved in meat fermentations. Meat Sci 89(3):303–309

Urso R, Comi G, Cocolin L (2006) Ecology of lactic acid bacteria in Italian fermented sausages: isolation, identification and molecular characterization. Syst Appl Microbiol 29(8):671–680

Wang Y (2006) Meat processing technology. China Environmental Science Publishing House, Beijing

Xiong T, Wei H, Qiao C (2013) Fermented food. China Quality Inspection Press, China Standards Press, Beijing

Xu Y (2011) Fermented food science. Zhengzhou University Press, Zhengzhou

Yang Z, Liu X (2008) Introduction and improvement prospect of Jinhua ham technology. Technol Mark (12):19–20

Zalacain I, Zapelena MJ, De Peña MP, Astiasarán I, Bello J (1997) Application of Lipozyme 10,000 L (from Rhizomucor miehei) in dry fermented sausage technology: study in a pilot plant and at the industrial level. J Agric Food Chem 45(5):1972–1976

Zhu S (2009) Ham processing principle and technology. China Light Industry Press, Beijing

Chapter 9
The Preparation Technology of Pharmaceutical Preparations of Lactic Acid Bacteria

Wei Chen and Linlin Wang

9.1 Application and Development of Lactic Acid Bacteria in Pharmaceutical Products

9.1.1 The History and Development of Lactic Acid Bacteria Used in Pharmaceutical Products

9.1.1.1 The Origin of Lactic Acid Bacteria Pharmaceutical Products

The application of lactic acid bacteria in medicine mainly originates from the efficacy of food. In 1906 Metchnikoff suggested that bacteria in yogurt are good for human health. Later, according to the phenomenon that yogurt can inhibit the growth of acid-tolerant bacteria, he assumed that similar phenomena will appear in the human intestine. The bacteria in the yogurt protect human health from spoilage bacteria by making some spoilage bacteria fail to proliferate and cannot produce toxic substances. This argument became a popular medical topic in Western Europe at the time. Following the first successful use of bacteria to treat diseases, Daviel Newman applied lactic acid bacteria to treat bladder infections and pointed out that the reason for the cure is that the lactic acid produced by the metabolism of lactic acid bacteria has anti-infective properties. However, because of the exhaustive information provided, the simple sample, and the lacking basic characteristics of the strain, the mechanism cannot be clarified. However, this finding laid the foundation for the clinical application of lactic acid bacteria. Subsequently, Rettger et al. made a lot of research on the role and application of *Lactobacillus acidophilus* in 1935, which made people to have a strong interest in yogurt once again. They agreed that yogurt has a mitigating effect on constipation, inflammatory bowel disease (IBD),

W. Chen (✉) · L. Wang
Jiangnan University, Wuxi, China
e-mail: chenwei66@jiangnan.edu.cn

© Springer Nature Singapore Pte Ltd. and Science Press 2019
W. Chen (ed.), *Lactic Acid Bacteria*,
https://doi.org/10.1007/978-981-13-7283-4_9

bacterial dysentery, obesity, and other diseases. So far, yogurt has prevailed in Europe and other places.

9.1.1.2 Status of Pharmaceutical Products of Lactic Acid Bacteria

The United States is one of the fastest-growing countries in the microecological product industry which started late but developed rapidly. Its probiotics are mainly *Lactobacillus acidophilus*. See Table 9.1 for details. Other developed countries, such as the United Kingdom, Germany, France, and Japan, have also developed a

Table 9.1 List of probiotics in the US market

Manufacturing company	Product name	Use bacteria and excipients
American Health	Chewable acidophilus with bifidus	*Lactobacillus acidophilus*, *Bifidobacterium* 50 mg each, live bacteria count 1 billion/grain
Futurebiotics	Longest Living Acidophilus Plus	Compound *Lactobacillus acidophilus*
General Nutrition Corporation (GNC)	Super Colon Cleanse with Herbs and Acidophilus	*Lactobacillus acidophilus*, herbal medicine
GNC	Acidophilus Plus	Compound *Lactobacillus acidophilus*, live bacteria number 5 billion/200 mg microcapsules
Wakunaga of American	KYO-Dophilus	*Lactobacillus acidophilus*, *Bifidobacterium longum*, *Bifidobacterium bifidum*, live bacteria count 1.5 billion/piece
West Caldwell	Acidophilus	*Lactobacillus acidophilus*: apple pectin (1:100)
Nutrition Now	Efficient probiotic	*Lactobacillus acidophilus*, *Lactobacillus casei*, *Lactobacillus plantarum*, *Lactobacillus thermophilus*, *Bifidobacterium bifidum*, *Bifidobacterium infantis*, *Streptococcus faecalis*, live bacteria number 14 billion/grain
GNC	Chewable acidophilus oral liquid	*Lactobacillus acidophilus*, *Bifidobacterium*, *Lactobacillus delbrueckii*, Bulgarian subspecies concentrated cherry juice, yucca juice, live bacteria number 5 billion/ml
Nature's Bounty	Chewable acidophilus oral liquid	*Lactobacillus acidophilus*, *Bifidobacterium bifidum*, live bacteria number 1 billion/100 mg
Pharm Assure	Chewable acidophilus oral liquid	*Lactobacillus acidophilus*, oligofructose, apple pectin, live bacteria count 1 billion/grain
Rite Aid pharmacists	Acidophilus	*Lactobacillus acidophilus*, *Bifidobacterium bifidum*, *Lactobacillus delbrueckii* subsp. Bulgarian, pectin, calcium gelatin, calcium stearate
Jarrow Formulas	Jarro-Dophilus+FOS	Six kinds of lactobacilli, live bacteria number 3.36 billion/grain

Table 9.2 List of probiotics in Japanese market

Manufacturing company	Strain	Description
Morinaga Milk Industry Co., Ltd	*Bifidobacterium longum*	Bifidobacterium powder BB536 (intestinal isolation in adults) with lactulose or lactic acid bacteria, live bacteria number 50 billion/g
Healthywal	*Bifidobacterium longum*	Strains with high physiological activity or stability, product nitrogen-filled packaging, live bacteria number 80 billion/g
Healthywal	*Bifidobacterium longum,* *Bifidobacterium bifidum,* *Lactobacillus acidophilus,* *Lactococcus lactis*	Strong stability, the number of viable bacteria per strain is 10 billion/g
Daiichi Sankyo Company, Limited	*Lactobacillus sporogenes*	Heat resistance, resistance, live bacteria 5 billion/g
Daiichi Sankyo Company, Limited	Acid-resistant *Bifidobacterium longum, Lactobacillus acidophilus, Enterococcus*	
Morishita Rendan Co., Ltd	*Bifidobacterium longum*	Raffinose, lactulose
Kyowa Hakko Kirin Co., Ltd	*Bifidobacterium longum*	Freeze-dried microcapsules, live bacteria number 50 billion/g

variety of microecological products and put them into production and use. In Japan, 150 of the 224 specific health foods are regulating intestinal health products, and the Japanese probiotic powders and products are shown in Table 9.2. In 2014, the Ministry of Health of China also announced ten probiotic strains that can be used in health foods, such as *Bifidobacterium infantis*, *Bifidobacterium bifidum*, *Bifidobacterium longum*, *Bifidobacterium adolescentis*, *Bifidobacterium breve*, and Germany. *Lactobacillus* subsp. *bulgaricus*, *Lactobacillus casei* subsp., *Lactobacillus acidophilus*, *Streptococcus thermophilus*, and *Lactobacillus reuteri*.

Developed countries such as Europe, America, and Japan have developed early in the microecological product industry, with large investment, mature technology, and advanced equipment. The product was rich in variety and stable in quality. In comparison, the domestic microecological product industry will be relatively backward, mainly in the form of single product category and lack of market competitiveness.

9.1.1.3 Application of Lactic Acid Bacteria Pharmaceutical Products

Recent studies have shown that nisin does not produce drug resistance as an antibiotic and is not easily destroyed by enzymes in the gastrointestinal tract. It has also been reported in the literature that nisin can treat multiple diarrhea and has superior superiority (Wang et al. 1995). Among them, the most popular products are "Intestinal

Health" and "Lizhu Dele," which have brought revolutionary research to gastrointestinal diseases (Bao et al. 1998). In the European and American markets, Enpac (mainly containing *Lactobacillus acidophilus*, a foreign drug name) is used to relieve the side effects caused by antibiotic treatment; Lactinex (containing *Lactobacillus acidophilus*, Y. gutta) is one of the treatments of stomatitis; and Infloran Berna (containing *Lactobacillus acidophilus*, *Bifidobacterium bifidum*) is used to treat gastric dysfunction (Zou and Yang 2007). In the domestic market, some lactic acid bacteria health-care foods or products such as Ang Li No. 1, Shuangqi Tianbao, Huangdi Lansheng Liquid, Wuzhuwang, Lactobacillus tablets, and Sanzhu oral liquids are prevalent which contain bifidobacteria *Lactobacillus acidophilus* (Yu et al. 1998).

Many studies have confirmed that the use of lactobacilli can prevent and treat bacterial and fungal vaginitis. It has been reported that the use of lactic acid bacteria in the treatment of vaginitis, while oral vitamins, can stimulate the growth and reproduction of lactobacilli and improve the efficacy (Yi et al. 2006).

In recent years, research, application, and development of lactic acid bacteria in the field of medicine have been unprecedentedly active, and a large number of research results have been obtained. For example, in order to meet the needs of special functional medical uses, it is particularly important to improve the genetic traits of lactic acid bacteria, and research on genetic engineering methods has received more attention (Xue and Yang 2008). Lactobacillus genetic engineering has achieved great development. The use of double-layer embedding technology makes the stability and re-workability of lactic acid bacteria far superior to that of common lactic acid bacteria and fully expands the application of probiotics (Chen and Weng 2005).

9.1.2 Lactic Acid Bacteria Commonly Found in Pharmaceutical Products

9.1.2.1 Lactobacillus

Lactobacillus is one of the probiotics in the human gut and is essential for regulating the intestinal microenvironment. The *Lactobacillus* product that is popular in production has been characterized of relatively simple production process, good oxygen resistance, and good curative effect. Many *Lactobacilli* can be used, including *Lactobacillus acidophilus*, *Lactobacillus casei*, *Lactobacillus paracasei*, *Lactobacillus delbrueckii* subsp. *bulgaricus*, *Lactobacillus brevis*, *Lactobacillus reuteri*, *Lactobacillus fermentum*, *Lactobacillus plantarum*, *Lactobacillus gasseri*, *Lactobacillus helveticus*, and *Lactobacillus salivarius*.

9.1.2.2 Lactococcus

The *Lactococcus* used in the lactic acid bacteria products is relatively few, and the species are relatively concentrated. *Lactococcus lactis* subsp. *cremoris*, *Lactococcus lactis* subsp. *lactis*, and *Lactococcus lactis* subsp. *diacetyl* are widely applied.

9.1.2.3 *Bifidobacterium*

The most widely used probiotics is *Bifidobacterium*. *Bifidobacterium* is one of the important probiotics in the human intestine, which plays a vital role in the ecological balance of the entire intestine. The bifidobacteria which can be currently available for medicine are *Bifidobacterium adolescentis*, *Bifidobacterium breve*, *Bifidobacterium bifidum*, *Bifidobacterium longum*, and *Bifidobacterium infantis*.

9.1.2.4 Other Lactic Acid Bacteria

Enterococcus faecalis and *Enterococcus faecium* are mainly applied in lactic acid bacteria products. The earliest lactic acid bacteria in China is Rumeisheng, which is mainly used to treat digestive disorders such as diarrhea, constipation, and indigestion.

9.1.3 The Medical Function of Lactic Acid Bacteria

9.1.3.1 Prevention and Treatment of Lactose Intolerance

Lactose intolerance occurs when lactose cannot be hydrolyzed into glucose and galactose due to the low secretion of lactase. Unabsorbed lactose is degraded by colonic microorganisms to produce lactic acid, hydrogen, carbon dioxide, and methane, which eventually lead to diarrhea, abdominal pain, abdominal distension, and other symptoms (Rabot and Rafter 2010). According to the survey data, 70% of the world's adult population has lactose intolerance in 2008 (Lomer et al. 2008). In China, the incidence of lactose intolerance in adults is as high as 86.7%, and the intolerance index is 0.9 (Yan et al. 1987). The study found that not all lactic acid bacteria have lactase activity. Gu et al. (2012) isolated *Lactobacillus plantarum* K2 from Xinjiang horse milk, lysed the bacteria solution, and collected the enzyme solution after centrifugation. The method was determined by Bradford method at 37 ° C and pH 6.5 (with human intestinal environmental conditions). The enzyme activity was as high as 6620 U/g, indicating that the β-galactosidase produced by *Lactobacillus plantarum* K2 has the effect of alleviating lactose intolerance. In addition, the dairy products fermented by lactic acid bacteria are semisolid milk products, and the dairy products slow down the gastric emptying rate and intestinal transit time and improve the hydrolysis rate of lactose, thereby achieving the purpose of improving lactose intolerance (Song et al. 2010). Vincent et al. (2012) cloned and expressed the β-galactosidase gene encoding *Lactococcus lactis* IL1403 in *E. coli*. The results showed that the lactase present the great activity at 15–55 °C and pH 6.0–7.5. The maximum activity can be obtained by adding 0.8 mmol/L Fe^{2+} and 1.6 mmol/L Mg^{2+}. The biotransformation rate of lactose is up to 98%, and the catalytic efficiency is 102 mmol/(L·s). Li et al. (2012) studied the alleviation of lactose intolerance in *L. lactis* MG1363/FGZW in mice and grouped

BALB/c mice aged 9–10 weeks after weaning, 20 rats in each group. The control group was intragastrically administered with 200 μL of normal saline for 4 weeks, and the activity of β-galactosidase, intestinal flora and stool frequency, and total fecal quality were used as evaluation indexes. The results showed that compared with the control group, the activity of β-galactosidase in the intestine of the mice was significantly increased and the proportion of bifidobacteria in the intestinal flora was significantly increased, in terms of diarrhea index, although the number of stools did not change significantly. Total mass loss and lactose intolerance symptoms are alleviated (Chen 2007). In treatment of pediatric secondary lactose intolerance with medicine Changlekang containing bifidobacteria, 144 patients were randomly divided into 2 groups (treatment group and control group): the treatment group was treated with Changlekang, and the control group was given daily vitamins. The treatment effect was evaluated by the number of days of treatment (<6 days effective, >6 days invalid), and the stool frequency, stool viscosity, and stool lactose content were considered as parameters. The results showed that the total effective rate was 98.6% in the treatment group and 87.5% in the control group; the efficacy was better in the treatment group. Kolars et al. (1984) divided 50 patients with lactose intolerance into 4 groups and 4 groups of dank lactose aqueous solution (containing 20 g lactose), milk (containing 18 g lactose), plain yogurt (containing 18 g lactose), and lactulose solution (including lactulose 10 g), respectively. Then the hydrogen level in the breath was measured after 8 h. The results showed that the hydrogen content in the yogurt group was one-third of that in the milk and lactulose solution group; the proportion of patients with diarrhea or flatulence in the yogurt group was only 20% and in the milk group was 80%; the lactase activity test showed that patients with lactose intolerance could degrade 50–100% of lactose by lactase in the duodenum after drinking yogurt for 4 h. Another study reported that lactase secreted by *Lactobacillus bulgaricus* in fermented yogurt can safely enter the small intestine (Pochart et al. 1989). The buffering capacity of yogurt and the ability of the *Lactobacillus delbrueckii* subsp. *bulgaricus* to maintain cell membrane integrity are the main reasons for lactase to be protected from gastric acid degradation (Savaiano 2014). In summary, the mechanism of lactic acid bacteria against lactose intolerance mainly includes using lactic acid bacteria-fermented yogurt with lactase activity to reduce the lactose content in raw milk; some lactic acid bacteria (such as *Lactobacillus delbrueckii* subsp. *bulgaricus*) can utilize the buffering ability of the yogurt and maintain the integrity of their cell membrane to protect the lactase released by themselves from being degraded by gastric acid and safely entering the small intestine, thereby degrading the lactose in the yogurt and improving the symptoms of sugar intolerance.

9.1.3.2 Prevention and Treatment of Diarrhea

Lactic acid bacteria can antagonize pathogenic bacteria by regulating normal flora. It's reported that the effective rate of Lizhu Changle in treating chronic diarrhea was 85.2% (Li et al. 1998), while the efficacy of Huichunsheng Capsule

(Huang and Li 1999) in the treatment of adult diarrhea, pediatric diarrhea (Huang 2000; Li et al. 1999), and neonatal diarrhea (Yang et al. 1999) was over 90%, and Golden bifidus (composed of *Bifidobacterium longum, Lactobacillus bulgaricus, and Streptococcus thermophilus*) was effective in the treatment of acute and chronic diarrhea (Wang et al. 2000).

9.1.3.3 Cholesterol Lowering

Cardiovascular diseases are risk factors for death. High cholesterol levels in the serum and excessive intake of food-borne cholesterol are considered to be the leading causes of such diseases. Therefore, how to reduce cholesterol intake and lower cholesterol levels is getting more and more attention from researchers.

The lactic acid bacteria can absorb part of the cholesterol and convert it into cholate to be discharged from the body (Ma and Zhuge 1999). In vitro tests have shown that *Lactobacillus plantarum* and *Enterococcus faecium* can remove cholesterol from cholesterol-containing mediators; *Lactobacillus acidophilus* can lower serum cholesterol by directly degrading cholesterol and dissociation of bile salts in early period: the principle of which is the coprecipitation of lactic acid bacteria with bile salts. In vivo tests showed that after feeding the mice with lactic acid bacteria-fermented whey for 6 weeks, the experimental group had the lowest serum cholesterol value and higher glutathione peroxidase (GSH-PX) activity than the control group, and the antioxidant enzymes (SOD and catalase) are the most active in red blood cells (Xu et al. 2011).

9.1.3.4 Antibacterial and Anti-infective

When lactic acid bacteria ferment lactose in the body, a large amount of lactic acid and acetic acid are produced, and the intestine is in an acidic environment. The decrease of pH can inhibit the growth of spoilage bacteria and other harmful bacteria such as *Listeria, Staphylococcus*, and *Salmonella*. Many lactic acid bacteria can produce bacteriocin, which has antagonistic effects on colicin and staphylococcus. Therefore, lactic acid bacteria can be used in health foods to enhance the body's resistance to pathogenic bacteria.

9.1.3.5 Increase Immunity

Bifidobacteria and its surface structure can enhance the production of various cytokines and antibodies, increase the activity of natural killer cells (NK cells) and macrophages, enhance local or systemic defense functions, and exert anti-infective and antitumor effects.

Compared to the control group, after feeding *L. casei* to mice with invasive ductal carcinoma, the levels of IL22 and IFN-γ in mice were increased in experimental group, the activity of NK cells in splenocytes was greatly improved, the growth rate of tumors in mice was significantly reduced, and the survival time of mice was significantly increased. The results indicated that daily intake of *Lactobacillus casei* increases the survival time of cancer mice (Soltan et al. 2012). Bleau et al. (2010) found that extracellular polysaccharides produced by *Lactobacillus rhamnosus* RW-9595M promoted the induction of anti-inflammatory factor IL-10 induced by macrophages in mouse peritoneal, thereby reducing inflammation. Dong et al. (2012) used four strains of *Lactobacillus*, including *Lactobacillus casei shirota*, *Lactobacillus rhamnosus GG*, *Lactobacillus plantarum* NCIMB8826, and *L. reuteri* NCIMB11951, and two strains of *Bifidobacterium longum* SP07/3 and *Bifidobacterium bifidum* MF20/5 and compared their immune regulation to the peripheral blood mononuclear cells in the bacteria. It showed that both groups can enhance the activity of lymphocytes, T lymphocytes, and NK cells. At the same time, the activities of cytokines such as IL-6, IL-10, and TNF-α were also increased to varying degrees, but *Lactobacillus* tended to regulate Th1 cytokines, while bifidobacteria preferred to produce anti-inflammatory cytokines to enhanced immune effects. In a study (Elmadfa et al. 2010), 33 young (22–29 years old) healthy women were divided into 2 groups and drank milk products and conventional yogurt which is fermented by *Lactobacillus delbrueckii*, *Streptococcus thermophilus*, and *Lactobacillus casei* DN114001, respectively. After 4 weeks, blood samples from the subjects were collected, and the activation of natural killer cells and mitogen-induced T lymphocytes was found, with the production of cytokines; the immune response was significantly enhanced. Further research found that *Lactobacillus delbrueckii* (yogurt fermentation agent) can promote secretion of pro-inflammatory cytokines IL-10; *Lactobacillus rhamnosus GG* can increase the expression of costimulatory molecules CD80, CD86, and CD54 and mature markers CD83 and induce dendritic cell maturation. Hiramatsu et al. gave mice oral administration of *Bifidobacterium breve* Bp JCM70411, and then Bp JCM70411 was surrounded by CD11c+ cells in the area of lymphoid aggregation and cecal patch, indicating that Bp JCM70411 induced CD11c+ cells to mediate immune responses directly. In addition, Bp JCM70411 significantly increased IL-10 and IL-12p40 production by Thy1.2 cells and bone marrow dendritic cells and induced the production of immunoglobulin A (IgA) by dendritic cells in the Peyer's nodule and cecal patches, indicating that Bp JCM70411 can enriched the immune function of dendritic cells in the Peyer's nodule and cecal patches. The above studies showed that lactic acid bacteria can interact with host cells by secreted metabolites and wall-related molecules to activate immune-related cell signaling, thereby regulating cytokine expression, antibody secretion, differentiation, and activity of immune cells, enhancing the body's immune function.

9.1.3.6 Prevention of Vaginal Diseases

Lactic acid bacteria (mainly *Lactobacillus*) have accumulated a wealth of information in the treatment of genitourinary infections. Lee et al. (2013) studied whether lactic acid bacteria mixture can prevent urinary tract infection in young children.

The young rats were used as experimental models. The incidence of pyelonephritis in the experimental group of bladder-infused lactic acid bacteria mixture was significantly lower than that in the control group. Infusion of lactic acid bacteria showed significant preventive effect on experimental pyelonephritis, indicating that mixed lactic acid bacteria can prevent urinary tract infection in young children. Chytilová et al. (2014) found that linseed oil (rich polyunsaturated fatty acids) combined with *Lactobacillus plantarum* can be used as an oral therapeutic agent for the treatment of neonatal gastrointestinal diseases. Prophylactic administration of *Lactobacillus* can provide anti-inflammatory properties and help regulate the inflammatory response caused by enterotoxic *Escherichia coli* (ETEC) by stimulating Th1 cell-mediated cellular immunity and phagocytosis.

In vivo tests have confirmed that in addition to co-coagulation, vaginal endogenous flora can also colonize the surface of the vaginal epithelium (Reid et al. 1987). Lactobacillus can co-agglomerate with other bacteria, and pathogens can be inhibited when co-coagulation and the metabolites of lactobacilli work together (Reid et al. 1990). In vitro tests have also shown that lactobacilli inhibited both aerobic and anaerobic bacteria in women's vagina (Skarin and Sylwan 1986).

9.1.3.7 Others

In addition to the above applications, lactic acid bacteria have an effect of controlling endotoxemia. More than 90% of endotoxin in the human gut is derived from the release and lysis of Gram-negative bacteria. It has been reported that *Bifidobacterium* inhibited the growth of harmful bacteria by regulating the pH value of intestinal tract. At the same time, *Bifidobacterium* prevented the colonization, invasion, and bacterial translocation of pathogenic bacteria by competing with pathogenic bacteria in order to reduce the production or absorption of endotoxin and the level of endotoxin in blood to alleviate the damage to other organs (Luo et al. 1999).

9.1.4 Standard for the Use of Medical Lactic Acid Bacteria

In general, intestinal microecology will prevent the invasion of foreign substances. However, when the human body is affected by external factors, such as changes in stress, disease, and diet, pathogens will take advantage of it (Macfarlane and Cummings 1999). In addition, since the flora in the host is not fixed (Mackie et al. 1999), in order to be healthy, people need to constantly and regularly ingest bacteria that are beneficial to the human body. Some characteristics of lactic acid bacteria should also be considered when supplementing lactic acid bacteria.

9.1.4.1 Stain Selection

In addition to the production requirements, the selected strains should have the characteristics of high adhesion, strong competitiveness, rapid adaptation to the environment, and rapid growth and should be nontoxic, be harmless, and have no side effects. LAB have to withstand the erosion of stomach acid and bile acid. If they cannot withstand the erosion of stomach acid and bile acid, they will not be able to reach the intestine safely. Thus high activity for LAB is necessary, but it is not enough. Besides that, the strain must be adhesive in order to survive and function in the intestine. Therefore, sufficient attention should be paid to the adhesion of the strain, because any kind of microflora, whether normal or abnormal, will have exclusive effects on foreign bacteria. If the strain used for treatment is weak in nature, it cannot be effectively colonized and propagated in the corresponding environment. Eventually, it is quickly excreted from the host and cannot fully exert its ecological effects.

9.1.4.2 Dose Selection

The efficacy of LAB preparation can exert a certain health-care effect on the body only after a certain amount of it is ingested. If the amount of LAB preparation is not enough, the dominant flora cannot be formed in the body, so it is difficult to play a probiotic role. In the preparations officially approved for production in China, the regulations on the number and dosage of live bacteria are as follows: *Bacillus* $\geq 5.0 \times 10^8$ cfu/g and *Lactobacillus* $\geq 1.0 \times 10^7$ cfu/g. If the microecological preparation is added to the feed, the target live bacteria content should be no less than 1.0×10^9 cfu/kg (Zhang 2007). Currently, there is no uniform standard for the dosage of LAB preparations. The main reason is that the theoretical research of LAB preparations lags behind its practical application, and its action mechanism is not completely clear. Most of the research only stays in the experimental use effect, and the results vary greatly. Therefore, it still requires in-depth research to evaluate probiotic effect.

9.1.4.3 Medical Course Selection

Different doses also affect the function of LAB preparations. Different tests were carried out using the same microecological preparations, one for continuous use and the other for single use. Although the minimum dose was unclear, the effects of the microecological formulation disappeared after the drug was stopped. Animal experiments show that after stopping the drug, the probiotic effect disappears. After 7 days of drug withdrawal, the LAB strain cannot be found in the intestinal tract. Therefore, it seems unlikely that the LAB strain can be permanently colonized in the host body (Shao and Zhu 2003).

9.2   Types and Preparation Processes of LAB Pharmaceutical Preparations

Currently, there are nearly 100 kinds of LAB preparations for human use. In terms of composition, some are simple live bacterial preparations, and some are compound preparations of probiotics and other ingredients besides LAB. In terms of the kind of bacteria, some are made up by single bacteria, and others are composed of various bacteria. According to the dosage form, there are tablets, capsules, granules, and liquid preparations, including dairy drinks, oral liquids, and so on.

9.2.1   Types of LAB Pharmaceutical Preparations

9.2.1.1   LAB Capsule

Capsule is a large dosage form of LAB products and refers to a solid preparation in which probiotic or a probiotic added with auxiliary material is filled in a hollow capsule or sealed in a soft capsule. LAB capsule has the characteristics of simple production process, convenient transportation, and high bioavailability. The opaque capsule can be used to reduce the exposure of probiotics to light, air, and heat. The disadvantage is that liquid and hygroscopic preparations should not be directly prepared into capsule. At the same time, capsules are not suitable for children.

9.2.1.2   LAB Tablets

LAB tablet is also a major dosage form of the preparation. It is prepared by compressing LAB and appropriate excipients. As can be seen from the concept of the tablet, the tablet is a solid preparation having higher density and smaller volume prepared by compressing the LAB powder and can adapt to various requirements for the treatment and prevention of drugs. By different preparation means, various types of tablets can be produced such as coated tablets, dispersible tablets, sustained-release and controlled-release tablets, multilayered tablets, and so on. Different tablets can exert different function, like quick action, long action, controlled-release, and enteric dissolution. There are two common types of lactic acid bacteria (LAB) tablets, one is oral tablet and another is topical tablet. Oral tablet (such as LAB milk slices, lactobacteriocin tablets, etc.) belongs to ordinary tablets which generally are required to have good disintegration and dissolution. It is best made after coating preparation in order to prevent and reduce oxygen damage. Another kind of oral tablet is enteric-coated tablet, which can be used to prevent the damage of stomach acid to bacteria as well as mitigate the impact of oxygen. Topical tablets are mainly used to treat vaginal diseases. Tablets have high degree of automation and mechanization in production process. The product has the advantages of having stable shape, accurate dosage, and stable quality (mainly due to the decrease of the contact surface between lactic acid bacteria and air, light, and water) and being convenient to use and carry, easy to identify, and low cost. The disadvantage is that infants and comatose patients are having difficulty of swallowing these tablets.

The active LAB agents are generally produced in capsule and less in tablet. The reason may be LABs have a relatively low survival rate and short effective period within tablet. If an effective protective reagent can be screened in a suitable tablet preparation process, the survival rate can be increased and the effective period can be extended. Thereby we can use tablet as a dosage form of the active microbial agents for cost reduction and easier application.

9.2.1.3 LAB Granules

Granules are also a major class of LAB products. LAB granules are dry granule preparations with a certain particle size, which are produced by mixing LAB powder with a suitable auxiliary material. The granule production process is the same as that of the tablet, which is relatively simple, does not require complicated equipment, and is convenient to carry and take. The granules are composed of lactobacillus powder, filler, stabilizer, and flavoring agent. According to dissolution, it can be divided into soluble granule, suspended granule, effervescent granule, enteric granule, sustained-release granule, controlled-release granules, etc., while suspended granule occupy most of LAB granules. The main ingredient of suspended granule is dry bacterial powder, and milk powder or starch often is the supplementary material. Meanwhile, based on product requirements, flavoring agents like sweeteners and flavors and nutritional supplements such as vitamins can also be added. For bifidobacteria granules, it may be fortified with bifidus promoter like oligosaccharide. LAB granules is made after fully mixing these ingredients and vacuum-packed with composite aluminum film. It is divided into different specifications according to the number of active bacteria contained per gram, because the number of active bacteria per gram can range from several hundreds of million to hundreds of billion. The main characteristics of LAB granules can be summarized for its relatively low ability in dispersion, adhesion, agglomeration, moisture absorption, etc. It can be directly swallowed, so can be made into granules of different color, aroma, and taste as needed. If necessary, coating is applied to protect its activity. According to the nature of the coating material, the particles can have a good resistance to moisture and slow-releasing or enteric property while dissolved in water to drink. The number of viable bacteria is the most significant and only indicator of LAB granules, so the water temperature should not exceed 50 °C during drinking.

9.2.1.4 LBS Liquid Preparation

The springing-up of liquid preparations in China has promoted the Chinese consumers' understanding of LAB preparations, as well as facilitated the research and development of LAB products in China. The deepening of basic research, product development, and scientific propaganda on LAB function eliminates the deviation of LAB, indicating that LAB products will exhibit bright prospects in China.

9.2.1.4.1 Types of LAB Liquid Preparation

According to dispersion system, liquid preparations can be classified into a homogeneous phase liquid preparation (solution) and a nonhomogeneous phase liquid preparation (sol agent, suspension, and emulsion), while LAB liquid preparation generally belongs to suspension.

According to manufacturing technology, LAB liquid preparations can be divided into two categories. In the first type, LAB are inoculated in the medium made of food grade raw materials for fermentation. The fermentation broth is directly sterilized under sterile conditions without sterilization. In this way, the product is obtained. In the second type, LAB powder or suspension is added in the mixed liquid of other supplementary materials.

In addition to the probiotic effect of the LAB itself, the metabolites of LAB also have a good beneficial effect on the human body. Those are the advantages of the first type product which contains fermentation broth.

The second type of liquid preparation can be prepared as needed and then mixed with LAB, which has the advantage that this liquid preparation can be prepared according to the physiological characteristics of LAB. To a certain extent, it's advantageous to prolong the shelf life of product. In order to ensure the number of viable bacteria in the product during its shelf life, the product is required to be stored at low temperatures and better to be sold nearby.

9.2.1.4.2 Characteristics of Probiotic Liquid Preparations

The probiotic liquid preparation is characterized by accurate dosage and is suitable for consumption by various people. The disadvantage is that it is bulky and inconvenient to carry and the number of live bacteria is more difficult to maintain. In addition, the sterility requirements for the production process are relatively high.

9.2.1.5 Lactic Acid Bacteria Microcapsule Preparation

Microcapsule technology, a technology that is developing rapidly and widely used in the world, is one of the most effective and practical methods to protect the vitality of bacteria. Microcapsule refers to a package having a polymeric wall shell and a microcontainer. Microcapsule granulation technology is the embedding of solid, liquid, or gas in a microcapsule. The substance in the microcapsule is isolated from the external environment and is minimally affected by the environment, so it remains stable only when under the appropriate conditions, the buried material can be released.

The advantages of microencapsulation of lactic acid bacteria are as follows: (1) they can separate lactic acid bacteria from the external environment, so as to protect them from harmful environment, (2) they are conducive to transportation and preservation, and (3) they are released at fixed point. Enteric-coated materials allow the lactic acid bacteria to be released only after reaching the intestines. The purpose is to allow more living bacteria to enter the intestine and colonize the intestinal mucosa, so as to truly play the role in health care (Liu et al. 2004).

However, regardless of the dosage form or the embedding method, in terms of the current industrial production technology and the characteristics of the live bacteria of the probiotics product, for the sake of caution, the preservation should be protected from light and low temperature (the ideal storage position is the constant temperature layer of the refrigerator), and try not to open the sealed package; once opened, they must be used in the shortest time. The number of live probiotics is reduced, and the activity is reduced rapidly. If exposed to the environment in contact with the surrounding air, the viable count or activity of most products is almost lost in 1 week.

9.2.2 The Preparation Process of Lactic Acid Bacteria Pharmaceutical Preparations

The preparation process of the lactic acid bacteria pharmaceutical preparation is shown in Fig. 9.1. The difference in the preparation process of the different dosage forms is mainly the step of preparing the semifinished products into different dosage forms, the tablets are mainly tableting, the capsules are mainly containing capsules, and the granules are granulation. Microcapsules are wall materials that are made of special materials. Only the liquid formulation is the simplest. After the suspension, the mixture is mixed with the flavoring agent and then packaged; that is the finished product.

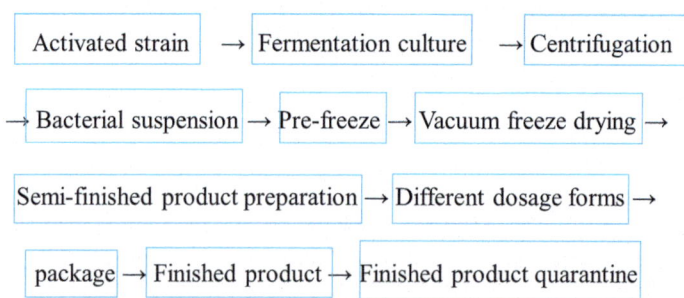

Fig. 9.1 Flow chart of preparation process of lactic acid bacteria pharmaceutical preparation

9.2.2.1   The Preparation of Lactic Acid Bacteria Capsules

9.2.2.1.1   Preparation Process of Capsule Preparation

Capsule preparations mainly include hard capsules and soft capsules. We will introduce the preparation process separately below.

The preparation process of the hard capsule mainly includes the preparation, filling, and sealing of the empty capsules and filling material and is now introduced as follows. The first is the preparation of empty capsules. The preparation process is mainly sol → blanking → drying → shelling → cutting → finishing. For ease of use, the empty capsule shell can also be printed with edible ink. After that, the empty capsules of the appropriate specifications are selected. After the preparation of the empty capsules is completed, the preparation of the filling materials is carried out, mainly by adding the auxiliary materials into the lactic acid bacteria suspension to form granules, and then filling them, and then filling the capsule caps after filling, currently more. The use of a lock-type capsule provides good airtightness and does not require sealing.

The difference between the preparation process of soft capsules and hard capsules is in the capsule material. Generally, soft capsules are mainly composed of gelatin, plasticizer, and water; the mass ratio is usually dry gelatin-dry plasticizer-water = 1: (0.4–0.6): 1. Soft capsule preparation methods are commonly used in the dropping method and the pressing method. The dropping method is completed by a dropping machine with a double-layer nozzle, and the soft capsule with mainly gelatin and the lactic acid bacteria suspension are, respectively, in the outer layer of the double-layer nozzle. The inner layer is sprayed at different speeds, so that the quantitative glue is wrapped with the quantitative lactic acid bacteria suspension, and then dropped into the cooling liquid which is incompatible with the gelatin, and gradually cooled and solidified into soft capsules. The pressing method is to make the glue into a thick and uniform film, and then place the lactic acid bacteria suspension between two films, and use a steel plate mold or a rotary mold to form a soft capsule.

9.2.2.1.2   Advantages and Disadvantages of the Preparation Process
of Capsule Preparation

The preparation process of hard capsules is mainly in the two steps of granulation and drying. The "one-step granulation method" is adopted, that is to say, the application of new equipment to mix materials, bond into granules, and dry the process is completed in one unit at a time. This method has high production efficiency, simplifies the process and equipment, and saves the plant and manpower. At the same time, the obtained particles are uniform in size, round in appearance, and good in fluidity. Artificial granulation has the disadvantages of being time-consuming and laborious and having uneven particles and low efficiency.

9.2.2.1.3   Representing the Application of the Product

Representative products of lactic acid bacteria capsule preparation are compound lactic acid bacteria capsules (polygram) that are mainly used for treating intestinal dysfunction caused by intestinal flora imbalance, such as acute diarrhea, chronic diarrhea, etc.; Lizhu Yule capsules (*Bifidobacterium* live bacteria capsule) that are used for the treatment of intestinal dysfunction caused by intestinal flora imbalance, such as acute diarrhea, chronic diarrhea, constipation, etc., and are used widely; Peifeikang capsules (*Bifidobacterium* triple live capsule) that are used for the treatment of intestinal tract acute diarrhea, chronic diarrhea, and constipation caused by dysbacteriosis and also used to treat mild to moderate acute diarrhea, chronic diarrhea, dyspepsia, and bloating and for adjuvant treatment of endotoxemia caused by intestinal flora imbalance; Befeida capsules (Bifidiary triple live capsule) that are used for the treatment of diarrhea and bloating caused by intestinal flora imbalance and can also be used for the treatment of mild to moderate acute diarrhea and chronic diarrhea; and Dingzhusheng capsules (*Lactobacillus* vaginal bacterium) that are commonly used preparation for vaginal administration for the treatment of bacterial vaginosis caused by bacterial flora disorder.

9.2.2.2   Tablets of Lactic Acid Bacteria

9.2.2.2.1   The Preparation Process of Tablets

The preparation process of the tablet mainly includes mixing, granulating, drying, and tableting. For probiotic preparations, good compressibility is essential; the better the compressibility, the lower the pressure required for tableting and the better the tablet can be made under moderate pressure because the tablet pressure directly affects the survival rate of lactic acid bacteria in the tablet. The preparation process of the tablet is as follows.

The lactic acid bacteria fermentation broth is thoroughly mixed with a suitable diluent and disintegrant to form a soft material, and the soft material is manually or mechanically extruded through a sieve, as a wet granule can be obtained by using a rocking granulator. The wet granules are granulated after drying (normal pressure box drying, fluidized bed drying, spray drying, infrared drying, microwave drying, and freeze-drying) in order to spread the agglomerated and adhered particles to obtain uniform size particles, then the lubricant is added to the particles, and the particles are placed in the mixing cylinder for "total mixing." If the lactic acid bacteria used are unstable to moisture and heat, the blank granules containing no lactic acid bacteria may be first prepared, then the lactic acid bacteria are added, and the tablets are sealed after several hours of sealing.

Most lactic acid bacteria preparations can be made into blank granules, then mixed with bacterial powder, mixed uniformly, and then compressed by tableting method to reduce the death of lactic acid bacteria during granulation.

9.2.2.2.2   The Advantages and Disadvantages of the Tablet Preparation Process

For the lactic acid bacteria preparation, evaluating the quality of the preparation process is mainly based on the survival rate of the bacteria, which can ensure that the survival rate of the live bacteria is the advantage of the tablet preparation process, and the two steps affecting the live bacteria are mainly dry and tablet. Among them, fluidized bed drying, spray drying, and freeze-drying have the advantages of high efficiency, high speed, short time, suitable for heat sensitive materials, and aseptic operation. Other drying methods have more or less disadvantages such as uneven drying of materials and large power consumption and influence on material stability.

The tableting is divided into two types: dry pressing and wet pressing. The majority of probiotic tablets are dry-pressed. For which have greet fluidity and compressibility, direct compression is generally used. For the liquidity is not good, it is necessary to first granulate, that is, after adding the binder, first make a large piece and then pulverize into granules to prepare the tablet. Wet method mainly includes two kinds, one is to make the mixture into soft material, granulate, dry, and then tablet, and the other is fluidized spray granulation. During the wet pressing process, some probiotics that are not resistant to heat and oxygen are greatly affected and should be used with caution.

9.2.2.2.3   The Application of the Representative Products

Representative products of tablets are *Lactobacillus* tablets (the mainly for abnormal fermentation in the intestine, indigestion, enteritis, and diarrhea in children), compound *Lactobacillus* acidophilus tablet (mainly for intestinal dysfunction caused by intestinal flora imbalance, such as transitional acute diarrhea), and Jin Shuangqi (live combined *Bifidobacterium* and *Lactobacillus* tablets, for diarrhea caused by intestinal flora imbalance, chronic diarrhea, antibiotic treatment, diarrhea, and constipation). Prebayer is a bifid and four-lived live bacteria tablet that can supplement normal physiological bacteria, regulate intestinal flora, stimulate immunity, form a strong biological barrier and chemical barrier in the intestine, synthesize vitamins, and promote food digestion and absorption of nutrients. Prebayer has good effects on acute and chronic diarrhea in adults and constipation and acute diarrhea in children. No adverse reactions such as in the blood, liver, and kidney were found.

9.2.2.3 Granules of Lactic Acid Bacteria

9.2.2.3.1 Preparation Process of Granules

The preparation process of the lactic acid bacteria granules is relatively simple and is similar to the preparation process of the tablets, but the tablets are directly loaded into the bag without the need of pressing into tablets. The general preparation process of the granules is as follows: the lactic acid bacteria fermentation liquid is thoroughly mixed with a suitable diluent and a disintegrating agent to form a soft material, and the soft material is manually or mechanically extruded through a sieve to obtain wet granules. After the wet granules are dried (box drying, fluidized bed drying), they are granulated and classified, and then bagged, and, finally, products were finished.

9.2.2.3.2 Advantages and Disadvantages of Granule Preparation Process

In the preparation process of granules, the preparation process is good or bad, mainly in the two steps of granulation and drying. The "one-step granulation method" is adopted, that is, the application of new equipment to mix materials, bond into granules, and dry. It is completed in one time in the unified equipment. This method has high production efficiency, simplifies the process and equipment, saves the plant and manpower, and at the same time produces the particle size; the appearance is round and the fluidity is good. Artificial granulation has the disadvantages of being time-consuming and laborious and having uneven particles and low efficiency. The particles obtained by the fluidized drying method have been dried and do not need to be dried again, and other drying methods are subjected to the step of drying and removing water.

9.2.2.3.3 The Application of the Representative Products

Representative products of granules are *Lactobacillus* granules (Jenocon) that are used for the treatment of abnormal fermentation in the intestine, indigestion, enteritis, and diarrhea in children; Peifeikang powder (*Bifidobacterium* triplet) that is used for the treatment of intestinal tract acute diarrhea, chronic diarrhea, and constipation caused by dysbacteriosis and is also used to treat mild to moderate acute diarrhea, chronic diarrhea and dyspepsia and abdominal distension and for the adjuvant treatment of endotoxemia caused by intestinal flora imbalance; and Lizhu Changle powder (*Bifidobacterium* live bacteria powder) that is used to treat intestinal dysfunction caused by intestinal flora imbalance, such as acute diarrhea, chronic diarrhea, constipation, etc.

9.2.2.4   Lactic Acid Bacteria Liquid Preparation

9.2.2.4.1   Preparation Process of Liquid Preparation

Lactic acid bacteria liquid preparations can be divided into two categories from the production process: one is made from food grade raw materials and then the other is inoculated with lactic acid bacteria for fermentation. The fermented fermentation is directly canned under sterile conditions, and the product is obtained; another is processed into a solution by a certain formula, and after being sterilized and cooled, the lactic acid bacteria powder or the bacterial suspension is mixed and canned to obtain a product.

9.2.2.4.2   Advantages and Disadvantages of the Preparation Process of Liquid Preparations

The advantage of the preparation process of the liquid preparation is that the lactic acid bacteria can be ensured in a large amount, and the production process is relatively simple, but the major problems are that it needs to be operated under aseptic conditions, the environment is relatively high, and the bacteria are easily infected. The conditions are high and the shelf life is also short.

9.2.2.4.3   The Application of the Representative Products

Representative product of liquid preparation include Jiajiain oral liquid (*Bifidobacterium* oral liquid) whose main effect is to improve the gastrointestinal function, immune regulation, growth promotion, and treatment of various common gastrointestinal diseases.

Kangbao (compound active lactic acid bacteria oral solution) is mainly used for the treatment of gastrointestinal diseases such as diarrhea, abdominal pain, and constipation; *Bifidobacterium* (*Bifidobacterium* oral liquid) is mainly used for the treatment of constipation.

9.2.2.5   Other Forms of Lactic Acid Bacteria Preparations

In addition, the forms of lactic acid bacteria preparation also include solid beverage (bacterial powder), microcapsule state, etc. Representative products include Lizhu Changle (live *Bifidobacterium*), which cure for the intestinal dysfunction caused by intestinal flora imbalance, such as acute diarrhea, chronic diarrhea, constipation, etc.; Peifeikang (triple viable *Bifidobacterium*), used for the treatment of acute diarrhea, chronic diarrhea, and constipation caused by intestinal flora imbalance as well as for the treatment of mild to moderate acute diarrhea, chronic diarrhea, dyspepsia, bloating, and endotoxemia caused by adjuvant treatment of intestinal flora disorders; and living preparation of *Lactobacillus* (live lactobacillus for vagina), used to treat bacterial vaginosis caused by bacterial flora disorder and which is currently a common preparation for vaginal medication.

9.3 Introduction of Well-Known Lactic Acid Bacteria Pharmaceutical Preparations at Home and Abroad

Now we will briefly introduce the development history, composition, function, and application of several well-known lactic acid bacteria pharmaceutical preparations at home and abroad.

9.3.1 Mamiai

9.3.1.1 Development History

Mamiai is a brand of Hanmi Pharmaceutical Group. Since its establishment in 1973, it has been a brand history of 40 years. In 1994, Mamiai appeared on the market in China. After more than 20 years of market and practice tests, Mamiai has become a basic standing medicine for children's intestinal health and is trusted by doctors and parents. From the aspect of raw material selection, the product adopts the bacteria that can be added to infant food recently promulgated by the Ministry of Health, and the number of live bacteria per pack of products is as high as 12 billion. High-quality strains and sufficient quantity ensure that probiotics can "live" to the intestine and colonize in the intestine. For children's health, this product does not add any preservatives and pigments so that infants can safely take it for a long time.

9.3.1.2 Ingredients

This product is a compound preparation. Its main components are 37.5 mg of live lyophilized powder per gram (1 bag) (1.35×10^8 cfu of *Enterococcus faecium*, 1.5×10^7 cfu of *Bacillus subtilis*), vitamin C 10 mg, vitamin B_1 0.5 mg, vitamin B_2 0.5 mg, vitamin B_6 0.5 mg, vitamin B_{12} 1.0 mg, nicotinamide 2.0 mg, calcium lactate 2.0 mg (corresponding to calcium 2.6 mg), and zinc oxide 1.2 mg (equivalent to zinc 1.0 mg).

9.3.1.3 Function and Application

Mamiai protects children from diarrhea by its *Enterococcus* colonization in the intestine and through its secretion (such as enterococcin, bacteriocin, lysozyme, etc.) which effectively inhibits harmful strains in the intestine and then helps the beneficial bacteria to proliferate. *Enterococcus faecalis* can settle in specific parts of the intestine after entering the body and then reproduce rapidly and play a barrier protection in the intestinal mucosa. It can effectively prevent the invasion of harmful microorganisms and has certain effects on infectious and noninfectious diarrhea. In addition, its prevention and treatment can also be expressed in the following aspects: diarrhea, intestinal infections, intestinal mucosal damage caused by the use of antibiotics, constipation, indigestion, abnormal fermentation in the intestines, malnutrition, enteritis, loss of appetite, abdominal distension, and other diseases.

9.3.2 Peifeikang

9.3.2.1 Development History

In the 1990s, Xinyi Pharmaceutical developed "Peifeikang," which pioneered China's microecological preparations. On October 5, 1998, Peifeikang passed the preliminary examination at the FDA Biological Products Review Center. This is the first international micro-pharmaceutical drug application accepted by the organization, and it is also the first Western medicine preparation in the mainland to be accepted by the FDA. In November 2001, Pfeiffer received a US patent.

9.3.2.2 Ingredients

This product is a compound preparation, and it contains *Bifidobacterium longum*, *Lactobacillus acidophilus*, and *Enterococcus faecalis*.

9.3.2.3 Function and Application

It can directly supplement the normal physiological bacteria of the human body, adjust the balance of intestinal flora, inhibit and eliminate the pathogenic bacteria in the intestinal tract, reduce the production of intestinal toxins, promote the digestion of nutrients by the body, synthesize the vitamins needed by the body, and stimulate the body immunity. It is mainly used in the following aspects: acute diarrhea and chronic diarrhea caused by intestinal flora imbalance, mild or moderate acute diarrhea and indigestion, constipation, and bloating.

9.3.3 Lizhu Changle

9.3.3.1 Development History

Lizhu Changle is a bifidobacterium preparation produced by Lizhu Pharmaceutical Group Inc. (Zhuhai). Its main component is *Bifidobacterium adolescentis*. It has been used clinically for more than 20 years and has a remarkable effect on the treatment of various digestive diseases. It is widely used for tumor surgery as an adjuvant therapy because it has no side effect. In the process of development and production, a series of technical problems have been solved by Lizhu Changle, and the product quality has reached the domestic advanced level. Firstly, anaerobic culture is used to ensure the stability of the obligate anaerobic *Bifidobacterium adolescentis* strain. Secondly, the appropriate medium and fermentation conditions are selected to ensure the viable count of *Bifidobacterium adolescentis* starter. Finally, a higher viable count of product during the storage period is ensured by the addition of protective agents and double-layer vacuum packaging. This product is a capsule that is easy to carry.

9.3.3.2 Ingredients

This product is a *Bifidobacterium* (*Bifidobacterium adolescentis*) powder, made by culture enrichment, freeze-drying, and mixing with lactose and magnesium stearate, each containing 0.5 billion live bacteria.

9.3.3.3 Function and Application

This product is a *Bifidobacterium* live bacterial preparation. Bifidobacteria together with other anaerobic bacteria occupy the surface of the intestinal mucosa, forming a biological barrier, preventing the colonization and invasion of the bacteria, and inhibiting the growth of pathogenic bacteria by producing lactic acid and acetic acid to reduce the pH in the intestinal tract. When people get sick or take antibiotics for a long time, they will cause the imbalance of flora and the proliferation of harmful bacteria, which will cause diarrhea. This product can rebuild the normal microecological system in the human intestinal tract and adjust the intestinal flora to stop diarrhea.

This product is mainly used to treat intestinal dysfunction caused by intestinal flora imbalance, such as acute diarrhea, chronic diarrhea, constipation, etc.

9.3.4 Golden Bifidus

9.3.4.1 Development History

Golden bifidus is the product of lactic acid bacteria produced by Inner Mongolia double Qi pharmaceutical Limited by Share Ltd. Golden bifidus is a bifidobacteria-based triple live bacterial preparation containing *Lactobacillus* and *Streptococcus thermophilus*. During the process of Golden bifidus's development and production, a series of technical problems have been solved, and the product quality has reached the domestic advanced level: firstly, ensuring the stability of the obligate anaerobic *Bifidobacterium* strain by anaerobic culture; secondly, producing the highest viable count in the *Bifidobacterium* starter through the appropriate medium and fermentation conditions; and finally, improving the effective amount of live bacteria in the product during the storage period and using of protective agent and double-layer vacuum packaging; thus the viable cell survival rate of the finished product is the highest among similar products in China. Golden bifidus products passed the identification of Inner Mongolia Autonomous Region in 1993. On August 10, 1994, the Inner Mongolia Autonomous Region Light Industry Research Institute in Hohhot held a result conference on the "Bifidobacteria preparation"; Professor Yu Ruomu, honorary director of the Chinese Nutrition Society, believes that Bifidobacteria tablet is better among the current microecological preparations, and the technology is at the leading level at the domestic while reaching the international advanced level. Since its launch in 1994, Golden bifidus has been welcomed by consumers. In the same year, it won the gold medal of first China's health-care boutique. In 1995, it won the first prize of China's light industry outstanding new products and meanwhile awarded the first prize for science and technology progress.

9.3.4.2 Ingredients

The product is a compound preparation containing *Bifidobacterium longum* (not less than 1.0×10^7 CFU per gram), *Lactobacillus bulgaricus* (not less than 1.0×10^6 CFU), and *Streptococcus thermophiles* (not less than 1.0×10^6 CFU). Excipients are skim milk powder, lactose, glucose, microcrystalline cellulose, sodium carboxymethyl starch, citric acid, sucrose, and magnesium stearate.

9.3.4.3 Function and Application

9.3.4.3.1 Supplement

It can quickly replenish more viable bifidobacteria, *Lactobacillus bulgaricus*, and *Streptococcus thermophiles* for the body. Golden bifidus contains bifidobacteria from the intestines of healthy humans, *Lactobacillus bulgaricus*, and *Streptococcus thermophiles*, which create environment for the reproduction of bifidobacteria in the intestine. It has a highly active, high colonization rate of intestinal probiotics combination which is quickly replenished into the intestines to restore the number of probiotics in the intestine to a healthy level.

9.3.4.3.2 Suppression

It can effectively inhibit harmful bacteria that damage intestinal function and induce various diseases. Golden bifidus inhibits the growth of harmful bacteria and restores to healthy intestinal microecological environment by capturing the oxygen required for the survival of harmful bacteria, producing lactic acid and occupying the adhesion site of harmful bacteria in the intestine.

9.3.4.3.3 Reconstruction

It can repair and rebuild the intestinal biological barrier, enhance immunity, and promote nutrient absorption. The probiotic combination can colonize the surface of the intestinal mucosa, repair and reconstruct the damaged intestinal biological barrier, prevent the invasion of pathogenic bacteria and toxins, induce various antibodies produced by the intestinal mucosa, enhance immunity, and truly maintain intestinal homeostasis.

Golden bifidus is mainly used to treat diarrhea caused by intestinal flora imbalance and chronic diarrhea caused by ineffective antibiotic treatment.

9.3.5   VSL#3

9.3.5.1   Development History

VSL#3 is developed and produced by VSL#3 company which is designed for intestinal inflammatory diseases. It has the advantage of high efficiency and no side effects.

Its main ingredients are derived from the beneficial bacteria in the intestine that are suitable for both young and old. The shape of the product is a tablet which is convenient to carry.

9.3.5.2   Ingredients

This product is a compound preparation, and its main components include *Bifidobacterium breve*, *Bifidobacterium longum*, *Bifidobacterium infantis*, *Lactobacillus acidophilus*, *Lactobacillus plantarum*, *Lactobacillus paracasei*,

Lactobacillus bulgaricus, and *Streptococcus thermophiles*. The gram of the product contains mixed live bacteria 2.25×10^{10} CFU), and other components mainly include microcrystalline cellulose, magnesium stearate, hydroxypropyl methylcellulose, and silica.

9.3.5.3   Function and Application

VSL#3 has a role in oxalate resistance. It has a certain effect on chronic irritable bowel syndrome (IBS) or ulcerative colitis (UC). The bacteria of VSL#3 are generally beneficial in the human intestine which are usually found in healthy digestive tracts. VSL#3 contains eight probiotics that bind to the digestive tract wall to form a barrier that protects the intestines and colonizes the gut to produce metabolites that inhibit inflammation. Simply speaking, taking VSL#3 can regulate the human intestinal flora to relieve or cure the disease.

In brief, taking VSL#3 can regulate the intestinal flora of the human to achieve the purpose of relieving or curing the disease.

9.3.6   Prohep

9.3.6.1   Development History

The research team led by Dr. Hani El-Nezami and Dr. Gianni Panagiotou of the Department of Biology of the University of Hong Kong, in collaboration with the Li Ka Cheng Medical College of the University of Hong Kong and the Department of Medicine of the University of East Finland, has developed a hybrid probiotic, Prohep, which, with potential therapeutic effects on hepatocellular carcinoma, has potential therapeutic effects on liver cancer. The research results were published in March 2016 in the *American Academy of Sciences* which is an internationally influential academic journal. Recently, the University of Hong Kong Technology Transfer Office and the University of Hong Kong Branch have applied for international patents for Prohep.

According to PhD El-Nezami, the composition of Prohep can alter the structure of the gastrointestinal flora and the interaction between bacteria and the host. These interactions are closely related to host metabolism, disease risk, and pathological development. Therefore, this new mixed probiotic may provide an alternative or complementary treatment for future liver cancer treatment.

9.3.6.2   Ingredients

This product is a compound preparation, the main components of which are *Lactobacillus rhamnosus* (LGG) (5 × 109 cfu/g), nonpathogenic *Escherichia coli* Nissle1917 (2.5–25 × 109 cfu/g), and heat-inactivated VSL#3 and the ratio of which is 1:1:1.

9.3.6.3   Function and Application

Prohep is a mixed preparation containing ten kinds of bacteria. Some of them are existing probiotics in the intestine. They can be quickly colonized in the human intestine and produce metabolites that inhibit inflammation. These inflammatory protease-inducing metabolites establish a circulatory pathway through the gut and tumor, modulating some of the pro-inflammatory cells that promote tumor development. In brief, taking Prohep can regulate the human intestinal flora to achieve the purpose of relieving or curing cancer, especially for liver cancer.

9.4   Existing Problems and Prospects of Pharmaceutical Preparations of Lactic Acid Bacteria

9.4.1   The Existing Problems of Pharmaceutical Preparations of Lactic Acid Bacteria

With the development of biotechnology, the research and development of lactic acid bacteria and their metabolites for medicine has become an important field in the pharmaceutical industry, and it has also brought new challenges to pharmaceutical preparations. Because biotechnology and chemical drugs are very different in terms of physical and chemical properties, biological properties, and drug properties, how to make such drugs into safe, effective, and stable preparations is a big problem. These substances are sensitive to enzymes, gastric acid, etc. Therefore, the choice of formulation needs careful consideration and measurement of various factors. At present, the problems of microecological preparations are as follows.

First, the strains of some domestic manufacturers are not strictly selected according to the standards of microorganisms. The source of the strains used for production is not standardized. The strains used are not the first generation of freeze-dried strains but the original species. It is not clear whether the strains that have been passaged many times have already produced mutations. Some manufacturers have simple equipment and many problems. The bacteria in the products seriously exceeded, and even the bacteria have an advantage, resulting in no obvious effect after using such products.

Second, the domestic reports on the efficacy of microorganisms are limited to the observation of the effects of product tests, and there are few systematic studies on the mechanism of microbial action and the safety evaluation of hosts. In fact, the strains used as feed additives should undergo strict pathological and toxicological tests to prove that they are nontoxic, non-teratogenic, drug-free, and drug-resistant. Therefore, the safety test or evaluation of the production strain should be carried out regularly.

Third, the selection of probiotics is blind, and the method of detecting probiotics is random and lacks uniform standards.

9.4.2 The Prospect of Pharmaceutical Preparations of Lactic Acid Bacteria

The research and development of biopharmaceutical drug which releases new technology and new dosage form is not only the task of modern pharmacy but also a hotspot of modern pharmacy research. At present, the research hotspots of lactic acid bacteria pharmaceutical preparations mainly include increasing the absorption and improving stability of biotechnological drugs; the preparation of microparticle carrier; embedding, dissolving, or adsorbing lactic acid bacteria substances in the microparticle carrier for delivery, thereby improving its stability; and having sustained release and targeting. Polymer materials not only guide the exploration of new materials in the research of particulate carrier pharmaceutical preparations but also guide the exploration of particle system preparation theory and technology; for the selection of high-efficiency and high-quality excipients, pharmaceutical excipients must be developed in terms of safety, functionality, applicability, efficiency, and economics; and apply multidisciplinary new theories, new technologies, and new methods to study important issues related to preparations.

At present, there are few types of pharmaceutical preparations for lactic acid bacteria in China, and there are not many clinical applications. However, we firmly believe that lactic acid bacteria preparations are still developing in China. The lactic acid live bacteria preparation will attract people to research and utilize it with its own characteristics and charm, making it a greater contribution to human health.

References

Bao X-h, Shen w-m, Shen y-f (1998) Experimental analysis and application of micro-ecological health products. Bull Sci Technol 27(2):61–64

Bleau C, Monges A, Rashidan K et al (2010) Intermediate chains of exopolysaccharides from *Lactobacillus rhamnosus* RW-9595M increase IL-10 production by macrophages. J Appl Microbiol 108(2):666–675

Chen Q (2007) *Bifidobacterium* and *Clostridium butyricum* combined treatment of infants with lactose intolerance in 72 cases. Tianjin Med J 35(3):237

Chen J-k, Weng W (2005) A technique of double layer embedment *Lactobacillus* and biological activity of embedded bacterium. Acta Agric Boreali-Occidentalis Sin 14(6):191–194

Chytilová M, Nemcová R, Gancarčíková S et al (2014) Flax-seed oil and *Lactobacillus plantarum* supplementation modulate TLR and NF-κB gene expression in enterotoxigenic *Escherichia coli* challenged gnotobiotic pigs. Acta Vet Hung 64(4):463–472

Dong H, Rowland I, Yaqoob P (2012) Comparative effects of six probiotic strains on immune function *in vitro*. Br J Nutr 108(3):459–470

Elmadfa I, Klein P, Meyer AL (2010) Immune-stimulating effects of lactic acid bacteria in vivo and in vitro. Proc Nutr Soc 69(3):416–420

Gu Y-h, Zhang J-s, An H-r et al (2012) HAO Yan-ling screening of β-galactosidase highly producing *Lactobacillus plantarum* and its enzymatic properties. Zhonggue Rupin Gongye 40(9):8–10

Hiramatsu Y, Hosono A, Konno T et al (2011) Orally administered *Bifidobacterium* triggers immune responses following capture by CD11c (+) cells in Peyer's patches and cecal patches. Cytotechnology 63(3):307–317

Huang Y (2000) Lizhuchangle treatment of 29 cases of chronic diarrhea in children. Chin J Microecol 12(6):321–321

Huang L, Li f-x (1999) Clinical observation of bifidobacteria live preparation for treating diarrhea. Chin J Microecol 11(2):96

Kolars JC, Levitt MD, Aouji M et al (1984) Yogurt—an autodigesting source of lactose. N Engl J Med 310(1):1–3

Lee JW, Lee JH, Sung SH et al (2013) Preventive effects of lactobacillus mixture on experimental *E. coli* urinary tract infection in infant rats. Yonsei Med J 54(2):489–493

Li Q-H, Tang G-H, Zhong M-M et al (1998) The short-term effect of oral *Bifidobacterium* Lizhuchang on chronic diarrhea. Chin J Micro-Ecol 10(1):45

Li Y-h, Ma L, Gu L-m et al (1999) Clinical observation of double fork milk for liver disease and intestinal endotoxemia. Chin J Micro-Ecol 11(3):153

Li J, Zhang W, Wang C et al (2012) *Lactococcus lactis* expressing food-grade beta-galactosidase alleviates lactose intolerance symptoms in post-weaning Balb/c mice. Appl Microbiol Biotechnol 96(6):1499–1506

Liu Y-h, Zhao J-b, Lv X-f (2004) Application of microencapsulation technology in probiotic products. Food Mach 20(2):58–60

Lomer MC, Parkes GC, Sanderson JD (2008) Review article: lactose intolerance in clinical practice – myths and realities. Aliment Pharmacol Ther 27(2):93–103

Luo J-x, Lin M-x, Xie J-P et al (1999) Clinical observation of Huichunsheng in treating pediatric diarrhea. Chin J Micro-Ecol 11(1):41

Ma M-r, Zhu G-j (1999) Research progress on microbial-derived cholesterol-lowering drugs. Chin J New Drug 8(9):603–605

Macfarlane GT, Cummings JH (1999) Probiotics and prebiotics: can regulating the activities of intestinal bacterial benefit health. Br Med J 318:999–1003

Mackie RI, Sghir A, Gaskins HR (1999) Developmental microbial ecology of the neonatal gastrointestinal tract. Am J Clin Nutr 69(suppl):1035S–1045S

Pochart P, Dewit O, Desjeux JF et al (1989) Viable starter culture, beta-galactosidase activity, and lactose in duodenum after yogurt ingestion in lactase-deficient humans. Am J Clin Nutr 49(5):828–831

Rabot S, Rafter JG (2010) Guidance for substantiating the evidence for beneficial effects of probiotics: impact of probiotics on digestive system metabolism. J Nutr 140(3):677S–689S

Reid G, Cook RL, Bruce AW (1987) Examination of strains of lactobacilli for properties that may influence bacterial interference in the urinary tract. J Urol 138(2):330–335

Reid G, Bruce AW, Mcgroarty JA et al (1990) Is there a role for lactobacilli in prevention of urogenital and intestinal infections. Clin Microbiol Rev 3(4):335

Savaiano DA (2014) Lactose digestion from yogurt: mechanism and relevance. Am J Clin Nutr 99(5 Suppl):1251S–1255S

Shao J-j, Zhu R-l (2003) The role of micro-ecological preparations and the problems that should be paid attention to in application. Sichuan Anim Vet Sci 30(9):45–46

Skarin A, Sylwan J (1986) Vaginal lactobacilli inhibiting growth of Gardnerella vaginalis, Mobiluncus and other bacterial species cultured from vaginal content of women with bacterial vaginosis. Acta Pathol Microbiol Immunol Scand B 94B(1–6):399–403

Soltan Dallal MM, Yazdi MH, Holakuyee M et al (2012) *Lactobacillus casei* ssp. *casei* induced Th1 cytokine profile and natural killer cells activity in invasive ductal carcinoma bearing mice. Iran J Allergy Asthma Immunol 11(2):183–189

Song Y-l, Li H-y, Liu C-j (2010) Research progress in the treatment of lactose intolerance by lactic acid bacteria. Chin J Micro-Ecol 22(8):751–753

Vincent V, Aghajari N, Pollet N et al (2012) The acid tolerant and cold-active β-galactosidase from *Lactococcus lactis* strain is an attractive biocatalyst for lactose hydrolysis. Antonie Van Leeuwenhoek 103(4):701–712

Wang Y, Cai F, Song G (1995) Therapeutic effect of lactic acid bacteria tablets in the treatment of 44 cases of diarrhea. Changzhi Med J (4):308

Wang J, Wang C, Li X-h (2000) Clinical observation of Golden bifidus in the treatment of acute and chronic diarrhea. Chin J Micro-Ecol 12(1):32

Xu J, Zou J, Yuan J (2011) Study on the effect of lactic acid bacteria on cholesterol lowering in high fat mice. Chin J Micro-Ecol 1:5–7

Xue Y, Yang Y (2008) New progress in lactic acid bacteria genomics and genetic engineering. Mod Food Sci Technol 24(6):617–620

Yan J, Li M, Zhong J et al (1987) Survey of lactose absorption and intolerance in milk and yogurt in normal adults. J Nutr 36(2):154–157

Yang Y, Li Y, Shi T (1999) Clinical observation of Huichunsheng in the treatment of neonatal diarrhea. Chin J Micro-Ecol 11(5):290

Yi X, Yang Y, Chen J et al (2006) Clinical observation of lactic acid bacteria vaginal capsule combined with metronidazole in the treatment of bacterial vaginosis. Cap Med 6(12):43

Yu M, Huo J, Yang J (1998) The role and development trend of micro-ecological regulators. Chin J Micro-Ecol 2:117–119

Zhang R (2007) Scientific use of microbial feed additives. Feed Livest 6:5–8

Zou J, Yang J (2007) Application and prospect of probiotic lactic acid bacteria in the field of medicine. Heilongjiang Med 20(1):28–30

Chapter 10
Lactic Acid Bacteria in Animal Breeding and Aquaculture

Gang Wang and Xing Jin

10.1 Application of Lactobacillus in Animal Husbandry

With the rapid development of social economy, animal husbandry, like agriculture, industry and handicraft industry have become one of the social industries. With the increase of mass consumption, the development of animal husbandry industry has reached a new height. However, a series of problems which restrain the development of animal husbandry industry have also entered people's vision with the continuous expansion of its scale. Animal diseases which always occur in livestock farming can lead to the death of livestock or poultry and result in serious economic losses. The common types of animal diseases can be grouped into the following three types.

1. Common diseases: medical, surgical, and obstetrical diseases of animals. The incidence of common diseases is high, and the clinical symptoms are also diversified.
2. Infectious diseases: this kind of disease is mainly caused by pathogenic microorganisms and has certain incubation period and clinical symptoms, epidemic and infectious. Viruses, bacteria, and fungi can be the pathogenic microorganisms which induce their occurrence. The clinical symptoms and pathological changes of animals have certain particularity, and these diseases are difficult to prevent.
3. Parasitic disease: caused by the invasion of parasites (arthropods, protozoa, and worms) into humans. The poultry and livestock eat soil, drinking water or feed containing worm eggs or larvae will sichen. Besides the detrimental factors of animals' growth or activity environment, a series of human factors may also cause animal diseases including the irrational use of drugs in breeding process, the imperfection of breeding equipment and construction, and the adverse management of breeding farms.

G. Wang (✉) · X. Jin
Jiangnan University, Wuxi, China
e-mail: wanggang@jiangnan.edu.cn

© Springer Nature Singapore Pte Ltd. and Science Press 2019
W. Chen (ed.), *Lactic Acid Bacteria*,
https://doi.org/10.1007/978-981-13-7283-4_10

10.1.1 Prevention and Treatment of Lactobacillus in Livestock Diseases

It has been a key issue for livestock farmers to prevent and control animal disease, and it plays an important role in improving the economic efficiency of the livestock industry with the development of farming. At present, the use of lactic acid bacteria preparation added to livestock feed increases year by year. The researches have reported that lactic acid bacteria can adjust the mammal intestinal microflora balance, strengthen the body's immunity and resistance, restrain the growth of pathogenic bacteria, keep animal intestinal flora balance, and adjust gastrointestinal digestion and absorption function. Adding a certain amount of lactic acid bacteria in animal feed formulation can compete ecological sites, inhibit the growth of pathogenic microorganisms, decrease the enterobacteriaceae bacteria in the gastrointestinal tract, and keep the balance of animal intestinal microecological bacteria when feed are eaten by animals.

With the rapid development of China's dairy industry, the dairy consumption market will continue to expand and become mature. Effectively expanding the efficiency of cow breeding, improving milk production, and reducing the economic loss caused by diseases also have become key issues to be urgently solved by farmers. Mastitis is a kind of multiple diseases in mammalian diseases and has the characteristic of wide range, high incidence, difficulty in curing, and easy to relapse. Once the cow is infected with mastitis and cannot be effectively controlled in time, it will bring huge economic loss to the farmers (Persson et al. 2011). At present, most antibiotics are used for the prevention and treatment of mastitis. But, it will bring many problems such as increased bacterial resistance, antibiotic residues, and environmental pollution caused by resistance genes. The use of probiotics as a substitute for antibiotics is attracting more attention from scientists. Yang (Yang et al. 2014) used the compound probiotics including *Lactobacillus casei* HM-09 and *Lactobacillus plantarum* HM-10 as 100–200 g per head daily dosage which had the living bacterium acuity 1.5×10^9 CFU/g of lactobacillus in Holstein cows from different areas of pasture. Meanwhile, CMT and DHI methods were adopted to test the cow mastitis inspection and the number of somatic cell detection to research the controlment of lactobacillus for dairy cow mastitis. The result showed that the average somatic cell number of cow decreased by 39.8–62.8% after 5–10 days. And more importantly, the cure rate of mild and severe mastitis was 71.21% and 33.00–74.19%, respectively, after 10–30 days of lactobacillus microecological preparation. It was showed that lactobacillus had a good therapeutic effect on mastitis.

Neonatal calf diarrhea is very harmful to the growth and development of calves. The symptoms are diarrhea and the feces is watery when the calf is sick. At the same time, dehydration and acidosis will also occur in calf which lead to the death of calf. Diarrhea is spreaded quickly and easily through the digestive tract that acts as the main transmission route and brings huge economic losses to cattle breeding industry. Less than 2 months calves have low resistance and less developed digestive system. So they are susceptible cattle. The categories of diarrhea can be divided into bacterial diarrhea and viral diarrhea. The average incidence of each other is 36.7% and 7.5%, respectively. The inhibitory activity of lactobacillus on *Salmonella*, *Shigella*, and *Escherichia coli* has been extensively studied, and the application of lactobacilli preparations in animal husbandry has great significance for promoting the sustainable and healthy development of the industry (Liao et al. 2012).Yao et al. (2014) used lactobacillus preparation to control the calf diarrhea. After 15 days prevention test, the rate of calf diarrhea in the lactobacillus preparation group was 14.29%, while that in the control group was more than twice as high (29.76%). Diarrhea calves in lactobacillus preparation group did not die, while the death rate of diarrhea calves in control group was 3.57%. The rate of diarrhea decreased from 100% to 3.70%, and the cure rate of diarrhea increased from 0% to 85.19% in lactobacillus preparation group. It was also found that early feeding of lactobacillus biologica to new calves would be more effective in preventing diarrhea.

In addition, cows will also suffer from a disease called cow endometritis after parturition. The incidence of cow endometritis is as high as 20–40% in China and has been a common frequently occurring disease. The extension of calving duration, decrease of milk production, and lesion of reproductive organs will happen when the cows are ill. At the same time, the cost of expensive treatment also brings great economic losses to farmers. Lactobacillus were the main bacteria which isolated from healthy cows, but *Staphylococcus aureus*, *Streptococcus agilis*, and *Escherichia coli* were the three most pathogenic bacteria isolated from the uterus of sick cows. Through staining microscopy, biochemical test, and PCR detection, the lactobacillus strain isolated from healthy cows was identified as *Lactobacillus acidophilus*. Safety animal studies determined that the three strains (*Lactobacillus acidophilus*, *Lactobacillus rhamnosus*, and *Bacillus natto*) did not cause death and other diseases in mice. Three strains of bacteria were cultured under the optimal culture conditions and then were mixed into a certain proportion of microecological preparations to treat the successfully modeled mice and rabbits. From the results of autopsy and pathological section, the treatment effect was certain. Some studies showed that lactobacillus fermentation was used to treat endometritis in cows. The results showed that the effective rate of endometritis in cows was 95.23% and the cure rate was 85.71% which indicated that the microecological preparation could be used as a treatment drug for endometritis in cows. It also provided a new approach to solve the problem of high incidence of endometritis in cows and antibiotic residues in milk.

Many studies had shown that compound probiotics as feed additives could improve intestinal development and promote animal health in young animals and have been widely used in pig breeding. Liu (Liu et al. 2012) studied that the fodder added probiotic preparations could significantly improve daily gain early weaning piglets and reduce the material weight ratio and diarrhea of weaning piglets. Bao (Bao et al. 2015) verified that probiotics could significantly improve conservation pig weight, improve daily gain, and reduce material weight ratio and morbidity rate. It was also reported that two different compound probiotics were added to piglet feed. The results showed that the daily weight gain of the group adding 0.1% probiotics increased by 6.53% and 4.02%, and the rates of diarrhea decreased by 15.55% and 4.44% compared with the control group.

In modern pig production, antibiotics are often used to prevent the occurrence of diseases in order to improve their resistance to diseases and their resistance to stress. Such as lincomycin, long-term use of antibiotics affects the intestinal microorganism, the intestinal microflora of pigs, and the digestion of feed. The use of microecological preparations can alleviate the side effects of antibiotics on intestinal microorganisms by supplementing beneficial microorganisms. The microecological preparation can inhibit the adhesion and growth of harmful microorganisms, enhance the elimination of harmful microorganisms, stimulate the differentiation and development of gastrointestinal tract, and improve the digestibility of nutrients by establishing a dominant microflora to maintain beneficial microorganisms in gastrointestinal tract. Harmful bacteria in the intestinal tract break down food residues and produce toxic compounds (such as ammonia, amines, indoles, skatole, and nitrite) which will be absorbed by host can reduce the animals' immunity and even lead to various diseases. Most *Escherichia coli* are normally harmless bacteria in the gut, but overmuch reproduction can cause digestive problems. It was reported that the addition of *saccharomyces cerevisiae* and lactobacillus lactobacilli into the feed could significantly reduce the amount of *Escherichia coli* in the feces of weaned piglets. And the number of *Escherichia coli* in the feces of weaned piglets decreased significantly from second week of the trial after feeding the related composite bacteria.

10.1.2 Improvement of Growth Performance on Livestock by Lactobacillus

With the acceleration of China's modern animal husbandry development and increasingly mature, small investment, short cycle and high yield of development mode has preliminary forming. It is important to promote the industrialized operation of agriculture and increase the income of farmers and herdsmen. It also has a big gap between our country and world's advanced level on breeding scale and management level of animal husbandry and a great potential development space on our country animal husbandry. The development of modern animal husbandry is based on the development of feed industry, the core of which is to vigorously develop green, safe, and efficient feed additives. As one of the main members of green feed additive, microecological preparation is more and more favored by farmers. Therefore, the application of microecological preparation in animal breeding and management becomes more and more important.

Animal microecological preparation is a kind of feed additive which can stimulate the growth of beneficial microorganism in the animal. Animal microecological preparations are mainly divided into probiotics, prebiotics, and biobiotics. At present, there are more than 40 kinds of microorganisms which can be directly fed in the world and 16 kinds of microorganisms which can be directly added to the feed announced by China's ministry of agriculture no. 658. It is reported that probiotics as growth-promoting feed additive substituting antibiotics had great potential. After adding probiotics to feed, it could improve the palatability of feed and improve livestock feed intake. At the same time, probiotics could not only degrade the complicated structure and large molecular weight protein to small molecular peptides and free amino acid, but also compound a variety of animal vitamins. What's more, the probiotics could promote the absorption of trace elements, benefit to animal gastrointestinal digestion and absorption, and improve feed conversion rate.

Through the breeding process on selecting probiotics, the healthy animals are used. The selected probiotics can produce acid and digestive enzymes, inhibit the pathogenic bacteria, and colonize the animal digestive tract. It also can tolerate animal gastric acid environment, bile salt, feed processing, and drug sensitivity at the same time. Probiotics can produce a certain number of bacteriostatic substances (such as antimicrobial peptides, bacteriocin) which can selectively inhibit the growth of some pathogenic bacteria. Meanwhile, a large number of probiotics which colonize in the animal's intestinal tract can form the advantages of bacterial flora and play a protective role for the animal's digestive tract (especially the intestinal tract). At the same time, the colonization of probiotics can maintain the facultative anaerobic environment of the animal digestive tract which can be conducive to the establishment of normal flora of animals and the restoration of animal health. Probiotics can selectively stimulate the rapid reproduction of native probiotics in animals' intestines; it plays an important role in the establishment and recovery of normal flora balance in animals. Therefore, in the process of animal breeding, the effect of promoting animal health and improving animal growth and production performance can be achieved by adding microecological preparations.

Probiotics are widely used in pig breeding, and there are different purposes at pig different stages (Chaucheyras-Durand and Durand 2010). In pregnant sows, probiotics are mainly used to improve the digestibility of feed, reduce constipation, and reduce stress. In lactating sows, the main purpose of adding lactobacillus is to improve the quality and yield of milk and increase the survival rate of piglets. For piglets, *Lactobacillus* is added mainly to increase body weight and reduce diarrhea. In the fattening stage, probiotics are mainly used to reduce diarrhea and improve feed utilization rate and meat quality. Weaning for piglets is a very complicated period which piglets need to be strongly separated from gilts, milk and delivery room to transited, plant feed and nursery. In addition, the digestive system of piglet is not well developed which is likely to cause stress, intestinal diseases, and economic losses.

Some studies have shown that probiotics could balance the intestinal flora, reduce the colonization of pathogenic bacteria, inhibit the proliferation of pathogenic bacteria, and improve the digestibility of nutrients and growth performance. Yang et al. (2009) added 350 mg/kg of compound probiotics containing lactobacillus, yeast, and bacillus subtilis to the feed of weaned pigs. The result showed that probiotics could improve the activity of protease, lipase, and amylase in the intestinal tract. Zeyner and Boldt (2006) showed that giving piglets oral fecal enterobacteria every day from birth to weaning could improve growth performance and reduce the incidence of diarrhea. Adding BioPlus 2B (*Bacillus subtilis*) to pig feed could reduce ammonia emissions, but it had no significant effect on growth performance. It is reported that adding 0.05% condensation bacillus on the basis of weaned piglet diet made weight significantly higher than the blank control group. In the experiment, feed conversion rate was observably enhanced, and diarrhea rate was decreased significantly. At the same time, lactobacillus and bifidobacterium increased, and *E. coli* decreased significantly in excrement and urine. Condensation bacillus preparation added in the growing pig feed can significantly increase the average daily gain of pigs and reduce feed costs. There was no significant difference between the bacillus and antibacterial dysentery bacteria (Zhou et al. 2012). Scientists believed that the addition of bacillus subtilis could promote the growth performance of the body, prevent the occurrence of diarrhea, increase the daily intake and daily weight gain of piglets, and decrease the rate of diarrhea and feed conversion. After feeding suckling pigs with *Bacillus coagulum* CNCMI-1061, Adami and Cavazzoni (1999) found that *Bacillus coagulum* could colonize the intestinal tract and inhibit the growth of strains such as *E. coli*. The results showed that adding probiotics into the feed could significantly improve its productive growth performance and promote the balance and stability of its intestinal flora. It also could improve the body immunity and prevent or treat diarrhea effectively.

In young ruminants, probiotics added into fodder mainly promote the perfection of rumen microflora, reduce the stress of weaning, and reduce the perniciousness of pathogenic microorganism. In milk production, probiotics are used to increase the milk quality and yield. In beef cattle production, the main purpose of probiotics is to increase body weight and improve feed conversion. Studies have shown that adding probiotics can regulate intestinal pH and reduce the risk of acidosis. Adams et al. (2008) found that *Propionibacterium jensenii* added into the diet could increase the body weight of calves and promote rumen development. The growth performance of beef cattle could be improved at the later stage, but the experimental results were often inconsistent. Qiao (Qiao and Shan 2007) found that the addition of bacillus licheniformis to the diets of Holstein cows significantly increased milk production. It is also reported that adding probiotics could improve the survival rate and weaning weight of lamb.

Although the digestive system of ruminants is complex, probiotics can improve the intestinal flora and inhibit pathogenic bacteria. Luan et al. (2008) found that *Bacillus natto* could significantly improve milk fat rate and increase milk protein rate and milk production. Zhang et al. (2008) believed that *Bacillus natto* could increase the daily weight gain and daily food intake of calves, reduce the feed weight ratio, and improve the ketone body rate. In addition, probiotics could promote the degradation metabolism of feed and the generation of volatile fatty acids in rumen. Other research results showed that feeding bacillus subtilis could significantly reduce the pH of chyme in duodenum and jejunum of calves and improve the quality of weaning early calves.

10.1.3 Preparation and Application of Lactobacillus Silage

According to the law of world agriculture development, the traditional agriculture which mainly produces grain has been transformed into modern agriculture which mainly produces animal husbandry. Since the beginning of twenty-first century, with the rapid development of China's economy and the gradual improvement of people's living standards, animal husbandry has maintained a rapid growth. But, conventional feed shortage has already turned into the main factors restricting the development of China's animal husbandry. Mining the potential of existing feed resources and adjusting measures to local conditions of land development and agricultural by-products (rice straw and green feed) have become the main ways to solve this challenge. It is also one of the world today the trend of the development of the feed at the same time.

With the development of science and technology, the fermentation method of pure seed inoculation is used to simplify the silage production process. This fermentation method guarantees the ecological security, is conducive to maintaining the stability of the product, and improves the nutritional quality of forage after fermentation. The simple and convenient use of microbiological preparations is conducive to realizing industrialized production, shortening the mature period of finished products, playing a normative role in the standardization and safety of product features, and greatly improving the market competitiveness of products. At present, more and more studies have been reported on the addition of biological agents to improve the quality of fermentation. Many countries have developed many additives specially used for silage and put them into the market as commodities.

By fermentation of lactobacillus, forage can improve quality essentially and solve food safety problems caused by food chain from the source. From the perspective of economy, lactobacillus can make use of almost all common plants and reduce the cost of farming. From the perspective of resources, the nutrients of the feed after fermentation by lactobacillus are more abundant than before fermentation, which improves the utilization rate of raw materials by animals and saves natural resources. The experience of the rapid development of animal husbandry at home and abroad proves that the development of silage is an important way of modern animal husbandry. With the development of modern biotechnology, silage processing technology has become more and more mature. New biological additives can significantly improve the quality of silage and make a qualitative leap in the feeding value of forage. Not only that, the safety of silage products with biological starter added has been greatly guaranteed. Therefore, the development of high-quality lactobacillus starter has become a key part of silage technology.

The mainly metabolites produced by lactobacillus are lactic acid, which can resultfully reduce the growth of plants' own pathogenic bacteria in the fermentation process and give the materials good flavor and texture. The material becomes soft and has the favorable fragrant sour taste and palatability after fermentation. It can stimulate the appetite of livestock, promote the secretion of digestive fluid, and increase the frequency of intestinal peristalsis. So, the fermented feed can enhance the digestive function of animals and prevent constipation. After fermentation, the nutritional characteristics of plant raw materials were improved. Generally, higher water and protein content will lead to lower fiber content. This will improved the digestibility of animal feed. These fodders are rich in vitamin and appropriate proportion of nitrogen and phosphorus. The artificial control of fermentation process makes it free from natural factors such as weather and reduces the probability of pollution and corruption. The metabolic products such as lactic acid can prevent the deterioration of the material, so as to achieve long-term preservation and meet the need of winter forage shortage.

Some researches showed that the content of pH, amino acid nitrogen, and cellulose of lactobacillus silage alfalfa was significantly reduced, and the number of living bacteria of lactobacillus was significantly increased. The digestibility of nutrients of fermentative alfalfa increased significantly with the extension of time, and the loss rate of dry matter did not exceed 12%. They also found that the contents of soluble carbohydrates, the digestibility of organics, the recovery rate of dry matter, the content of lactic acid, and the ratio of lactic acid or acetic acid were increased in the silage with lactobacillus addition. The content of neutral washing fiber, acidic washing fiber, pH, ammonia nitrogen/total nitrogen, and volatile fatty acid decreased.

In recent years, some studies have carried on the preliminary fermentation juice (previously fermented juices, PFJ). PFJ is similar to lactobacillus preparation. It is prepared by fermenting the inherent lactobacillus on the plant to achieve the goal of rapid proliferation of lactobacillus. The addition of PFJ can rapidly increase the content of lactic acid in the stock, reduce pH, inhibit the activity of *Clostridium difficile*, and protect the decomposition loss of protein. Its addition effect is not affected by the growth period of silage material, DM content, and silage conditions. It is more stable than lactobacillus in improving silage fermentation quality. PFJ enables the proliferation of lactobacillus on the surface of forage grass as a promoter of lactic acid fermentation. What's more, it is also a good substitute for lactobacillus preparation.

A large number of experimental studies have confirmed that probiotics can improve production efficiency and improve animal health in animal husbandry, but there are still some results that show that probiotics have no promoting effect on animal husbandry production and even have negative side effects. The main factors influencing the effect of probiotics include the species and physiological state of target animals, the preparation and storage conditions of feed, the adding mode of probiotics, and the characteristics and dosage of probiotics.

10.1.3.1 Characteristics and Dosage of Probiotics

Probiotics (such as bifidobacterium, lactobacillus, and bacillus) have benefits in gastrointestinal tract function mainly in the form of living bacteria. For forming the stomach competitive exclusion engraftment on harmful bacteria and producing antibacterial substances to inhibit the growth of harmful bacteria, the probiotic bacteria must be able to pass the acidic environment of the stomach and intestinal bile secretion. According to statistical analysis by Jadamus et al. (2002), one of the important reasons for the inefficacy of probiotics in production was that the probiotics used were difficult to survive in the animal intestines. Therefore, the intestinal normal bacteria should be selected as much as possible to improve the tolerance of probiotics to the intestinal environment. The synergism and antagonism between different strains should be fully considered. The effects of probiotics are closely related to the number of live bacteria. If the number of live bacteria ingested by animals is too small, it will be difficult to colonize and grow in the intestinal tract. When the number of live bacteria is too large, it is easy to cause waste and even produce side effects. Because the proliferation of microorganisms requires energy consumption.

10.1.3.2   The Adding Mode of Probiotics

Timmerman et al. (2005) emphasized that the feeding mode and adding time of probiotics were important factors influencing the effect of probiotics. The main ways to add probiotics are through feed, water, spray, or irrigation. In production practice, it is often used in feed and drinking water. Studies had shown that probiotics were more effective when added to water than when added to feed. Because of drinking water sanitation, interference, and the need for special quantitative adding equipment, probiotics are mainly added to feed in actual production (Karimi Torshizi et al. 2010). Because probiotics cannot be permanently colonized in the gastrointestinal tract, the addition of probiotics needs to be sustained to ensure its effectiveness.

10.1.3.3   The Preparation and Storage Conditions of Fodder

Some ingredients in feed can also affect the use of probiotics, such as antibiotics and pesticides added to feed that reduce the effect of probiotics. Wheat and barley would increase the viscosity of intestinal chyme and also had adverse effects on probiotics (Choct et al. 1996). High temperature of machine may lead to inactivation of probiotics during the production of feed. In addition, direct friction and extrusion of feed ingredients also affect the activity of probiotics during the mixing of feed. When the storage temperature exceeds 30 °C,, chemical substance and enzyme reaction in majority of probiotic bacteria (such as lactobacillus and bifidobacterium) will be quickened. It may cause the probiotic inactivation. For fear of influencing the activity of bacteria, probiotic preparations should generally be refrigerated in 2–8 °C environment temperature and prevented oxidation reaction.

10.1.3.4 The Species and Physiological Stage of Target Animals

Probiotics play a role on the gastrointestinal tract through colonization. Different probiotics have different colonization abilities in the intestinal tracts of different animals. Therefore, it is better for the probiotics to come from the normal intestinal flora of target animals to improve the colonization and growth capacity. Early addition of probiotics is important for animals because probiotics can regulate the expression of genes in intestinal epithelial cells and create a better environment for the colonization of probiotics. In addition, some special physiological states (such as weaning, lactation, and stress) may affect the effect of probiotics.

To sum up, probiotics must remain active at the action site to realize function. Therefore, the tolerance of strains to temperature, gastric acid, and bile in gastrointestinal tract should be fully considered in the selection of strains. Probiotics can be embedded in microcapsules to maintain sufficient activity when reaching the action site. In addition, probiotics can be combined with additives (such as oligosaccharides, enzyme preparations, and acidifier) to maximize the effect of probiotics.

10.1.4 The Other Application of Lactobacillus

10.1.4.1 The Microcapsules Technology of Probiotics

Microencapsulation technology refers to cover solid, gas, or liquid with natural or artificial polymers continuous thin film. The embedded object is isolated from the external environment. When embedded objects encounter certain external stimuli or can be gradually released in a specific environment. PH, mechanical extrusion, temperature, enzyme activity, osmotic pressure, infiltration of water molecules, the presence of certain chemicals, and changes in storage time are all possible factors which stimulate the release of membrane objects. Many tests had shown that microencapsulation could keep probiotics alive in acidic, alkaline, or bile environments. Microencapsulation can improve the survival rate of probiotics in low-temperature drying, high-temperature processing, and the presence of antibiotics. Both lactobacillus and bifidobacterium are highly sensitive to pH and oxygen in the environment. Sufficient number of live bacteria can be guaranteed when it reaches the action site through microcapsule embedding technology.

10.1.4.2 The Combination Between Probiotics and Oligosaccharides

Fructo-Oligosaccharides, also known as oligosaccharides, are compounds formed by two to ten glycosidic bond polymerizations. Due to the lack of enzymes capable of degrading oligosaccharides in animals, oligosaccharides are difficult to be digested and utilized by passive objects. So it enters the animal's posterior intestine directly. The animal's hindgut is teeming with microbes which can use oligosaccharides. Oligosaccharides can be used as nutrients for beneficial bacteria such as lactobacillus and bifidobacterium, but cannot be used by pathogenic bacteria such as *Escherichia coli*, *Salmonella*, and *Clostridium perfringens*. Oligosaccharides can be the energy source of probiotics to improve the competitive advantage of probiotics. In addition, the digestive enzymes secreted by probiotics can degrade oligosaccharides into monosaccharides which can be used by the body. The effect of probiotics can be improved by adding oligosaccharides. The results showed that the growth performance of weaned piglets was improved by the combination of probiotics and oligosaccharides. It is also reported that the addition of both oligosaccharides and probiotics can significantly improve the growth performance of broilers and increase the amount of lactobacillus and bifidobacterium in the intestinal tract.

10.1.4.3   The Combination Between Probiotics and Enzyme Preparations

Enzyme preparation is a biochemical product with high efficiency and specific catalysis. Enzyme preparations are mainly used to supplement the endogenous enzymes of animals which are not enough to improve the utilization rate of nutrients in animals so as to improve the production efficiency. At present, the enzymatic preparation used in animal production mainly includes phytase, protease, lipase, xylanase, and various complex enzymes. There is a synergistic relationship between enzymatic preparations and probiotics. The catalyzed nutrients are not only used for animals but also provide nutrients for probiotics. In addition, probiotics can produce digestive enzymes such as amylase and protease and improve the activity of enzyme preparations. Lin qian et al. (2012) found that combined use of probiotics and complex enzyme preparations could yield better daily weight gain and metabolic energy than using singly.

10.1.4.4   The Combination Between Probiotics and Acidulant

Acidifier can be divided into organic acid, inorganic acid, and compound acidifier. Acidifier can improve the environment of the intestinal tract by reducing the pH of the gastrointestinal tract and inhibiting the growth of harmful bacteria such as *Escherichia coli* and *Salmonella*. The digestive tract function of young livestock and poultry is not sound; the secretion of gastric acid and digestive enzymes cannot meet the demand. Acidifier can activate the secretion of digestive enzymes in the gastrointestinal tract. In addition, acidifier can mask some bad odors in the feed, improve the palatability of the feed, stimulate taste bud cells, increase the secretion of digestive enzymes in the mouth, and enhance animal appetite. 0.1% fumaric acid, lactic acid, and probiotics were added into the feed of 12-day-old yellow-feathered broiler chickens. The results showed that the combination of organic acids and probiotics increased the daily weight gain of broiler chickens and improved the feed conversion rate.

10.2 Application of Lactobacillus in Poultry Production

10.2.1 Prevention and Treatment of Lactobacillus in Poultry Disease

How to maintain the balance of the intestinal flora of animals, promote the development of their immune organs, inhibit the production of harmful bacteria, and improve the body immunity and speed up the production capacity of poultry has become the common goal of every farmer in the poultry farming industry. With the expansion of the scale of farming, it is urgent to reduce the poultry disease probability and control the spread of disease. At present, antibiotics added in the feed are the main way to solve the poultry disease from the source. But now, the use of antibiotics has been more and more excluded by people, and its residues in food have caused wide public concern. A growing number of antibiotics substitutes are researched and developed by the people.

Lactobacillus is reported to be able to effectively antagonize pathogenic bacteria in poultry. The mechanisms involved can be divided into the following categories:

1. Theory of dominant flora: excluding pathogenic bacteria through competitive inhibition. There are tens of thousands of microorganisms in poultry intestinal tract. Beneficial bacteria in the intestinal tract can form competitive adhesion to intestinal epithelial cells and compete for the adsorption sites and nutrients of pathogenic bacteria in the intestinal tract.
2. Biological oxygen capture theory: the pathogenic bacteria are mostly aerobic bacteria. The anaerobic bacteria are predominant in the intestinal tract of poultry. Some probiotics which belong to aerobic bacteria grow and colonized in the animal body to consume oxygen and create an anaerobic environment. It inhibits the reproduction and growth of aerobic bacteria and facultative anaerobic pathogenic bacteria.
3. Bacteria barrier theory: metabolic products of lactic acid bacteria, such as organic acids and bacteriocins, can kill pathogenic bacteria to a certain extent.

Aiba et al. (Yuji Aiba and Nobuyuki Suzuki 1998) found that lactobacilli could produce 15–156 mmol/L of lactic acid with it growing. It could significantly inhibit the urease activity of *Helicobacter pylori*, reduce the activity of *H. pylori*, and antagonize the growth of *H. pylori* co-cultured with *H. pylori*. For some Gram-negative bacteria, bacteriocin contains bioactive protein components and has antibacterial or bactericidal effects. It is reported that coccus bacteriocins had a significant effect on inhibiting *H. pylori*. A variety of pathogenic bacteria will colonize in poultry intestinal tract including *Salmonella*. When people eat poultry with undevitalized *Salmonella*, it will cause a series of complications such as nausea, vomiting, diarrhea, fever, and abdominal cramps. Infants, the elderly or immunocompromised symptoms are usually more serious. *Salmonella* infection also causes other serious complications by bacteremia. Therefore, it is imminent to avoid poultry infected with *Salmonella* strictly. Several types of lactobacillus have been found to be effective in preventing or reducing salmonella infection in broiler chickens. Jin et al. (1996) found that lactobacillus inhibited the growth of salmonella enteritis in chickens and different species had different mechanisms. Pascual et al. (1999) thoroughly cleared the 1.0×10^6 CFU *Salmonella* typhimurium through intragastric administration of both 1.0×10^8 CFU *Lactobacillus salivarius* CTC219 in 21 days chicken experiment. Inoculation of *Lactobacillus reuteri* on hatching eggs could reduce the colonization of newly hatched chicks of *Salmonella* and *Escherichia coli* to reduce the poultry death. *Lactobacillus plantarum* is a sensitive receptor of mannose, which can compete with pathogenic bacteria for the binding site of mannose in chicken intestines to inhibit the colonization of pathogenic bacteria in chicken intestines. Scientists also isolated *Enterococcus faecium* J96 from fecal. It could colonize in poultry intestinal tract and secrete lactic acid and bacteriocin to inhibit *Salmonella* in chickens. Some probiotics compound feed have been gradually developed and become a commercial product. Mountzouris (Mountzouris et al. 2009) studied the treatment result in *Salmonella*-infected chicken of probiotics and antibiotics added into chicks feed and drinking water. They used 6.0×10^5 CFU of *Salmonella* to infect 5 days of age chicks then fed chicken with probiotics and antibiotics added feed, respectively. The probiotic group was 2.0×10^9 CFU/kg of probiotics BP5S feed additives. Probiotics BP5S feed additives included *Lactobacillus reuteri* isolated from healthy adult chicken crop, *Enterococcus faecium* isolated from jejunum, bifidobacterium isolated from ileum, and *Lactobacillus salivarius* isolated from cecum and lactobacillus salivary. The tests found that probiotics significantly reduced the prevalence of salmonella in broilers infected with enteritis and the amount of *salmonella* in broilers with typhoid caecum. The therapeutic effect was essentially effective in adding antibiotics to feed.

Campylobacter *jejuni* is also a kind of pathogenic bacteria colonized in the poultry intestinal. The poultry itself have no obvious clinical symptoms when they carry germs. However, people will have diarrhea, fever, and acute enteritis by *Campylobacter jejuni* infection when they eat undercooked chicken or duck. Therefore, controlment of *C. jejuni* colonization in poultry is one of the necessary problems for farmers. Chaveerach et al. found that organic acids could be effectively saved and inhibited *C. jejuni* growth effectively in low pH. Lactobacillus P30 isolated from poultry had good tolerance in the artificial simulation of the gastrointestinal tract. It also could effectively reduce the *Campylobacter jejuni* in chickens' in vivo engraftment. Another key factor that lactobacillus could effectively inhibit the growth of *C. jejuni* was its secretion of bacteriocin OR-7 from *Lactobacillus salivarius* NRULB-30514. Bacteriocin OR-7 could effectively antagonize *C. jejuni* and inhibit pathogenic bacteria colonization in chickens (Stern et al. 2006).

10.2.2   Improvement of Growth Performance on Poultry by Lactobacillus

In a normal, healthy, and stress-free poultry gut, the microecological environment is balanced, and the poultry can maximize their growth. Beneficial bacteria, especially lactic acid bacteria, will decrease, and harmful bacteria begin to grow in the case of external stimuli. Poultries in subhealth state are prone to diarrhea, and production efficiency or feed utilization will decline. The balance of gut flora will be broken by factors such as feed and environment. Under natural conditions, chicks can get a complete microbial system from the hen's feces to avoid bacterial infection. However, under commercial incubation conditions, chicks are usually incubated in an incubator without normal gut microbes which is completely isolated from hens or other adult chickens. So, the use of probiotics in the pre-growth phase of poultry is more important and more useful than in other animals.

In broiler production, probiotics can increase body weight and feed conversion, inhibit the colonization and growth of harmful bacteria, and improve meat quality and immune function. Molnar et al. (2012) added *Bacillus subtilis* to the feed of broilers; the results showed that probiotics significantly improved body weight gain, feed conversion rate, and immune response. Adding 1000 mg/kg of lactobacilli into the feed of broiler chickens increased body weight, improved feed conversion rate, and decreased the deposition of belly fat on 8 days chicken (Kalavathy et al. 2003). Mountzouris et al. (2007) added probiotics (*Lactobacillus reuteri, Enterococcus faecium, Bifidobacterium*, and *Lactobacillus salivarius*) to drinking water of chicken to improve the body weight gain, food intake, feed conversion rate, and digestive enzyme activity of broilers. It is reported that lactobacillus increased the pectoral muscle rate of broilers, the content of flavoring amino acids (such as glycine and alanine), water, and intermuscular fat.

In the production of laying hens, the main purpose of adding probiotics is to improve the utilization rate of feed, the amount of eggs, and the quality of eggs. Mikulski (Mikulski et al. 2012) pointed out that the addition of 100 mg/kg *Lactococcus lactis* into the diet of laying hens increased the weight of eggs, the thickness of shells, the relative mass of shells, and feed conversion rate. At the same time, it could also reduce the rate of broken eggs, unshelled eggs, and the cholesterol content of the yolk. Two thousand layers were used to carry out the fermentation feed feeding test of multiple strains. The result showed that the material egg ratio of the experimental group was significantly lower than the control group, and the egg production rate and weight of the experimental group were higher. At the same time, the feed cost was declined by fermentation. Wang et al. (Wang and Zhang 2013) showed that the addition of 150 g/kg bacillus coagulum preparation $(8.0 \times 10^9 \text{CFU/g})$ in the feed for laying ducks could improve the egg production rate, daily output, average weight of eggs, and feed conversion rate. The microbial preparation also could reduce the intake of eggs and improve the color of egg yolk, protein content, and quality of eggs.

10.3   Application of Lactobacillus in Aquaculture

With the gradual development of China's aquaculture industry, large-scale intensive farming mode gradually replaced the traditional extensive cultivation mode. But, the competition of feed and survival space and the accumulation of waste residue lead to the breeding pollution and deterioration of water quality and the growth of harmful microbes. It causes frequent disease of aquatic animal and product quality decline, which hinders the rapid development of aquaculture. In the face of all kinds of disease problem, it is main to use a variety of broad-spectrum antibiotics and chemical drugs for controlling the occurrence and spread of the disease. But, it causes new problems such as bacteria resistant, water quality pollution, microbial flora, and ecological balance destruction with long-term use. Meanwhile, the chemical drugs and antibiotics left in aquatic products will seriously threaten human health and do not meet the basic requirements of food safety production. Therefore, it is especially important to find a safe, nontoxic, non-residue, and nonresistant alternative to achieve sustainable development of aquaculture.

As an important probiotics, lactobacillus has been widely used in the fields of medicine, light industry, and food and animal husbandry. Lactobacillus plays a very important role in aquaculture as a microecological preparation. It is reported that lactobacillus applied in aquaculture could purify water body, improve water environment, colonize in host body, inhibit the growth of harmful bacteria, provide nutrients, and maintain ecological balance. Moreover, lactic acid bacteria in the fermentation process would produce a series of metabolites such as organic acids, special enzymes and bacteria on the surface of the active ingredients to stimulate tissue growth, body's physiological function, immune response, nutritional status, drug effect, and stress reaction. It could also be applied in fish, crustaceans, and trepang farming.

10.3.1   The Probiotic Function of Lactobacillus on Aquatic Animals

Lactobacillus preparation or feed used in aquaculture has the basic function of regulating host intestinal flora, maintaining microecological balance, and supplementing nutrients. The role of lactobacillus as probiotics in aquaculture is mainly reflected in two aspects: (1) the effects on water microecological balance and water environment and (2) the effects on host gastrointestinal microecological balance. Therefore, it can give full play to the probiotics of lactic acid bacteria, such as regulating the balance of intestinal flora, purifying water, improving water quality, pretreating diseases, promoting growth, and improving host body immunity.

10.3.1.1 Regulating the Balance of Intestinal Flora

Lactic acid bacteria can produce a variety of metabolites such as organic acids, hydrogen peroxide, and bacteriocins during fermentation. Hydrogen peroxide secreted by probiotics could activate the catalase system which had antibacterial effect and could inhibit the growth and reproduction of pathogenic bacteria such as *Escherichia coli* and *Salmonella* (Zheng et al. 2005). Acidic substances such as lactic acid reduced the pH of the environment and antagonized harmful microorganisms with weak acid tolerance. This environment could inhibit the growth of harmful bacteria and regulate the microecological balance of intestinal flora (Lin 2012). The secreted bacteriocins such as nisin have a good germicidal effect. Through the destruction of the cell membrane of harmful microorganisms and membrane potential changes, bacteriocins improve the permeability of the cell membrane to lose a large amount of ATP and ions in the cell membrane and lead to the death of pathogenic bacteria at last. At the same time, the bacteriocin has a broad spectrum of antibacterial activity and a good inhibitory effect on the closely related strains. Some studies have found that the metabolites produced by lactobacillus after fermentation have an inhibitory effect on harmful microorganisms which played a key role in the inhibitory effect.

Lactobacillus is a kind of facultative anaerobic microorganism. After entering the intestinal tract of aquatic animals, it conducts aerobic metabolism, consumes the oxygen in the intestinal tract, and reduces its concentration to inhibit the growth of harmful aerobic bacteria. By competing for oxygen with pathogenic bacteria, it inhibits the pathogenic bacteria and maintains the microecological balance of the intestinal flora.

Lactic acid bacteria are a kind of normal flora in fish intestinal tract and have a good colonization ability to become the advantage bacterium group. It can contend the nutrients from living space and living environment with pathogenic bacteria and make the pathogens rarely get adequate nutrition and space to grow. This domino effect can inhibit the growth of harmful bacteria and maintain the balance of intestinal flora. Therefore, lactobacillus mainly inhibits the growth of harmful bacteria through the effect of metabolites and the competition with harmful bacteria for oxygen, space, and nutrients.

10.3.1.2 Improving the Water Quality of Aquaculture

With the continuous development of aquaculture, intensive aquaculture mode is continuously improved. Overstocking and concentrated feeding will lead to a large number of residual bait, excretion waste of aquaculture animals, and accumulation of animal or plant at the bottom of the pond. Especially under the condition of insufficient dissolved oxygen in water, the decaying and decomposition of organics will produce ammonia nitrogen, nitrite, hydrogen sulfide, and other toxic substances. These harmful substances will pollute the water environment and seriously damage the entire ecological environment of aquaculture to bring huge economic losses to the aquaculture industry (Hall and Holby 1992). Studies have shown that lactobacillus could effectively remove nitrites in culture water, such as the application of *Lactobacillus brevis* A1.558, which had a good removal effect in complex water environment (Wei 2007). The removal of nitrite by lactobacillus mainly has two stages (enzyme removal and acid removal). When the environmental pH is >4.5, enzyme removal is the main part, and nitrite reductase is generated during the fermentation process to degrade harmful nitrites into nontoxic ammonia. When environmental pH <4.0, acid removal is the main part. In this process, through oxidation, ammonification, nitrification, denitrification, and nitrogen fixation effect, the metabolism organic acid decomposed the accumulated residual bait, waste, and other harmful substances into carbon dioxide, nitrate, phosphate, and nontoxic harmless beneficial substances. These effects achieved the purpose of purifying aquaculture water and improving water quality (Gong et al. 2011; Qifang 2002). Li (Chang and Wang 2010) studied the removal capability of several lactic acid bacteria to nitrite and found that acid may be the main factor to remove nitrite. The lower the pH, the faster the removal speed. Lactobacillus could also inhibit the growth and reproduction of pathogenic bacteria and corrupt bacteria in aquaculture water to decompose harmful substances in water and avoid eutrophication in water. Some special enzymes could also be produced to reduce ammonia nitrogen, chemical oxygen consumption, and nitrite in the water to improve the water quality (Feng and Chen 2005). Zhang et al. (Zhang and Li 1999) studied the impact of bacillus lactobacilli on the water environment. The results showed that the content of ammonia nitrogen and nitrite in the water environment of the feeding bacteria were relatively low compared with that of the water body of tilapia lactobacillus.

10.3.1.3 Preventing and Treating the Disease

Lactobacillus is the most widely used type of probiotics at present. After entering the host intestinal colonization, lactobacillus regulates the microecological structure in the host body through the growth, reproduction, and metabolites of the bacteria. So it can show the optimal physiological state and growth rate under the balance of the microecosystem. In the process of aquaculture, the most serious diseases are infectious diseases. The pathogenic microorganisms mainly include viruses, bacteria, and fungi such as intestinal point-like aeromonas and mild aeromonas which belong to opportunistic pathogenic bacteria. A large number of harmful substances (such as ammonia and nitrite) were the main reason for aquatic animal disease outbreak. The harmful substance destroyed the normal microbiota on animal skin mucous membrane which causes pathogenic bacteria to enter the body. At the same time, these substances increased secretion of bacteria endotoxin activity and broke microbial ecological balance. It is reported that lactic acid bacteria could prevent and cure diarrhea or enteritis caused by intestinal bacteria. Lactic acid bacteria produced strong adhesion force through the proteins and fat teichoic acids secreted by bacteria on intestinal mucosa cells. This adhesion ability could increase the lactic acid bacteria colonization and make them become the advantage bacterium group. The pH was low through bacteriocin, organic acid, and hydrogen peroxide to inhibit the growth of harmful bacteria and make lactic acid bacteria become the preponderant bacterium group.

10.3.1.4 Improving the Host Immunity

Disease hazards and antibiotic residues have been the main problems in aquaculture industry. The farmers not only restrain the growth of harmful pathogenic bacteria and reduce the probability of the occurrence of diseases but also improve the immunity ability of aquatic animals. Fish and crustaceans are lower vertebrates, and their immune systems are relatively weak compared with mammals. Immune organs and tissues mainly include the spleen, thymus, and anterior kidney; blood lymphoid and alimentary tract lymphoid tissues rely on the non-specific immune system to improve the body's resistance to external antigens. Lactobacillus can stimulate the development of aquatic animals' immune organs and enhance their activity against interferon, macrophages, complement, and lysozyme. Fish are more dependent on non-specific immune regulation than higher vertebrates. Crustaceans are mainly non-specific and have no specific immune response. Probiotics improve the immune capacity of the body mainly by enhancing the activity of various immune enzymes such as lysozyme, catalase, superoxide dismutase, and phosphatase in animal body fluids or walls.

In addition, lactobacillus can improve the immune function of aquaculture animals. Lactobacilli and their metabolites are immune to intestinal mucosa. In the aspect of humoral immunity, the production of cytokines such as IL-1, IL-5, and IL-6 can be increased by activating macrophages, NK cells, and B lymphocytes. In terms of cellular immunity, lactobacillus can enable antigens to enter ileal junction through M cells, activate Th2 cells, produce factor IgA, and enhance the secretion of SIgA antibody. SIgA can effectively prevent virus adhesion, neutralize bacterial toxins, inhibit bacterial growth, and maintain the ecological balance of normal intestinal flora. It plays an important role in local anti-infection process.

10.3.1.5 Promoting Growth

Lactobacillus produces growth-promoting factors such as folic acid, biotin, vitamin B6, vitamin K, and amino acids. It provides the nutrients needed by animals directly and participates in their metabolism to promote their development and growth. In addition, the metabolism of lactobacillus can produce a large number of acidic substances to improve the speed of enzymatic reaction which is conducive to the digestion and utilization of nutrients by enzymes. The acidic substances can accelerate the intestinal peristalsis and improve the digestibility of intestinal feed. Lactobacillus also can produce a variety of digestive enzymes or secretory enzymes (such as protease, amylase) to help the body to digest protein, sugar, and other nutrients. It also helps to digest substances which are not available to the animals themselves and promote the dissociation of certain mineral elements. Lactic acid bacteria fermentation, such as amino acids and vitamins, is also beneficial to the health of animals.

10.3.2 Application of Lactobacillus in Fish Farming

China is a big country in aquaculture, but there is a certain gap between its overall farming efficiency and that of developed countries. The problems of poor benefit and high incidence of aquaculture generally exist which seriously affect the development of aquaculture. In the intensive and high-density breeding process, a large amount of residual bait and fecal excreta enter the water environment. Because of the limit for automatic purification function of water, the waste will lead to the increase of harmful microorganisms and the growth of pathogenic bacteria. It brings great economic losses of the breeding industry. Therefore, people are constantly looking for safe, nontoxic, and effective prevention and treatment methods. Green, healthy, and safe aquaculture has become a new trend in the future. Since the 1980s, lactobacillus as an ideal probiotics in aquaculture has been applied in aquaculture as a biological additive. The primary condition for screening probiotics is whether the lactobacillus strain can survive in the water environment and give play to the effect of superior flora, whether it can have a certain tolerance to the gastrointestinal environment, and whether it can promote the growth of aquaculture animals and the body immunity. At present, lactic acid bacteria strains with good physiological and biochemical functions have been isolated from aquaculture animals and their water environment. Various microbial preparations have been developed which have been fully utilized in fish, crustaceans, and shellfish breeding and can play its beneficial effects.

In the process of fish breeding, the pathogenic bacteria, caused by disease hazards that mainly include salmon biocidium and vibrio, can respectively induce fish furuncle disease and vibrio disease. It brings huge economic losses to the aquaculture industry. Pathogenic bacteria enter the body mainly through the skin, gills, and gastrointestinal system of fish. Therefore, the key to fish disease is whether the pathogenic bacteria successfully colonize and grow in these tissues. The mucosa of the gastrointestinal tract of fish is composed of epithelial cells, and different locations have different structure and function. Pathogenic bacteria can enter the body through epithelial cells and cause diseases. The probiotic lactobacillus isolated from healthy fish intestinal tract or water environment has significant bacteriostatic effect on its metabolites such as competing with pathogenic bacteria for adhesion sites, preventing pathogenic bacteria from colonizing in fish sensitive locations and reducing the release of pathogenic toxin. As a pathogenic bacteria in the intestinal tract, *Escherichia coli* could adhere to the intestinal epithelial cells through the mannose receptor. *Lactobacillus plantarum* could metabolize the specific adhesion of mannose to inhibit the colonization of the intestinal epithelial cells of *E. coli*. Salina et al. (2008) found that *Carnobacterium divergens* which isolated from salmon had a certain inhibitory effect on colonization and reproduction of pathogenic bacteria in the host.

As a natural active microecological preparation, lactobacillus preparation could colonize slickly in the fish intestinal tract without obvious host specificity. It also could effectively regulate the intestinal flora to maintain the microecological balance and enhance the immune capacity of fish or the barrier layer of epithelial cells (Villamil and Tafalla 2002). Liu found that the superoxide dismutase, alkaline phosphatase, and acidic phosphatase activity in serum were significantly increased when *Lactobacillus acidophilus* was added to the fish fodder. At the same time, the weight gain rate, protein efficiency, and growth rate were also increased. Suzer et al. (2008) found that the total number of bacteria in the intestinal and water bodies of gingival bream was significantly different from that of the control group when lactobacillus was fed on *Sparus aurata*. The total number of bacteria in the water bodies of the control group was significantly higher than that of the experimental group. In addition, the activity, survival rate, and specific growth rate of digestive enzymes in larva were also significantly improved.

However, lactobacillus is not dominant in fish gut flora. Lactobacillus usually is regarded as the normal intestinal flora in fish and has good ability to colonize in intestinal tract without apparent host specificity. It is a certain relationship between lactobacillus quantity and environmental factors or nutrition. But, the number of lactic acid bacteria can maintain at a high level through artificial feeding (Ring 1998). Grouper is one of the most important commercial farmed fish in Southeast Asia. However, the pollution and deterioration of water environment increase the incidence of these diseases, and the pathogenic virus can induce nerve necrosis and other diseases of post-ovulation grouper (Fukuda et al. 1996). Some pathogenic bacteria (such as pseudomonas and vibrio) can also make grouper pathogenic, which brings certain difficulties to grouper breeding. Some scientists added different concentrations of *Lactobacillus plantarum* into fish fodder and found the average weight of grouper, feed conversion rate, and survival rate were significantly improved than the control group. Especially in the 1.0×10^{10} CFU/kg lactobacillus group, the average weight and feed conversion rate increased by 404.6% and 1.26 times, and the fish survival rate increased by 36.7% compared with the control group. In addition, the grouper hind gut lysozyme activity and glutathione peroxidase activity compared with the control group increased by 136.6% and 113.3%, respectively. The phagocyte cells in the immune system activity index, index and head kidney leukocytes breath also increased at the same time. The results confirmed that the plant lactobacillus could promote the growth of grouper and improve immune function and resistance to disease.

Grass carp enteritis is a common infectious disease caused by pathogenic bacteria. In order to control the occurrence of diseases, antibiotics, synthetic drugs, and pesticides are used in large quantities. It also results in the emergence of drug-resistant strains, water environment pollution, and drug residue problems of aquatic products which pose a threat to the safety of aquatic products. In order to find antibiotics substitute, Xia et al. (Xia and Fang 2013) adopted modern biological fermentation technology and unique freeze-drying process to make composite living bacteria preparations. The results of clinical trials showed that compound lactobacillus could improve the water transparency, dissolved oxygen, nitrite, ammonia nitrogen, and other water quality indexes of aquaculture water, promote the growth of cultured grass carp, reduce the feed coefficient, and increase the output of fish. Lactic acid bacteria can inhibit the reproduction of spoilage organisms and pathogenic bacteria in water, decompose harmful material to avoid eutrophication of water body, produce some special enzyme system, and reduce the ammonia nitrogen in water, chemical oxygen consumption, and nitrite content.

In grass carp aquaculture, Xu (2007) studied the influence between *Lactobacillus pentosus* and fish breeding water or intestinal flora. In this experiment, after adding *Lactobacillus pentosaceus* R1 to fish, the number of lactobacillus in the intestinal tract increased from the original 1.0×10^2 to 1.0×10^6 CFU/g. When the juvenile grass carps were fed, part of R1 is colonized in the gastrointestinal tract and a large part is excreted into the water through feces. Only part of the water in the tank was exchanged every day, so the amount of R1 in the water was increased. While the number of *Lactobacillus pentosaceus* R1 increased, the number of vibrio presented a downward trend. This suggest that *Lactobacillus pentosaceus* R1 could colonize in the intestinal tract of juvenile grass carps and control the amount of vibrio in the intestinal tract. In the first 30 days of the experiment, *Lactobacillus pentosaceus* R1 colonized in the intestinal tract of grass carps larvae and antagonized the proliferation of vibrio. This domino effect was achieved through the adjustment of intestinal flora by R1. But it had no effect on the aerobic heterotrophic bacteria in the intestinal tract. Although R1 was not the dominant microbiome in the intestinal tract and its composition in the intestinal microbiome was very small even after a large amount of feeding, the stable colonization of R1 in the intestinal tract inhibited the increase of the number of vibrio in the intestinal tract. At the same time, vibrio in water was also controlled to some extent by lactobacillus in the excretion of feces. According to the test results of each indicator, it was found that the effect of feeding R1 was similar to that of antibiotics.

In the application of fish feed, lactobacillus preparation is also gradually affecting the change of feed situation. Scientists conducted the application test and analysis of lactobacillus preparation in pond mud fish culture. They found that the use of lactobacillus preparations reduced the levels of ammonia nitrogen, nitrite, and hydrogen sulfide in aquaculture water. This effect improved the water aquaculture environment and the bottom of ponds, inhibited pathogenic microorganisms, enhanced the immune defense ability of cultured objects, reduced the use of fishing drugs, provided a good growth environment for the mud carp breeding, and promoted the healthy growth of fish ecology. The water amount in the breeding production of dace fish had been reduced by 50–60%. At the same time, the discharge of waste water had reached the qualified standard which played an ecological protection role to the surrounding environment. The use of lactobacillus preparation not only improved the breeding environment but also reduced the occurrence of blight. The feed coefficient of breeding production was reduced by 0.08–0.15. Although the use cost of lactobacillus preparation in the test ponds in 2011 and 2012 was 24 yuan/667 m², the production cost of dace fish per kg decreased by 0.71–0.78 yuan/kg. At the same time, the production increased by 93.5–125 kg/667 m².

10.3.3 Application of Lactobacillus in Crustaciculture

A large number of studies have shown that probiotics have been widely used in the cultivation of crustaceans. The probiotics can promote the growth of shrimp and crabs, improve the activity of digestive enzymes, and enhance immunity. Lactic acid bacteria can carry out metabolic activities in shrimp and crab aquatic animals and provide them with necessary nutrients such as amino acids, multivitamins, and growth-promoting factors. It even increases the biological activity of mineral elements to enhance the nutrient metabolism and promote the reproduction and growth of shrimp and crab. Adding lactobacillus into the feed can improve the activity of protease and amylase in the liver, pancreas, and intestinal tract. The immune system of crustacean breeding animals is not perfect, which mainly relies on non-specific immunity to improve disease resistance. An important indicator of immune ability is the level of lysozyme and peroxidase in serum. Including lysozyme can destroy and clean up into the host for the foreign body, and peroxidase has the ability of foreign body recognition and defense capabilities. Lactic acid bacteria can activate crustaceans peroxidase in the original system, increase the activity of peroxidase, improve the recognition of foreign bodies and host defense function, and reduce free radical damage to normal cells and removal of reactive oxygen species in the process of cell metabolism.

The giant prawns, also known as Malaysian prawns and freshwater longarm prawns, are the largest freshwater prawns in the world. It is widely distributed in the tropical and subtropical freshwater waters of the west Pacific region of India. It has the advantages of fast growth; large individual size; wide edible, delicious taste; rich nutrition; easy domestication; strong adaptability; and short production cycle. It is a kind of breeding variety with high economic efficiency and one of the main shrimp species in freshwater aquaculture in China. When the feeding density of *Macrobrachium rosenbergii* is relatively high, the overfeeding feed will enter the water body in the form of residual feed. The ammonium nitrogen, nitrite, hydrogen sulfide, and total nitrogen produced by the decomposition of residual feed and shrimp dung will exceed the standard which leads to deterioration of water quality and higher incidence of *Macrobrachium rosenbergii*. Especially in the mid and late stage of cultivation, the amount of organic matter such as baits, excreta, and animal carcasses in the breeding pond increased. This matter will result in the decrease of dissolved oxygen in the pond and the sharp increase of ammonium nitrogen and nitrite nitrogen which lead to the overgrowth of cyanobacteria. Therefore, low-density cultivation of *Macrobrachium rosenbergii* shrimp has become a major problem to restrain its reproduction and sale to adapt to the market. Tan et al. (Yang and Zhang 2013) added lactobacilli preparation in the feed through high-density macrobrachium rosenbloom shrimp cultivation. In the breeding cycle of *Macrobrachium rosenbergii*, the overall change of water quality in the breeding pond was as follows: (1) The pH was relatively stable in the tolerance range of *Macrobrachium rosenbergii*. (2) The concentration of ammonia nitrogen showed a trend of accumulation over time, which was higher than the environmental quality of surface water. (3) Total phosphorus concentration exceeded the environmental quality of surface water in five categories. (4) The concentration of total nitrogen increased with time and exceeded five standards of environmental quality of surface water. (5) The treatment effect of lactobacilli preparation on ammonia nitrogen and total nitrogen was obvious. Cyanobacteria and cyanobacteria also proliferated in the high-density aquaculture ponds in the late stage of aquaculture, but there was no "water bloom." Under the treatment of lactobacillus, the high-density *Macrobrachium rosenbergii* pond showed a similar pattern to the low-density pond which had a significant effect of increasing yield. It is reported that complex microbial preparations including yeast, bacillus, and lactobacillus had been used into the feed of *Macrobrachium rosenbergii*. The results showed that this kind of compound microbial preparation not only promoted growth but also significantly increased the number of beneficial microorganisms in the intestinal flora compared with the control group.

The river crab breeding industry in China has developed rapidly. With the continuous expansion of the breeding scale and the deterioration of the breeding environment, the occurrence of river crab diseases is frequent, which seriously restricts the development of the river crab breeding industry. The prevention and treatment of river crabs has been using antibiotics and other chemical drugs to kill pathogenic microorganisms. This method not only causes the endogenous and secondary infections of animals to develop resistance and reduce the resistance but also directly affects human health and ecological environment. There are a number of research applying probiotic preparations to open duck and conventional breeding through the mechanism of action of probiotics and the application of probiotic preparations for reference in other aquaculture experience. The scientists also explore the probiotic preparations of river crab breeding ecological environment improvement, prevention, and growth-promoting role and influence on breeding efficiency. And it provides a scientific basis for the popularization and application of probiotic preparations in river crab breeding. It is studied the application of probiotics in river crab breeding. They found that probiotics significantly improved the ecological environment of river crab breeding in the monitoring results of water samples and the observation of breeding process. It was mainly reflected in the effective control of chemical oxygen demand, ammonia nitrogen, nitrite nitrogen, and sulfide in the pond. The dissolved oxygen content gradually increased, while the control pond was the opposite. The results of bacterial count monitoring showed that the total number of bacteria decreased significantly after probiotics were applied to the test tank, while the control tank showed an increasing trend. No disinfectant or drug was applied to the test tank, and no deaths were found in river crabs. The average survival rate was 77.9%. However, after using disinfectant and drugs in the control pool, the morbidity and mortality were still very serious with a survival rate of only 60.6%. The results indicated that probiotic preparation had good preventive effect on pond raising river crab. In addition, it could be seen from the observation and record in the process of breeding that the river crabs in the test pool had a greater appetite and growth than those in the control pool. It could be seen that probiotics preparation has a remarkable growth-promoting effect on river crabs.

10.3.4 Application of Lactobacillus in Mollusk Breeding

Apostichopus japonicus is a kind of highly commercial mariculture animal in China which belongs to the family phylum Echinodermata. It is also a marine organism with the same origin of medicine and food. Because of its higher drug value, economic value, and nutritional value, *Apostichopus japonicas* has become the largest mariculture species in North China. However, in the process of *Apostichopus japonicas* culture, the problems of skin diseases such as skin ulcer syndrome and dermatitis caused by vibrio seriously lead to water pollution. It reduces the host immunity and increases the mortality of *Apostichopus japonicas*. It has become a bottleneck problem in the development of the *Apostichopus japonicas* industry (Huan et al. 2009; Xiong and Yao 2006). Lactic acid bacteria can grow under the condition of pH 6.0–8.5, 40 °C. The temperature of trepang feeding is 5–28 °C, and water for aquaculture environment pH ranges 7.5–8.6 (Xiong and Yao 2006). It can be seen that the conditions of the cultivation of ginseng can make the normal growth and reproduction of lactobacillus. In addition, as an anaerobic microorganism, lactobacillus can also survive and grow under the condition of oxygen deficiency and play a probiotic effect. Successful colonization of lactobacillus japonicus in the body will inhibit the reproduction of pathogenic bacteria.

It is reported that two strains of *Lactobacillus acidophilus* RS-1 and RS-2 from the intestinal and aquatic breeding environment of stichopus japonicus by certain screening methods and analyzed their probiotics effect on the radix salviae japonicus. The results showed that lactobacillus could secrete inhibitory substances and significantly inhibit vibrio growth by way of competing nutrients. At the same time, *Bacillus subtilis* YB-1 and *Bacillus cereus* YB-2 were also isolated. The effects of compound bacteria on immunity and growth of *Salvia miltiorrhiza* were investigated. The results showed that the compound microecological preparation could stimulate the activity of coeliac immune cells and intestinal digestive enzymes of the ginseng, promote its growth, inhibit the infection of vibrio, and reduce the mortality rate of the ginseng by more than 70%. In addition, the compound bacillus could also play the role of purifying the water in the cultivation of the ginseng. Wang (2009) study the addition of lactic acid bacteria and bacillus mixed formulation into the trepang water for aquaculture environment. It was found that probiotics mix not only made the trepang's growth rate increase but also reduced the harmful substances in aquaculture water such as ammonia nitrogen, nitrite, and phosphate content to enhance the purification capacity of water body. Zhang et al. (Zhang 2009) isolated *Lactobacillus* L-2 and *Bacillus* K-3 and J-9 from the intestinal tract of healthy *Codonopsis pilosula* and found that the activity of codonopsis protease and amylase could be improved and the immune function could be enhanced. Lactobacillus preparations and their metabolites could also increase the growth rate of ginseng and enhance the activity of alkaline phosphatase, acidic phosphatase, and lysozyme in the body cavity cells. Compared with the control group, the number of intestine vomit in the experimental group after anhydrous stress was lower. At the same time, the number of heterobacteria, vibrio, and *Escherichia coli* in the intestinal flora and aquaculture water was also significantly reduced, while the number of lactic acid bacteria was significantly increased. The results showed that lactobacillus preparation and its metabolites could regulate the intestinal flora of salvia japonicus.

References

Adami A, Cavazzoni V (1999) Occurrence of selected bacterial groups in the feces of piglets fed with *Bacillus coagulans* as probiotic. Basic Microbiol 39(1):3–9

Adams MC, Luo J, Rayward D (2008) Selection of a novel direct-fed microbial to enhance weight gain in intensively reared calves. Anim Feed Sci Technol 145(1–4):41–52

Bao J, Zhang Y, Ping F (2015) Effect of probiotics on production performance and the incidence rate in nursery pigs. Feed Ind 36(2):39–40

Chang Y, Wang L (2010) The effect on the amino butyric acid of black bean in the different cultivation condition. Food Ind 4:7–9

Chaucheyras-Durand F, Durand H (2010) Probiotics in animal nutrition and health. Benefic Microbes 1(1):3–9

Choct M, Hughes RJ, Wang J (1996) Increased small intestinal fermentation is partly responsible for the anti-nutritive activity of non-starch polysaccharides in chickens. Br Poult Sci 37(3):609–621

Feng J, Chen Y (2005) Influence of microecologicson the water quality. Trans Oceanol Limnol 4(15):104–108

Fukuda Y, Nguyen HD, Furuhashi M (1996) Mass mortality of cultured sevenband grouper, *Epinephelus septemfasciatus*, associated with viral nervous necrosis. Fish Pathol 31(3):165–170

Gong G, Lu Y, Guan S (2011) Preparation and properties of nitrite reductase from lactobacillus. China Brew 1:58–60

Huan D, Chongbo H, Zhou Z (2009) Isolation and pathogenicity of pathogens from skin ulceration disease and viscera ejection syndrome of the sea cucumber *Apostichopus japonicus*. Aquaculture 287:18–27

Jadamus A, Vahjen W, Schafer K (2002) Influence of the probiotic strain *Bacillus cereus* var. toyoi on the development of enterobacterial growth and on selected parameters of bacterial metabolism in digesta samples of piglets. J Anim Physiol Anim Nutr 86:42–54

Jin LZ, Ho YW, Ali MA (1996) Effect of adherent Lactobacillus spp. on in vitro adherence of salmonellae to the intestinal epithelial cells of chicken. J Appl Bacteriol 81:201–206

Kalavathy R, Abdullah N, Jalaludin S (2003) Effects of Lactobacillus cultures on growth performance, abdominal fat deposition, serum lipids and weight of organs of broiler chickens. Br Poult Sci 44(1):139–144

Karimi Torshizi MA, Moghaddam AR, Rahimi S (2010) Assessing the effect of administering probiotics in water or as a feed supplement on broiler performance and immune response. Br Poult Sci 51(2):178–184

Liao D-j, Ye Y-g, Wang M (2012) Investigation and measures to solve the prevalence of diarrhea disease in calves. Dairy Health 16:34–35

Lin SU (2012) Lactic acid bacillus mechanism of action and application prospects in aquaculture. Jilin Anim Husb Vet 5:25–29

Lin Q, Dai Q, Bin S (2012) Probiotics and enzyme preparation: effects on growth performance of yellow- feathered broilers and its mechanism. Chin J Anim Nutr 24(10):1955–1965

Liu H, Zhang M, Feng J (2012) Effects of probiotic preparation on growth performance and immune indices of early weaner piglets. Chin J Anim Nutr 24(6):1124–1131

Luan G, Wang J, Bu D (2008) Effects of supplementation of Bacillus subtilis Natto on milk production and quality of lactating cow. J Northeast Agric Univ 39(9):58–61

Mikulski D, Jankowski J, Naczmanski J et al (2012) Effects of dietary probiotic (Pediococcus acidilactici) supplementation on performance, nutrient digestibility, egg traits, egg yolk cholesterol, and fatty acid profile in laying hens. Poult Sci 91(10):2691–2700

Molnár AK, Podmaniczky B, Kürti P (2012) Effect of different concentrations of Bacillus subtilis on growth performance, carcase quality, gut microflora and immune response of broiler chickens. Br Poult Sci 20(43):1–9

Mountzouris KC, Tsirtsikos P, Kalamara E (2007) Evaluation of the efficacy of a probiotic containing lactobacillus, bifidobacterium, enterococcus, and pediococcus strains in promoting broiler performance and modulating cecal microflora composition and metabolic activities. Poult Sci 86:309–317

Mountzouris KC, Balaskas C, Xanthakos I (2009) Effects of a multi-species probiotic on biomarkers of competitive exclusion efficacy in broilers challenged with Salmonella enteritidis. Br Poult Sci 50(4):467–478

Pascual MN, Hugas M, Badiola JI (1999) Lactobacillus salivarius CTC2197 prevents Salmonella enteritidis colonization in chickens. Appl Environ Microbiol 65(11):4981–4986

Per Hall OJ, Holby O (1992) Chemical fluxes and mass balances in a marine fish cage farm. IV. Nitrogen. Mar Ecol Prog Ser 89:81–91

Persson Y, Nyman AK, Gronlund-Andersson U (2011) Etiology and antimicrobial susceptibility of udder pathogens from cases of subclinical mastitis in dairy cows in Sweden. Acta Vet Scand 53(36):1–8

Qiao G, Shan A (2007) Study of the effect of probiotics on performance and rumen fermentation in dairy cattle. Chin Dairy Cattle 3:10–14

Qifang Z (2002) The study on mechanism of nitrite degradation by lactic acid Bacteria. Food Ferment Ind 28(8):28–31

Ring E (1998) Lactic acid bacteria in fish: a review. Aquaculture 160:177–203

Salinas I, Myklebust R, Esteban MA (2008) In vitro studies of Lactobacillus delbrueckii subsp. lactis in Atlantic salmon (Salmo salar L.) foregut: tissue responses and evidence of protection against Aeromonas salmonicida subsp. salmonicida epithelial damage. Vet Microbiol 128(1–2):167–177

Stern NJ, Svetoch EA, Eruslanov BV (2006) Isolation of a Lactobacillus salivarius strain and purification of its bacteriocin, which is inhibitory to Campylobacter jejuni in the chicken gastrointestinal system. Antimicrob Agents Chemother 50(9):3111–3116

Suzer C, Çoban D, Kamaci HO (2008) Lactobacillus spp. bacteria as probiotics in gilthead sea bream (Sparus aurata, L.) larvae: effects on growth performance and digestive enzyme activities. Aquaculture 280(4):140–145

Timmerman HM, Mulder L, Everts H (2005) Health and growth of veal calves fed milk replacers with or without probiotics. J Dairy Sci 88:2154–2165

Villamil L, Tafalla C (2002) Evaluation of immunomodulatory effects of lactic acid bacteria in turbot (Scophthalmus maximus). Clin Vaccine Immunol 9(6):1318–1323

Wang L (2009) Effects of microecological preparation on the culture of Apostichopus japonicus sclenkain a closed recalculating aquaculture system. Chin J Microecol 21(6):497–504

Wang X-r, Zhang X (2013) Effects of Bacillus coagulans on laying performance, egg quality and serum biochemical indices of laying ducks. Acta Ecol Anim Domastici 34(2):69–74

Wei W (2007) Removal of nitrite from aquacultural water with lactobacillus brevis. J Ecol Rural Environ 23(4):37–40

Xia L, Fang M (2013) Study on the safety evaluation of grass carp and the effect evaluation of control bacterial enteritis from multiple-lactobacillus. Fish Inf Strategy 28(1):44–49

Xiong S, Yao X (2006) Effects of different temperatures and pH on the growth of lactic acid bacteria. Xinjiang Agric Sci 43(6):533–538

Xu T (2007) Effects of Lactobacillus pentosaceus R1 on the microflora in water and larvae guts of grass carp (*Ctenopharyngodon idella*). J Anhui Agri Sci 35(34):11112–11114

Yang M, Zhang Y (2013) Application test of EM bacteria in *Macrobrachium Rosenbergii* aquaculture. Pop Sci Technol 15(165):152–154

Yang H-y, Yang Z-b, Yang W-r (2009) Effect of probiotics and xylo -oligosaccharide on performance, digestive enzyme activities, blood index and intestinal microflora of early weaned piglets. Chin J Vet Sci 29(7):914–919

Yang H-j, Zhang S-t, Li J, Cui J-l (2014) The application of lactobacillus probiotics in prevention and treatment of cow subclinical mastitis. Dairy Health 17(2):51–54

Yao G-q, Zhao S-p, Gao P-f (2014) Applied research of lactobacillus probiotics in the prevention and treatment of calf diarrhea. Dairy Health 17:55–58

Yuji Aiba MT, Suzuki N (1998) Lactic acid-mediated suppression of helicobacter pylori by the oral administration of lactobacillus salivarius as a probiotic in a gnotobiotic murine model. Am J Gastroenterol 93(11):2098–2101

Zeyner A, Boldt E (2006) Effects of a probiotic *Enterococcus faecium* strain supplemented from birth to weaning on diarrhoea patterns and performance of piglets. J Anim Physiol Anim Nutr (Berl) 90(1–2):25–31

Zhang T (2009) Effect of different combinations of probiotics on digestibility and immunity index in sea cucumber *apostichopus japonicus*. J DaLian Fish Univ 24:64–68

Zhang Q, Li Z (1999) The effects of microbiological compound on ecological factors in culture waters. J Shanghai Fish Univ 8(1):43–47

Zhang H-t, Wang J-q, Bu D-p (2008) Effect of Bacillus subtilis Natto on growth performance of weaned calves. Chin J Anim Nutr 20(2):158–162

Zheng H, Shi Q-q, Shi B-h (2005) Comparison of Bacillus sp. on depuration of aquaculture waterbody. J Microbiol 25(6):41–44

Zhou Y, Zhou X, Wu S (2012) Effect of different level of Bacillus subtilis on the growth performance and diarrhea in weaned piglets. Feed Rev 4:29–31

Printed by Printforce, the Netherlands